Second Order Equations With Nonnegative Characteristic Form

Second Order Equations With Nonnegative Characteristic Form

O. A. Oleĭnik
Moscow State University
Moscow, USSR

and

E. V. Radkevič
Institute for Problems of Mechanics
Academy of Sciences of the USSR
Moscow, USSR

Translated from Russian by
Paul C. Fife
University of Arizona
Tucson, Arizona

AMERICAN MATHEMATICAL SOCIETY, PROVIDENCE, RHODE ISLAND
PLENUM PRESS • NEW YORK-LONDON

Library of Congress Cataloging in Publication Data

Oleĭnik, O. A.
Second order equations with nonnegative characteristic form.

Translation of Uravneniía vtorogo poríadka s neotrítsatel'noĭ kharakteristicheskoĭ formoĭ.
Bibliography: p.
1. Differential equations, Partial. 2. Boundary value problems. I. Radkevich, E. V., joint author. II. Title.
QA377.04313 515'.353 73-16453
ISBN 978-1-4684-8967-5 ISBN 978-1-4684-8965-1 (eBook)
DOI 10.1007/978-1-4684-8965-1

The original Russian text was published for the All-Union Institute of Scientific and Technical Information in Moscow in 1971 as a volume of Itogi Nauki–Seriya Matematika

The present translation is published under an agreement with Mezhdunarodnaya Kniga, the Soviet book export agency.

URAVNENIYA VTOROGO PORYADKA S NEOTRITSATEL'NOI KHARAKTERISTICHESKOI FORMOI
O. A. Oleĭnik and E. V. Radkevič

Уравнения второго порядка с неотрицательной характеристической формой
О. А. Олейник, Е. В. Радкевич

Softcover reprint of the hardcover 1st edition 1973

A Division of Plenum Publishing Corporation
227 West 17th Street, New York, N.Y. 10011

United Kingdom edition published by Plenum Press, London
A Division of Plenum Publishing Company, Ltd.
Davis House (4th Floor), 8 Scrubs Lane, Harlesden, London, NW10 6SE, England

Dedicated to I. G. Petrovskiĭ

CONTENTS

EDITOR'S NOTE

For reasons of economics, most displayed formulas in this translation have been inserted from the original Russian. This means that certain letters unfortunately have a different appearance in formulas from their counterparts in the text. The principal instances are summarized in the following table:

Displayed formulas	*Text*
$\varkappa$	κ
φ	ϕ
ε	ϵ

INTRODUCTION

Second order equations with nonnegative characteristic form constitute a new branch of the theory of partial differential equations, having arisen within the last 20 years, and having undergone a particularly intensive development in recent years.

An equation of the form

$$L(u) \equiv a^{kj}(x)\, u_{x_k x_j} + b^k(x)\, u_{x_k} + c(x)\, u = f(x) \qquad (1)$$

is termed an equation of second order with nonnegative characteristic form on a set G, if at each point x belonging to G we have $a^{kj}(x)\xi_k\xi_j \geqslant 0$ for any vector $\xi = (\xi_1, \cdots, \xi_m)$. In equation (1) it is assumed that repeated indices are summed from 1 to m, and $x = (x_1, \cdots, x_m)$. Such equations are sometimes also called degenerating elliptic equations or elliptic-parabolic equations.

This class of equations includes those of elliptic and parabolic types, first order equations, ultraparabolic equations, the equations of Brownian motion, and others.

The foundation of a general theory of second order equations with nonnegative characteristic form has now been established, and the purpose of this book is to present this foundation.

Special classes of equations of the form (1), not coinciding with the well-studied equations of elliptic or parabolic type, were investigated long ago, particularly in the paper of Picone [105], published some 60 years ago.

The memoir by Tricomi [131], as well as subsequent investigations of equations of mixed type, elicited interest in the general study of elliptic equations degenerating on the boundary of the domain, i.e. equations of the form (1) with the condition that $a^{kj}(x)\xi_k\xi_j > 0$ for $\xi \neq 0$ at points of the domain Ω and $a^{kj}(x)\xi_k\xi_j \geqslant 0$ for all ξ at points on its boundary.

The 1951 paper of M. V. Keldyš [63], initiating a long series of papers, played a significant role in the development of the theory. It was this paper of Keldyš that first brought to light the fact that in the case of elliptic equations degenerating on the boundary, under definite assumptions a portion of the boundary may be free from the prescription of boundary conditions. Many investigations following the paper of Keldyš were devoted to detailed studies of boundary problems for elliptic equations of arbitrary order degenerating on the boundary. These studies employed various methods which had previously been applied in the theory of elliptic equations and in other branches of the theory of partial differential equations (a survey of some of these papers is found in the book [123]). In connection with functional analysis approaches to the study of elliptic equations degenerating on the boundary, there arose a theory of function spaces with weighted norms, for which imbedding theorems analogous to those of S. L. Sobolev (see [71, 84, 133] and others) have been obtained. Studies were recently made of pseudodifferential equations degenerating on submanifolds lying inside the domain or on its boundary. (This is the concern of the papers [24, 77, 78, 134–136] and others.) The methods of the theory of pseudodifferential equations in many cases led to nearly definitive results for elliptic equations degenerating on the boundary, and in particular has led to significant advances in the study of the classical Poincaré oblique derivative problem (see [24, 77, 78]).

We shall investigate general equations of the type (1) without imposing any restrictions on the set of points x where the characteristic form $a^{kj}(x)\xi_k\xi_j$ may vanish for $\xi \neq 0$. The theory developed in the present book contains, in particular, results for elliptic equations of second order degenerating on the boundary.

The paper of Fichera [29], published in 1956, was an important step in the development of a general theory of second order equations with nonnegative characteristic form. In that paper boundary value problems analogous to the Dirichlet and Neumann problems for elliptic equations were posed for general second order equations with nonnegative characteristic form. The first boundary value problem (or Dirichlet problem) for the equation (1) in a domain Ω with boundary Σ was formulated as follows: to find a function $u(x)$ such that

$$L(u) = f \text{ in } \Omega \tag{2}$$

and

$$u = g \text{ on } \Sigma_2 \cup \Sigma_3, \tag{3}$$

where f and g are functions defined on Ω and on $\Sigma_2 \cup \Sigma_3$ respectively, the latter being a subset of Σ defined below. The entire boundary Σ of the domain Ω is separated into the sets $\Sigma_0, \Sigma_1, \Sigma_2, \Sigma_3$. Suppose $\bar{n} = (n_1, \cdots, n_m)$ is the inward normal vector to the boundary Σ. We denote by Σ_3 the noncharacteristic part of the boundary Σ, i.e. Σ_3 is the set of points of Σ where the condition $a^{kj}n_k n_j > 0$ holds. On the set $\Sigma \backslash \Sigma_3$, where $a^{kj}n_k n_j = 0$, one examines the Fichera function $b \equiv (b^k - a^{kj}_{x_j})n_k$. One denotes by Σ_0, Σ_1, Σ_2 the subsets of $\Sigma \backslash \Sigma_3$ where $b = 0$, $b > 0$, and $b < 0$ respectively. (In the case of elliptic equations degenerating only on the boundary Σ, problem (2), (3) coincides with the problem of M. V. Keldyš [63].)

We note that the separation of the boundary Σ into the parts Σ_0, Σ_1, Σ_2, Σ_3 is invariant with respect to changes of independent variables.

With regard to the problem (2), (3), the following questions arise: under what conditions on the coefficients of the equation and on the boundary of the domain will a solution of (2), (3) exist, be unique, and possess a definite degree of smoothness? (Simple examples show that problems (2), (3) might not have smooth solutions (see Chapter I, § 8).)

In the papers of Fichera [29, 30], a priori estimates for smooth solutions of (2), (3) were obtained in the spaces $L_p(\Omega)$. In particular it was shown that if $c < 0$ and $c^* < 0$ in $\Omega \cup \Sigma$, then for all functions u of class $C^{(2)}(\Omega \cup \Sigma)$ satisfying the condition $u = 0$ on $\Sigma_2 \cup \Sigma_3$ and for all $p \geqslant 1$ the estimate

$$\| u \|_{\mathscr{L}_p(\Omega)} \leqslant \frac{p}{\min[-c^* + (1-p)\, c]} \| L(u) \|_{\mathscr{L}_p(\Omega)} \tag{4}$$

holds. Here $c^* = a^{kj}_{x_k x_j} - b^k_{x_k} + c$, and by $C^{(k)}(\Omega \cup \Sigma)$ we denote the class of functions whose derivatives up to order k inclusive are continuous in $\Omega \cup \Sigma$ (see Chapter I, §2).

Using estimate (4) and the theorem on the representation of a linear functional in the space $\mathfrak{L}_p(\Omega)$, Fichera obtained an existence theorem for a weak solution of (2), (3) in the space $\mathfrak{L}_p(\Omega)$.

A weak solution of (2), (3) for $g = 0$ is defined as a function $u(x)$ in the space $\mathfrak{L}_p(\Omega)$ which, for any function $v(x)$ in the class $C^{(2)}(\Omega \cup \Sigma)$, equal to zero on $\Sigma_3 \cup \Sigma_1$, satisfies the integral identity

$$\int_\Omega u L^*(v)\, dx = \int_\Omega v f\, dx. \tag{5}$$

Here

$$L^*(v) \equiv a^{kj} v_{x_k x_j} + b^{*k} v_{x_k} + c^* v, \quad b^{*k} \equiv 2a^{kj}_{x_j} - b^k$$

(see Chapter I, §3).

Fichera also proved the existence of a solution of (2), (3) in the Hilbert space $\mathcal{H}$ with scalar product

$$(u, v)_{\mathcal{H}} = \int_{\Omega} (a^{kj} u_{x_k} v_{x_j} + uv)\, dx + \int_{\Sigma_1 \cup \Sigma_2} |b|\, uv\, d\sigma.$$

This proof is based on the Riesz representation theorem.

Existence theorems for weak solutions of (2), (3) in the classes $\mathcal{H}$ and $\mathcal{L}_p(\Omega)$, or in the class of bounded functions, may be proved by the method of elliptic regularization (see [91, 93] and Chapter I, §5). In this procedure the weak solution of (2), (3) for smooth f and g is obtained as the limit as $\epsilon \to 0$ of solutions of the Dirichlet problem for the elliptic equations

$$\epsilon \Delta u + L(u) = f \text{ in } \Omega, \quad \epsilon = \text{const} > 0, \tag{6}$$

with boundary condition

$$u = g_1 \text{ on } \Sigma, \tag{7}$$

where g_1 is a smooth function coinciding with g on $\Sigma_2 \cup \Sigma_3$. In order to prove the existence of a limit as $\epsilon \to 0$ of solutions of (6), (7), uniform (in ϵ) estimates are established for the norms of these solutions in a suitable space, as well as for the derivatives of these solutions on the boundary of the domain.

In his papers, Fichera posed the problem of the uniqueness of weak solutions of (2), (3) constructed in the spaces $\mathcal{L}_p(\Omega)$ and $\mathcal{H}$.

Questions of uniqueness of weak solutions of (2), (3) were first investigated in [91] (see also [93] and Chapter I, §6). Uniqueness theorems for such generalized solutions were proved by the method of elliptic regularization. Under broad assumptions on the coefficients of equation (2) and on the boundary Σ of the domain Ω, it was established that the weak solution in the sense of (5) is unique in the class $\mathcal{L}_p(\Omega)$ for $p \geqslant 3$. In this result it is assumed that the boundary Γ of the set Σ_2 in Σ has $(m-1)$-dimensional measure zero.

In Chapter I, §6, examples are given of problems (2), (3) for which all the assumptions of the uniqueness theorem for weak solutions in the class $\mathcal{L}_p(\Omega)$ for $p \geqslant 3$ are fulfilled, but uniqueness does not hold within the class $\mathcal{L}_p(\Omega)$ for $p < 3$. These examples show that the fundamental theorem of Chapter I, §6, on uniqueness of weak solutions is in a certain sense the best possible result.

Phillips and Sarason [102] gave an example of a problem (2), (3) in which Γ has positive $(m-1)$-dimensional measure and the weak solution of this problem is not unique, even in the class of bounded measurable functions.

They also prove in [102] a uniqueness theorem for weak solutions of (2), (3) in the space $\mathcal{H}$ by reducing (2) to a symmetric system and applying the theory of extensions of symmetric operators. A similar theorem is proved in Chapter I, §6. In this latter theorem requirements are imposed on the weak solution which are weaker than existence of a finite norm in the space $\mathcal{H}$. However, the coefficients of the highest order derivatives in (2) are required to be continuable to a neighborhood of Σ_2 with retention of the nonnegativity of the characteristic form, and also required to have bounded second order derivatives. In this theorem the continuability condition on the coefficients may be replaced by the requirement that the weak solution assumes the given values on Σ_2 in a certain weak sense.

The uniqueness theorems for solutions of (2), (3) are established also for piecewise smooth domains.

The question of existence of smooth solutions of (2), (3) for a special class of second order equations with nonnegative characteristic form was considered in the dissertation of A. M. Il'in (see [57, 58, 59]). He found that within this class, for the existence of a smooth solution of (2), (3), the coefficients of (2) and their derivatives should satisfy certain inequalities.

In [92, 93] broad sufficient conditions for the existence of a smooth solution of (2), (3) are given. Theorems on smoothness of weak solutions of this problem are also proved in these papers. The proofs of these theorems are based on the following lemma (see [93], and also Chapter I, §7).

Suppose $a^{kj}(x)\xi_k\xi_j \geqslant 0$ for all x in R^m and all $\xi = (\xi_1, \cdots, \xi_m)$, and suppose $a^{kj}(x)$ belong to class $C_{(2)}(R^m)$. Then every function $v \in C_{(2)}(R^m)$ satisfies the inequality

$$(a^{kj}_{x_p} v_{x_k x_j})^2 \leqslant M a^{kj} v_{x_k x_s} v_{x_j x_s}, \tag{8}$$

where the constant M depends only on the second derivatives of a^{kj}. We denote by $C_{(k)}(\Omega)$ the class of functions with derivatives up to order k inclusive bounded in Ω. This lemma also has significant applications in the study of local smoothness of weak solutions of equation (1) (see Chapter II, §6).

For example, a sufficient condition for the existence of a solution of (2), (3) in the class $C_{(s)}(\Omega)$ is that the coefficients of the equation and the function f belong to the class $C_{(s)}(\Omega)$, that the boundary of the domain Ω be sufficiently smooth, that the inequality $c \leqslant -c_0 < 0$ be satisfied, where c_0 is a number depending on s, and that the coefficients a^{kj} be continuable into a neighborhood of $\Sigma_0 \cup \Sigma_1 \cup \Sigma_2$, in which neighborhood the a^{kj} belong to class $C_{(s)}$ and the characteristic form $a^{ki}\xi_k\xi_i$ is nonnegative.

Moreover it is assumed that the sets $\bar{\Sigma}_3$, $\bar{\Sigma}_2$, and $\Sigma_0 \cup \Sigma_1$ have no points in common (see [92, 93, 147]). The question of the existence of a smooth solution of (2), (3) (without the assumption that one can continue the coefficients a^{kj} through the boundary with the properties indicated) was considered in the paper of Kohn and Nirenberg [64] (see Chapter I, §9). They obtained conditions for the existence of solutions of (2), (3) in the Sobolev spaces $W_2^N(\Omega)$. In these results it is assumed that the intersection of $\Sigma_0 \cup \Sigma_1$ and $\overline{\Sigma_2 \cup \Sigma_3}$ is empty, that in Ω the coefficient $c \leqslant -c_0 < 0$, and on Σ_2 that $b \leqslant -b_0 < 0$, where the constants c_0 and b_0 are sufficiently large. For simplicity the coefficients of equation (2), the boundary of Ω, and the functions f and g are all assumed infinitely differentiable. In the paper referred to, the solution of (2), (3) is also obtained by the method of elliptic regularization. Thus in a neighborhood of $\Sigma_0 \cup \Sigma_1$ the left side of (2) is supplemented by an elliptic operator multiplied by ϵ, the order of this operator depending on N, and outside a neighborhood of $\Sigma_0 \cup \Sigma_1$ in Ω it is supplemented with a second order elliptic operator multiplied by ϵ. A priori estimates, uniform in the parameter ϵ, for solutions of the corresponding problem for the elliptic equation are established in the space $W_2^N(\Omega)$. Analogous theorems are obtained by us by other means in Chapter I, §8, where conditions are given under which the solution of the problem (2), (3) belongs to class $C_{(k)}(\Omega)$ in the case that the a^{kj} are not continuable through the boundary with the properties indicated above. It is shown that in this case the solution of (2), (3) may be obtained by regularizing (2) with the aid of a second order elliptic operator with small parameter ϵ. We note that the method of elliptic and parabolic regularization has been widely used by Lions (see, for example, [75]), and also in the study of discontinuous solutions of nonlinear hyperbolic equations (see [96], the method of vanishing viscosity). The second boundary problem for (1), analogous to the Neumann problem for elliptic equations, was investigated in [27, 110].

There is great interest in the study of local smoothness of the solutions of (2), (3), and in the related question of hypoellipticity of second order equations. The concept of hypoellipticity of an operator was introduced in Schwartz's book [120] (see also [50, 51]).

A linear differential operator P with infinitely differentiable coefficients defined in a domain Ω is called hypoelliptic in Ω if, for any distribution $u(x)$ in $D'(\Omega)$ and any domain Ω_1 contained in Ω, the condition that Pu is infinitely differentiable in Ω_1 implies that $u(x)$ is also infinitely differentiable in Ω_1.

For equations and systems with constant coefficients, necessary and sufficient conditions for hypoellipticity have been found [50, 51]. For differential equations and systems with variable coefficients, and also for pseudodifferential operators, various sufficient conditions for hypoellipticity have been found.

L. Hörmander proved in [55] that second order hypoelliptic equations have non-negative characteristic form at each point of the domain Ω, after possible multiplication by -1 (see also Chapter II, §3).

In [55] Hörmander gave a sufficient condition for hypoellipticity for second order equations of the form

$$Pu \equiv -\sum_{j=1}^{r} X_j^2 u + iX_0 u + cu = f, \tag{9}$$

where X_j $(j = 0, 1, \cdots, r)$ are first order differential operators with infinitely differentiable real coefficients

$$X_j \equiv \sum_{k=1}^{m} a_j^k(x) D_k; \; D_k = -i\frac{\partial}{\partial x_k}.$$

The condition of Hörmander for the operator (9) is a condition on the Lie algebra of the operators X_j $(j = 0, 1, \cdots, r)$. For hypoellipticity of the operator (9) in the domain Ω it is sufficient that at each point of Ω, among the operators X_j $(j = 0, 1, \cdots, r)$ and the commutators generated by these operators, there exist m linearly independent operators. This condition of Hörmander is also a necessary condition for hypoellipticity of the operator P in the domain Ω, if one restricts attention to the class of operators of the type (9) for which, at every point x, among the operators X_j and the commutators generated by them there exist exactly μ linearly independent operators, where $\mu \leqslant m$ is independent of x. In this case we say that the system $\{X_0, \cdots, X_r\}$ has rank μ in Ω. Hörmander's proof uses the theory of Lie algebras and certain special function spaces.

Another proof of Hörmander's theorem on the hypoellipticity of the operator (9) is given in [112] and a more general theorem in Chapter II, §5. This proof is based on the theory of pseudodifferential operators. With the aid of pseudodifferential operators a priori estimates for solutions of (9) are found in the norm of the spaces $\mathcal{H}_s$; and these estimates imply the hypoellipticity of P, using the theorem proved in Chapter II, §4. The a priori estimates established in Chapter II, §5, also imply local smoothness of the generalized solution of (9).

In Chapter II, §5, we exhibit a class of hypoelliptic operators of the form (9) for which the condition of Hörmander may be violated on a certain set M of points of Ω. It is proved that the operator P is hypoelliptic in Ω, if: 1) in $\Omega \setminus M$ the system X_j $(j = 0, 1, \cdots, r)$ has rank m, where M is a bounded set of points lying on a finite number of $(m-1)$-dimensional smooth manifolds $\mathfrak{M}$ with closure in Ω, and 2) at each point x in M, either for some $j = 1, \cdots, r$

$$\sum_{k=1}^{m} a_j^k \Phi_{x_k} \neq 0,$$

or, in case

$$\sum_{j=1}^{r} \left| \sum_{k=1}^{m} a_j^k \Phi_{x_k} \right| = 0,$$

the condition

$$\sum_{j=1}^{r} - X_j^2 \Phi + i X_0 \Phi \neq 0$$

is fulfilled, where $\Phi(x_1, \cdots, x_m) = 0$ is the equation for $\mathfrak{M}$ in a neighborhood of the point x, and $\operatorname{grad} \Phi \neq 0$.

In the case when the set M consists of a single point x^0, the operator (9) is hypoelliptic in Ω if at this point one of the coefficients a_j^k $(k = 1, \cdots, m; j = 0, \cdots, r)$ is different from zero. Certain hypoelliptic operators of the form (9) for which the condition of Hörmander is violated were also found by V. S. Fediĭ [28].

Not every equation (1) with infinitely differentiable coefficients may be reduced to the form of equation (9). Hilbert (see [48]) constructed an example of a nonnegative polynomial $P(x, y)$ of two independent variables x and y of degree six which is not representable as a finite sum of squares of polynomials. It is easily shown that an operator of the form

$$L(u) \equiv z^6 P\left(\frac{x}{z}, \frac{y}{z}\right) \Delta u + Tu,$$

where T is any operator of first order with infinitely differentiable coefficients and $\Delta u = u_{xx} + u_{yy} + u_{zz}$ is not representable in the form (9) in any neighborhood of the origin.

For general second order equations of the form (1) with nonnegative characteristic form, a condition for hypoellipticity is given in [113] and a more general theorem in Chapter II, §6.

Let $L^0(x, \xi) = a^{kj}(x) \xi_k \xi_j$. We denote by L^0 the pseudodifferential operator with symbol $L^0(x, \xi)$ (see Chapter II, §2). For any pseudodifferential operator A we denote by $A^{(j)}$ and $A_{(j)}$ the pseudodifferential operators corresponding to the symbols $\partial A(x, \xi)/\partial \xi_j$ and $D_j A(x, \xi)$ respectively, where $A(x, \xi)$ is the symbol of the operator A and $D_j = -i\partial/\partial x_j$. We write the operator $L(u)$ in the form

$$L(u) \equiv -D_j(a^{kj}D_k u) + iQu + cu,$$

where $Qu \equiv (b^k - a^{kj}_{x_j})D_k u$. We consider the system of operators $\{Q_0, Q_1, \cdots, Q_{2m}\}$ where $Q_0 = Q$, $Q_j = L^{0(j)}$ for $j = 1, \cdots, m$, $Q_j = E_{-1}L^0_{(j-m)}$ for $j = m+1, \cdots, 2m$, E_{-1} is a pseudodifferential operator with symbol $\phi(x)(1+|\xi|^2)^{-1/2}$ and $\phi(x) \in C_0^\infty(\Omega)$. For any multi-index $I = (\alpha_1, \cdots, \alpha_k)$, where $\alpha_l = 0, \cdots, 2m$ for $l = 1, \cdots, k$, we set $|I| = \Sigma_1^k \lambda_l$, where $\lambda_l = 1$ for $\alpha_l = 1, \cdots, 2m$, and $\lambda_l = 2$ for $\alpha_l = 0$. To each multi-index I we associate the operator

$$Q_I = \operatorname{ad} Q_{\alpha_1} \cdots \operatorname{ad} Q_{\alpha_{k-1}} Q_{\alpha_k},$$

where $\operatorname{ad} AB = AB - BA$ for any operators A and B.

We consider the operators Q_I generated by the system of operators $\{Q_0, \cdots, Q_{2m}\}$. The operator Q_I may be represented in the form $Q_I = Q_I^0 + T_I$, where the operator T_I has order at most zero, and Q_I^0 is a pseudodifferential operator with symbol $q_I^0(x, \xi)$. We shall say that the system of operators on the compact set K has rank m if there exists a number $R(K)$ such that

$$1 + \sum_{|I| \leqslant R(K)} |q_I^0(x, \xi)|^2 \geqslant C_1(1 + |\xi|^2), \quad C_1 = \text{const} > 0 \tag{10}$$

for all $x \in K$ and all ξ in R^m.

If the rank of the system of operators $\{Q_0, \cdots, Q_{2m}\}$ is equal to m on every compact set K belonging to Ω, then the operator L of form (1) is hypoelliptic in Ω. The operator L is also hypoelliptic in Ω if the system of operators $\{Q_0, \cdots, Q_{2m}\}$ has rank m on every compact set K in $\Omega \backslash M$, and at the points of M,

$$a^{kj}\Phi_{x_k}\Phi_{x_j} + |a^{kj}\Phi_{x_k x_j} + b^k\Phi_{x_k}| > 0$$

(the set M has the properties indicated before). In this, as in the analogous theorem for operators of the type (9), the domain Ω_1 figuring in the definition of hypoellipticity either contains the set M, or does not intersect it. Under these conditions equation (1) has the property of local smoothness of generalized solutions, and a priori estimates in the spaces $\mathcal{H}_s$ may be established which are analogous to the known Schauder interior estimates [118].

If the set M consists of a single point, then for hypoellipticity of (1) in Ω it is sufficient that the inequality

$$\sum_{j=1}^{m} a^{jj} + |b^j| > 0$$

be satisfied at that point. An investigation of the hypoellipticity of operators of the form (1) is also carried out on the basis of the theory of pseudodifferential operators in Chapter II, §6.

For equations (1) and (9) with analytic coefficients, necessary and sufficient conditions for hypoellipticity have been found (see [20], [98] and Chapter II, § 8).

It is easy to verify that the equation of Brownian motion satisfies the conditions of hypoellipticity formulated above. Classes of equations analogous to the Brownian motion equation were considered in papers by T. G. Genčev [43], Hörmander [55], A. M. Il'in [57], and others. Fundamental solutions for such equations were constructed in [55] and [57].

For hypoelliptic equations satisfying the Hörmander condition or condition (10), a solution of the first boundary value problem in a nonsmooth domain may be constructed by a procedure analogous to that used by M. V. Keldyš [63] for second order elliptic equations degenerating on the boundary of Ω (see Chapter II, §7). In this procedure the domain Ω is approximated by domains $\Omega_1, \cdots, \Omega_n, \cdots$, such that $\Omega_n \subset \Omega_{n+1}$, the continuous boundary function g is continued inside the domain Ω, and in Ω_j one considers the solution u_j^ϵ of the elliptic equation

$$\epsilon\Delta u + L(u) = f$$

with boundary condition $u_j^\epsilon = g$ on the boundary of Ω_j. On the basis of a priori estimates proved in Chapter II, §§5 and 6, it is established that at interior points the sequence $u_j^\epsilon(x) \to u(x)$ as $\epsilon \to 0$ and $j \to \infty$. The boundary continuity of solutions $u(x)$ of problem (2), (3) is proved by means of barriers.

Qualitative properties of solutions of second order equations with nonnegative characteristic form have been studied. The maximum principle for smooth solutions of (1) was proved by Fichera in [29], and this same principle for generalized solutions was proved in [91] and in Chapter I, §5 (see also Chapter I, §1). We note the difference between the maximum principle proved in Chapter I, §1, and the theorems on the maximum principle proved in [29] and in Chapter I, §2. An investigation of a strong maximum principle for general second order equations with nonnegative characteristic form was undertaken in papers by Pucci [108, 109] and A. D. Aleksandrov [1–5], and for equations of the type (9) in the papers of Bony [15–17] (see Chapter III, §1).

At each point x of G we consider the characteristic vectors corresponding to positive characteristic values of the matrix $\|a^{kj}\|$, and let $E(x)$ denote the linear space spanned by these vectors. A curve l is called a line of ellipticity for equation (1) if in a neighborhood of each of its points there exists a vector field $\overline{Y} = (Y_1(x), \cdots, Y_m(x))$ such that $Y_j \in C^{(1)}$, at each point x of this neighborhood the vector $(Y_1(x), \cdots, Y_m(x))$ lies in the plane $E(x)$, $a^{kj}(x)Y_k(x)Y_j(x) \geqslant \text{const} > 0$, and the curve l is a trajectory of the vector field $\overline{Y}$. A set $\mathfrak{M}$ is called a set of elliptic connectivity for equation (1) if any two points of $\mathfrak{M}$ may be joined by a curve consisting of a finite number of arcs of lines of ellipticity, and if there exists no set properly containing $\mathfrak{M}$ and possessing this same property. If the entire domain Ω is a set of elliptic connectivity, then equation (1) is called elliptically connected in Ω. It is not difficult to find examples of equations which are elliptically connected but not elliptic in Ω.

For equation (1) the strong maximum principle has the following form (see [1–5] and Chapter III, §1): suppose $L(u) \geqslant 0$ in the domain Ω and the coefficient $c(x)$ and $M = \sup_\Omega u$ satisfy $Mc \leqslant 0$ in Ω. If $u(x^0) = M$ for some $x^0 \in \Omega$, then, on the set of elliptic connectivity containing the point x^0, either $u \equiv 0$, or $u \equiv M$ and $c \equiv 0$. For equation (1), A. D. Aleksandrov [1–5] also proved a theorem analogous to the strong maximum principle for parabolic equations. For equations of the form

$$Pu = -\sum_{j=1}^{r} X_j^2 u + iX_0 u + cu = f$$

theorems on the strong maximum principle were proved by Bony [15–17]. Suppose $\overline{X}_j(x)$ denotes the vector $(a_j^1(x), \cdots, a_j^m(x))$. Suppose F denotes the set of points of Ω at which $u(x) = M = \sup_\Omega u$. We shall assume that $c \leqslant 0$ and $M \geqslant 0$.

It was proved [15] that if $P(u) \geqslant 0$, $u \in C^{(2)}(\Omega)$ and a trajectory $x(t)$ of the vector field $\overline{X}_j(x)$ $(j = 1, \cdots, r)$ contains a point $x(t_0)$ of the set F, then the entire trajectory $x(t)$ belongs to F. In the case when the system of operators $\{X_1, \cdots, X_r\}$ has rank m in Ω, the strong maximum principle for such an equation has the same form as for the Laplace equation: if $P(u) \geqslant 0, c \equiv 0$ in $\Omega, u \in C^{(2)}(\Omega)$ and $u(x)$ assumes a greatest positive value M at a point x^0 belonging to Ω, then $u \equiv M$ in Ω.

In the case when the rank of the system of operators $\{X_1, \cdots, X_r\}$ is equal to $\mu < m$ at each point of Ω, the strong maximum principle for equation (9) has the same form as for the heat equation. Namely, suppose that $P(u) \geqslant 0$ in Ω, $u \in C^{(2)}(\Omega)$, and that $u(x)$ takes on its largest positive value M at a point x^0 belonging to Ω. Then $u \equiv M$ at each point of Ω with the following properties: (1) It may be joined to the point x^0 by some curve consisting of a finite number of arcs of trajectories of the vector fields $\bar{X}_j(x)$, $j = 0, \cdots, r$. (2) When this curve is followed away from the point x^0, any portion of a trajectory of the field $\bar{X}_0(x)$ lying on the curve must be followed in the direction of the vector $\bar{X}_0(x)$.

In the papers of Bony [15–17] uniqueness theorems for the Cauchy problem for equations of the form (9) with analytic coefficients are proved, as well as Harnack theorems. A. D. Aleksandrov [1–5] studied the level set $u = M$ of a function $u(x)$ satisfying in the domain Ω the relation $L(u) \leqslant 0$, under the condition that $u \geqslant M$ at all points of the domain considered.

A large number of papers have been devoted to the study of degenerating parabolic equations of the form

$$u_t = a^{kj}(x, t)\, u_{x_k x_j} + b^k(x, t)\, u_{x_k} + c(x, t)\, u + f(x, t), \tag{11}$$

where $a^{kj}(x, t)\xi_k\xi_j \geqslant 0$ at all points of the domain considered. Clearly (11) is a particular case of (1). The Cauchy problem for (1) has been studied by various methods in a series of papers (see, for example, [92, 124] and others). We note that the Cauchy problem for (11) with initial condition at $t = 0$ may also be studied by the methods of Chapters I and II (see [92]).

In connection with the study of the Tricomi problem for equations of mixed type, interest has arisen in studying hyperbolic equations degenerating on the boundary. In [88, 89] the Cauchy problem was studied for second order equations such that at each point of the domain considered, the characteristic form has one negative characteristic value, the remaining being positive or zero. The Cauchy problem for such equations can be investigated by methods similar to those used in the study of second order equations with nonnegative characteristic form. In [88, 89] and in Chapter III, §2, the solution of the Cauchy problem for the equation

$$u_{tt} = L(u) + f \tag{12}$$

with initial conditions

$$u|_{t=0} = \varphi(x), \quad u_t|_{t=0} = \psi(x), \tag{13}$$

where $L(u) \equiv a^{kj}u_{x_k x_j} + b^k u_{x_k} + cu$ is a second order operator with nonnegative characteristic form and with coefficients depending on x and t, is obtained by the method of hyperbolic regularization. Inequality (8) plays a significant role in establishing a priori estimates for solutions of this problem. In particular it has been shown (see Chapter III, §2) that the Cauchy problem (12), (13) has a unique solution in the class W_2^s, and that an energy estimate holds, if the coefficients in (12) and the functions f, ϕ and ψ are sufficiently smooth and either

$$\alpha t\,(b^k\xi_k)^2 \leqslant Aa^{kj}\xi_k\xi_j + a_t^{kj}\xi_k\xi_j \tag{14}$$

or

$$\alpha\,(T-t)\,(b^k\xi_k)^2 \leqslant Aa^{kj}\xi_k\xi_j - a_t^{kj}\xi_k\xi_j + \frac{a^{kj}\xi_k\xi_j}{\alpha\,(T-t)} \tag{15}$$

holds in the domain considered $\{0 \leqslant t \leqslant T\}$, for any ξ. Here α and A are positive constants with α depending on the degree of smoothness of the coefficients of (12), of the function f, and of the initial functions (13).

Condition (14) contains as particular cases many criteria for correctness of the Cauchy problem known earlier only for hyperbolic equations of second order with two independent variables degenerating on the line carrying the initial data (see [10, 12, 106], and others). A survey of some of these results is found in the book [123].

Some necessary conditions for correctness of the Cauchy problem (12), (13) are given in [62, 149] (see also Chapter III, §3).

Many problems in hydrodynamics (boundary layer theory), the theory of filtration, physical problems connected with the study of Brownian motion, problems from probability theory (Markov processes), and problems from other areas lead to second order equations with nonnegative characteristic form. The study of quasilinear and nonlinear second order equations with nonnegative characteristic form (see [26, 36, 90, 96, 140]) is significant for applications in gas dynamics and boundary layer theory, and in other branches of mechanics.

Problem (2), (3) has also been studied by methods of probability theory based on the stochastic equations of K. Itô. In this situation a generalized solution of (2), (3) is considered in the class of bounded measurable functions and is defined in terms of the theory of Markov processes (see [32–36], and others).

There are many interesting unsolved problems connected with second order equations with nonnegative characteristic form, as well as with the analogous equations of higher order. (We note that the boundary value problem (2), (3) has not been studied completely, even for the simplest equation of heat conduction, $u_t = u_{xx}$. This question is discussed in detail in [64].) Among the unsolved problems we indicate the question of the spectrum of the problem (2), (3). For elliptic equations degenerating at the boundary, questions having to do with the character of the spectrum of the first boundary value problem were considered in [80, 128, 129].

There is interest in the further study of questions concerning smoothness conditions and nonsmoothness conditions for generalized solutions of (2), (3). It would be interesting to study the nature of the singularities of nonsmooth generalized solutions of (2), (3) and to clarify the conditions under which they arise.

In connection with problems of geometry in the large, there is interest in describing the class of equations of the form (1) with analytic coefficients and analytic function f, for which all sufficiently smooth solutions are analytic (some results on this question may be found in [148]).

The problem of describing all correct boundary value problems for (1) is unsolved. For elliptic differential and pseudodifferential equations degenerating only on the boundary, this problem was studied in [44, 134–136]. A parametrix was recently constructed, i.e. the principal part of the Green's function was found, for degenerating quasielliptic equations (not only of second order) (see, for example, [134–136]). In these papers conditions for normal solvability of boundary value problems for the indicated equations were found, analogous to the known conditions for normal solvability of boundary value problems for elliptic operators.

It would be interesting to delineate the class of equations of the form (12) for which the Cauchy problem is not correct. This question is part of the larger problem of studying hyperbolic equations with multiple characteristics. In recent years significant progress has been made on this problem by means of the theory of pseudodifferential operators and asymptotic solutions, (see [82, [127], [62], [116], [149], Chapter III, §3, and others).

We shall be pleased if the appearance of this book draws attention to the problems mentioned above.

CHAPTER I

THE FIRST BOUNDARY VALUE PROBLEM

§1. Notation. Auxiliary results. Formulation of the first boundary value problem

We shall denote by Ω a bounded domain in Euclidean space R^m, and by $x = (x_1, \cdots, x_m)$ a point in this space. All functions considered in Chapter I are assumed real unless otherwise specified. As usual, the symbol $A \subset B$ means that the set A is contained in B, and $\bar{A}$ denotes the closure of the set A.

We shall say that the function $u(x)$ belongs to the class $C^{(k)}(G)$ if it has continuous derivatives up to order k inclusive at all points of the set $G \subset R^m$. We shall denote by $C_{(k)}(G)$ the class of functions whose weak derivatives up to order k inclusive are bounded in G. The class $C^{(0)}$ consists of functions continuous in G, and $C_{(0)}$ contains all measurable functions bounded in G.

We shall denote by $C^{(k,\beta)}(G)$ the class of functions with derivatives in G up to order k inclusive satisfying a Hölder condition with exponent $\beta, 0 < \beta < 1$.

We shall say that the domain Ω with boundary Σ belongs to class $A^{(k)}, A_{(k)}$, or $A^{(k,\beta)}$ $(k \geqslant 1)$ if in some neighborhood Q of each of its points P, the boundary Σ is representable in the form

$$x_l = f_l(x_1, \ldots, x_{l-1}, x_{l+1}, \ldots, x_m)$$

for some l, where the functions f_l belong to the corresponding class $C^{(k)}(Q_l)$, $C_{(k)}(Q_l)$ or $C^{(k,\beta)}(Q_l)$ and Q_l is the projection of $Q \cap \Sigma$ on the plane $x_l = 0$.

We shall also have to do with domains Ω with piecewise smooth boundaries Σ. Classes of such domains, denoted by $B^{(k)}, B_{(k)}$ and $B^{(k,\beta)}$, are defined by induction on the dimension of the domain. A one-dimensional domain $B^{(k)}, B_{(k)}$ or $B^{(k,\beta)}$ is an interval. Furthermore we say that the domain Ω with boundary Σ belongs to class $B^{(k)}, B_{(k)}$ or $B^{(k,\beta)}$ if Σ may be separated into a finite number of pieces S^j, each homeomorphic to an $(m-1)$-ball, which intersect, if at all, only at boundary points, and such that S^j may be represented in the form

$$x_l = f_l(x_1, \ldots, x_{l-1}, x_{l+1}, \ldots, x_m)$$

for some l, where the function f_l is given in some $(m-1)$-dimensional closed region of class $B^{(k)}$, $B_{(k)}$ or $B^{(k,\beta)}$ respectively on the hyperplane $x_l = 0$, and in this region belongs to the class $C^{(k)}$, $C_{(k)}$ or $C^{(k,\beta)}$ respectively.

We denote by $\bar{n} = (n_1, \cdots, n_m)$ the interior normal vector at boundary points of Ω; $\varnothing$ denotes the empty set; $|x| = (\Sigma_1^m x_j^2)^{1/2}$.

Suppose $\alpha = (\alpha_1, \cdots, \alpha_m)$ is a multi-index, where the α_j are nonnegative integers, and $|\alpha| = \Sigma_1^m \alpha_j$.

In Chapter I we shall sometimes use the following notation for derivatives:

$$D^\alpha u = \left(\frac{\partial}{\partial x_1}\right)^{\alpha_1} \cdots \left(\frac{\partial}{\partial x_m}\right)^{\alpha_m} u.$$

Usually (see, for example, [56]), one uses the notation $D^\alpha u = D_1^{\alpha_1} \cdots D_m^{\alpha_m} u$, where $D_j = -i\partial/\partial x_j$, $i = \sqrt{-1}$. It will be convenient for us to use this latter notation in Chapter II. We set

$$\|u\|_{C_{(k)}(G)} = \sum_{|\alpha| \leqslant k} \sup_G |D^\alpha u|.$$

As a rule, constants will be denoted by C_j, and the enumeration of them by means of the index j is retained only within the confines of the proof of a given theorem.

We denote by $C_0^\infty(\Omega)$ or $C_0^\infty(R^m)$ the class of infinitely differentiable functions which may differ from zero only on a compact set belonging to the domain Ω or to R^m, respectively.

A equation of second order of the form

$$L(u) \equiv a^{kj}(x) u_{x_k x_j} + b^k(x) u_{x_k} + c(x) u = f(x) \tag{1.1.1}$$

with the condition

$$a^{kj}(x)\xi_k\xi_j \geqslant 0 \tag{1.1.2}$$

for any real vector $\xi = (\xi_1, \cdots, \xi_m)$ and any point $x \in \Omega$ will be termed a second order equation with nonnegative characteristic form in Ω.

Here and throughout it is assumed that repeated indices are summed from 1 to m.

It is clear that the class of equations with nonnegative characteristic form includes equations of elliptic and parabolic types, first order equations (the case $a^{kj}\xi_k\xi_j \equiv 0$), ultraparabolic equations, the equations of Brownian motion, Tricomi's equation in the upper halfplane, the equation of M. V. Keldyš, and others.

We consider the first boundary value problem for equation (1.1.1) in a domain Ω; this problem was first set up in its general form by Fichera in [29]. We assume that condition (1.1.2) is fulfilled for all points x in $\Omega \cup \Sigma$ and all $\xi \in R^m$, that Ω belongs to class $A_{(1)}$, and $a^{kj} \in C_{(2)}(\Omega)$, $b^k \in C_{(1)}(\Omega)$, $c \in C_{(0)}(\Omega)$. We denote by Σ^0 the set of points of Σ where $a^{kj}(x)n_k n_j = 0$. At the points of Σ^0 we consider the function

$$b(x) \equiv (b^k - a^{kj}_{x_j})n_k, \tag{1.1.3}$$

which we shall call the Fichera function for the equation (1.1.1). We denote by Σ_1 the set of points of Σ^0 where $b > 0$, by Σ_2 the set of points of Σ^0 where $b < 0$, and by Σ_0 those points of the set Σ^0 where $b = 0$. The set $\Sigma \backslash \Sigma^0$ will be denoted by Σ_3.

If the boundary Σ of Ω belongs to class $B^{(k)}$, $B_{(k)}$, or $B^{(k,\beta)}$, then Σ_0, Σ_1, Σ_2, Σ_3 are defined analogously, for this purpose considering only interior points of S^j.

The first boundary value problem for equation (1.1.1) consists in the following: to find a function u in $\Omega \cup \Sigma$ such that

$$L(u) = f \text{ in } \Omega, \tag{1.1.4}$$

$$u = g \text{ on } \Sigma_2 \cup \Sigma_3, \tag{1.1.5}$$

where f is a given function in Ω and g is a given function on $\Sigma_2 \cup \Sigma_3$. Clearly if (1.1.1) is elliptic, then problem (1.1.4), (1.1.5) is the Dirichlet problem. For parabolic equations in a cylindrical domain, (1.1.4), (1.1.5) constitutes the mixed problem, or, as it is sometimes called, the first boundary value problem for a parabolic equation.

As another example, for the first order equation $-u_{x_1} = f(x)$ the boundary Σ of the rectangle $\Omega = \{0 < x_1 < X_1, 0 < x_2 < 1\}$ contains the set Σ_0 lying on the lines $x_2 = 0, x_2 = 1$, the set Σ_2 lying on the line $x_1 = 0$, and Σ_1 lying on the line $x_1 = X_1$. In this case the problem (1.1.4), (1.1.5) coincides with the Cauchy problem.

For the equation of M. V. Keldyš

$$y^\gamma \frac{\partial^2 u}{\partial y^2} + \frac{\partial^2 u}{\partial x^2} + a(x, y)\frac{\partial u}{\partial x} + b(x, y)\frac{\partial u}{\partial y} + c(x, y)\, u = 0, \tag{1.1.6}$$
$$\gamma = \text{const} > 0,$$

considered in a domain Ω of the (x, y) plane bounded by a segment of the x axis and a curve Γ lying in the upper halfplane $y \geqslant 0$, the problem (1.1.4), (1.1.5) with $\gamma \geqslant 1$ coincides with the problem E or D studied in [63].

We introduce the notation

$$L^*(v) \equiv (a^{kj}v)_{x_k x_j} - (b^k v)_{x_k} + cv$$
$$= a^{kj}v_{x_k x_j} + b^{*k}v_{x_k} + c^*v, \tag{1.17}$$

where

$$b^{*k} = 2a^{kj}_{x_j} - b^k, \quad c^* = a^{kj}_{x_k x_j} - b^k_{x_k} + c.$$

It is easy to see that the Fichera function b^* for the operator $L^*(v)$ is equal to $-b$, where b is the Fichera function (1.1.3) for the operator $L(u)$.

We shall show that the decomposition of the boundary Σ of the region Ω for the operator $L(u)$ into subsets $\Sigma_0, \Sigma_1, \Sigma_2, \Sigma_3$ is invariant relative to smooth nondegenerate changes of independent variables, and that therefore the problem (1.1.4), (1.1.5) is invariant with respect to such coordinate transformations.

LEMMA 1.1.1. *The sign of the function*

$$b \equiv (b^k - a^{kj}_{x_j})n_k$$

at points Σ^0 *of the boundary* Σ *of the region* Ω *does not change under smooth nondegenerate changes of independent variables in* (1.1.1).

PROOF. In equation (1.1.1) we effect a change of variables

$$y_l = F^l(x_1, \cdots, x_m), \quad l = 1, \cdots, m. \tag{1.1.8}$$

Suppose, in a neighborhood of a point of the boundary Σ under consideration, that Σ is given by the equation

$$F(x_1, \cdots, x_m) \equiv \Phi(y_1, \cdots, y_m) = 0, \tag{1.1.9}$$

and suppose that $\operatorname{grad} F$ and $\operatorname{grad} \Phi$ have the same direction as the interior normal. In the new variables equation (1.1.1) has the form

$$a^{kj}F^l_{x_k}F^s_{x_j}u_{y_l y_s} + b^k F^l_{x_k}u_{y_l} + a^{kj}F^l_{x_k x_j}u_{y_l} + cu = f. \tag{1.1.10}$$

The function b for (1.1.1) may be written in the form

$$b \equiv \left(b^k - a^{kj}_{x_j}\right) n_k = \left(b^k - a^{kj}_{y_l}F^l_{x_j}\right) \Phi_{y_s}F^s_{x_k}\left(F_{x_\rho}F_{x_\rho}\right)^{-1/2}.$$

We shall now calculate the Fichera function $\widetilde{b}$ for (1.1.10). We have

$$\begin{aligned}\tilde{b} &= [b^k F^l_{x_k} + a^{kj} F^l_{x_k x_j} - (a^{kj} F^l_{x_k} F^s_{x_j})_{y_s}] \Phi_{y_l} (\Phi_{y_\rho}\Phi_{y_\rho})^{-1/2} \\ &= [b^k F^l_{x_k} - a^{kj}_{y_s} F^l_{x_k} F^s_{x_j}] \Phi_{y_l} (\Phi_{y_\rho}\Phi_{y_\rho})^{-1/2} \\ &\quad + [a^{kj} F^l_{x_k y_s} F^s_{x_j} - a^{kj} F^l_{x_k y_s} F^s_{x_j} - a^{kj} F^l_{x_k} F^s_{x_j y_s}] \Phi_{y_l} (\Phi_{y_\rho}\Phi_{y_\rho})^{-\frac{1}{2}} \\ &= b (F_{x_k} F_{x_k})^{\frac{1}{2}} (\Phi_{y_\rho}\Phi_{y_\rho})^{-\frac{1}{2}} - a^{kj} F^l_{x_k} F^s_{x_j y_s} \Phi_{y_l} (\Phi_{y_\rho}\Phi_{y_\rho})^{-\frac{1}{2}}. \end{aligned} \tag{1.1.11}$$

The last term in (1.1.11) is equal to zero, since

$$a^{kj} F^l_{x_k} F^s_{x_j y_s} \Phi_{y_l} \equiv a^{kj} F_{x_k} F^s_{x_j y_s} \quad \text{and} \quad a^{kj} F_{x_k} = 0$$

on Σ^0. Consequently

$$\tilde{b} = b (F_{x_k} F_{x_k})^{\frac{1}{2}} (\Phi_{y_\rho}\Phi_{y_\rho})^{-\frac{1}{2}}$$

and the lemma is proved.

THEOREM 1.1.1. *The subsets* $\Sigma_0, \Sigma_1, \Sigma_2, \Sigma_3$ *of the boundary* Σ, *defined for the operator* $L(u)$, *remain invariant under smooth nonsingular changes of independent variables in* (1.1.1).

PROOF. The invariance of the set Σ_3 under a change of independent variables of the form (1.1.8) follows from the equation

$$\begin{aligned} a^{kj} n_k n_j &= a^{kj} \Phi_{y_l} F^l_{x_k} \Phi_{y_s} F^s_{x_j} (F_{x_\rho} F_{x_\rho})^{-1} \\ &= a^{kj} F^l_{x_k} F^s_{x_j} n'_l n'_s (F_{x_\rho} F_{x_\rho})^{-1} (\Phi_{y_\nu} \Phi_{y_\nu}), \end{aligned}$$

if in a neighborhood of a point under consideration the boundary Σ is given by (1.1.9), and $\vec{n}' = (n'_1, \cdots, n'_m)$ denotes the internal normal vector to Σ in the space $(y_1, \cdots, y_m)$. The invariance of the sets $\Sigma_0, \Sigma_1, \Sigma_2$ follows from the invariance of Σ_3 and from Lemma 1.1.1. The theorem is proved.

We now obtain a Green's formula for the operator $L(u)$. Suppose u and v are functions in the class $C^{(2)}(\Omega \cup \Sigma)$, and that the domain Ω belongs to class $B^{(1)}$. The operator $L(u)$ may be written in the form

$$L(u) \equiv (a^{kj} u_{x_k})_{x_j} + (b^k - a^{kj}_{x_j}) u_{x_k} + cu. \tag{1.1.12}$$

Setting $b^k - a^{kj}_{x_j} = l^k$, we have

$$L^*(v) \equiv (a^{kj}v_{x_k})_{x_j} - (l^k v)_{x_k} + cv,$$
$$L(u)v - L^*(v)u \equiv (a^{kj}vu_{x_k} - a^{kj}uv_{x_k})_{x_j} + (l^k uv)_{x_k}.$$

Integrating the latter identity over the domain Ω and applying Ostrogradskiĭ's formula, we obtain

$$\iint_\Omega (L(u)v - L^*(v)u)\,dx = -\int_\Sigma [(a^{kj}u_{x_k}v - a^{kj}v_{x_k}u)\,n_j + l^k uvn_k]\,d\sigma, \qquad (1.1.13)$$

where $\bar{n}$ is the interior normal vector to Σ and $d\sigma$ is the surface element of Σ.

It is clear that on Σ^0 the equation $a^{kj}n_j = 0$ is satisfied for $k = 1, \cdots, m$, and on Σ_3 the length of the vector $\bar{\nu} = (\nu_1, \cdots, \nu_m)$, where $\nu_k = a^{kj}n_j$, is different from zero. By definition $l^k n_k = b$. Therefore from (1.1.13) it follows that

$$\int_\Omega (L(u)v - L^*(v)u)\,dx = -\int_{\Sigma_3}\left\{v\frac{\partial u}{\partial \bar{\nu}} - u\frac{\partial v}{\partial \bar{\nu}}\right\}d\sigma - \int_\Sigma buv\,d\sigma, \qquad (1.1.14)$$

where $\partial/\partial\bar{\nu} \equiv a^{kj}n_j\partial/\partial x_k$. Formula (1.1.14) is called the Green's formula for (1.1.1).

Suppose, in a neighborhood of some point P of the boundary Σ, the equation for Σ has the form $F(x_1, \cdots, x_m) = 0$, with $\operatorname{grad} F \neq 0$, and $F > 0$ in Ω. On Σ we consider the function $\beta \equiv L(F)$. It is easy to see that β is invariant relative to changes of independent variables in (1.1.1). We denote by Σ_0', Σ_1', Σ_2' the set of points of Σ^0 where $\beta = 0, \beta > 0, \beta < 0$ respectively. Let $\sigma^* = \Sigma_0' \cup \Sigma_1'$.

LEMMA 1.1.2. *Suppose $u \in C^{(2)}(\Omega \cup \sigma^*)$ and $u \in C^{(0)}(\Omega \cup \Sigma)$, and that the domain Ω belongs to class $A_{(2)}$. Then if either $L(u) \leqslant 0$ and $c < 0$ in $\Omega \cup \sigma^*$, or $L(u) < 0$ and $c \leqslant 0$ in $\Omega \cup \sigma^*$, it follows that $u(x)$ may assume a least negative value only on $\Sigma_2' \cup \Sigma_3$. If the domain Ω belongs to class $B_{(2)}$, then in place of $\Sigma_2' \cup \Sigma_3$, one should consider $\Sigma\backslash\sigma^*$.*

PROOF. Suppose the function $u(x)$ assumes a least negative value at an interior point P of the domain Ω.

We perform a change of independent variables $x = Ay$, where A is a constant matrix and $\det \|A\| \neq 0$. We choose A so that the operator $L(u)$ assumes at the point P in the new coordinate system a canonical form

$$L(u) = \alpha^{kj}u_{y_k y_j} + \beta^k u_{y_k} + cu. \qquad (1.1.15)$$

This means that at P the coefficients $\alpha^{kj}=0$ for $k\neq j$, and that α^{kk} are equal to zero or one ($k=1,\cdots,m$). Since by assumption the function u has a local minimum at P, it follows that at that point P we have $u_{y_k}=0$ and $u_{y_k y_k}\geqslant 0$. Therefore $L(u)>0$ at the point P if $c<0$, and $L(u)\geqslant 0$ if $c\leqslant 0$, which contradicts the hypothesis of the lemma.

We show now that a least negative value of u may not be taken at a point of σ^* either. Suppose the contrary, and that u assumes a least negative value at a point P belonging to σ^*. In a neighborhood of P we introduce new coordinates $y_1,\cdots$, y_m so that the boundary Σ lies on the plane $y_m=0$ and the interior normal to Σ at P coincides with the direction of the y_m axis. In the new variables the operator $L(u)$ takes the form (1.1.15).

Clearly, at the points of σ^* in some neighborhood of P we have $\alpha^{mj}=0$ for $j=1,\cdots,m$, and $\beta^m=\beta\geqslant 0$.

By performing, if necessary, another coordinate transformation which leaves the plane $y_m=\text{const}$ invariant, we may suppose that the operator (1.1.15) has canonical form at the point P.

Since the function u assumes a negative minimum value at the point P, it follows that at that point

$$u_{y_k}=0,\ u_{y_k y_k}\geqslant 0\ (k=1,\ \ldots,\ m-1);\ u_{y_m}\geqslant 0.$$

Hence at P the inequality $L(u)>0$ holds if $c<0$, whereas $L(u)\geqslant 0$ if $c\leqslant 0$. But this contradicts the hypothesis of the lemma. Therefore a least negative value of the function u may be assumed only at a point of $\Sigma\backslash\sigma^*$.

THEOREM 1.1.2. (MAXIMUM PRINCIPLE.) *Suppose that* $u\in C^{(2)}(\Omega\cup\sigma^*)$, *that* $u\in C^{(0)}(\Omega\cup\Sigma)$, *and that the domain* Ω *belongs to class* $A_{(2)}$. *Assume that* $c<0$ *on* $\Omega\cup\Sigma$ *and that* $L(u)=f$ *in* $\Omega\cup\sigma^*$. *Then*

$$|u|\leqslant\max\left\{\sup_{\Omega}\left|\frac{f}{c}\right|,\ \max_{\Sigma_2'\cup\Sigma_3}|u|\right\}=M. \tag{1.1.16}$$

If $c<0$ *on* $\Omega\cup\Sigma$ *and* $L(u)=0$ *in* Ω, *then at points of* Ω

$$|u|\leqslant\max_{\Sigma_2'\cup\Sigma_3}|u|. \tag{1.1.17}$$

If the domain is in class $B_{(2)}$, *then in place of the set* $\Sigma_2'\cup\Sigma_3$ *one should consider* $\Sigma\backslash\sigma^*$. *(In case* $\Sigma_2'\cup\Sigma_3=\varnothing$, *set* $\max_{\Sigma_2'\cup\Sigma_3}|u|=0$.)

PROOF. Consider the function

$$v_{\pm} = M \pm u.$$

Then in $\Omega \cup \sigma^*$ we have

$$L(v_{\pm}) = cM \pm f \leqslant 0,$$

since $M \geqslant \sup |f/c|$. Therefore according to Lemma 1.1.2 the function $v_{\pm}$ may assume a least negative value only at a point of $\Sigma \backslash \sigma^*$. But on $\Sigma \backslash \sigma^*$ the function $v_{\pm} = M \pm u \geqslant 0$. Consequently $v_{\pm} \geqslant 0$ in $\Omega \cup \Sigma$ and $|u| \leqslant M$, i.e. (1.1.16) holds.

Inequality (1.1.17) is a consequence of (1.1.16). The theorem is proved.

A theorem close to Theorem 1.1.2 was obtained by a different method in [30] (see also Chapter I, §2). It is easy to prove that interior points of $\Sigma_0 \cup \Sigma_1$ are contained in the set $\Sigma_0' \cup \Sigma_1'$.

In Chapter I, §5, we shall prove a theorem in some sense more general than Theorem 1.1.2; namely, we prove a maximum principle for generalized solutions of (1.1.1).

We note that in the proof of Theorem 1.1.2, only that portion of Σ which lies in $\Sigma_0' \cup \Sigma_1'$ was required to be smooth. Theorem 1.1.2 contains the known maximum principles for elliptic and parabolic equations.

We also remark that the assumption $c < 0$ in Theorem 1.1.2 is essential and may not be replaced by the condition $c \leqslant 0$. In fact, for the equation

$$L(u) \equiv x_1 \frac{\partial u}{\partial x_2} - x_2 \frac{\partial u}{\partial x_1} = 0$$

in the annulus Ω of the form $\frac{1}{2} < x_1^2 + x_2^2 < 1$ the entire boundary belongs to Σ_0; however, the equation is satisfied by any function $u = \text{const} \neq 0$, and (1.1.17) does not then hold.

§2. A priori estimates in the spaces $\mathfrak{L}_p(\Omega)$

Here we obtain a priori estimates for smooth solutions of the problem

$$L(u) \equiv a^{kj} u_{x_k x_j} + b^k u_{x_k} + cu = f \text{ in } \Omega,$$
$$u = 0 \text{ on } \Sigma_2 \cup \Sigma_3, \tag{1.2.1}$$

which will be used later in proving existence theorems for this problem. We assume for the operator $L(u)$ and the domain Ω that Green's formula (1.1.14) is valid. We introduce the notation

$$\|u\|_{\mathcal{L}_p(\Omega)} = \left(\int_\Omega |u|^p dx\right)^{\frac{1}{p}}, \quad \|u\|_{\mathcal{L}_p(\Sigma_j)} = \left(\int_{\Sigma_j} |u|^p d\sigma\right)^{\frac{1}{p}}.$$

The results given in this and the next two sections were obtained by Fichera [30].

LEMMA 1.2.1. *Suppose* $u \in C^{(2)}(\Omega \cup \Sigma)$, $u = 0$ *on* $\Sigma_2 \cup \Sigma_3$, *and* p *is any number such that* $1 \leqslant p < \infty$. *Suppose there exists a function* $w \in C^{(2)}(\Omega \cup \Sigma)$ *satisfying the condition*

$$w \leqslant 0,\ L^*(w) + c(p-1)w > 0 \ \text{ in } \ \Omega \cup \Sigma. \tag{1.2.2}$$

Then

$$\|u\|_{\mathcal{L}_p(\Omega)} \leqslant \frac{\max\limits_{\Omega \cup \Sigma} p\,|w|}{\min\limits_{\Omega \cup \Sigma}[L^*(w) + (p-1)cw]} \|Lu\|_{\mathcal{L}_p(\Omega)}. \tag{1.2.3}$$

PROOF. We apply the Green's formula (1.1.14) to the functions w and $(u^2 + \delta)^{p/2}$, where $\delta > 0$ is arbitrary. We have

$$\begin{aligned}\int_\Omega L^*(w)(u^2+\delta)^{p/2}dx &- \int_\Omega wL((u^2+\delta)^{p/2})\,dx \\ &= \int_\Sigma bw(u^2+\delta)^{p/2}d\sigma - \int_{\Sigma_3}(u^2+\delta)^{p/2}\frac{\partial w}{\partial \bar{\nu}}\,d\sigma,\end{aligned} \tag{1.2.4}$$

bearing in mind that

$$\frac{\partial (u^2+\delta)^{p/2}}{\partial \bar{\nu}} = p(u^2+\delta)^{\frac{p}{2}-1} u a^{kj} u_{x_k} n_j = 0$$

on Σ_3. It is easy to see that

$$\begin{aligned}L((u^2+\delta)^{p/2}) = p(u^2+\delta)^{\frac{p}{2}-1} uL(u) &+ c(u^2+\delta)^{\frac{p}{2}-1}[(1-p)u^2+\delta] \\ &+ a^{kj}u_{x_k}u_{x_j}p(u^2+\delta)^{\frac{p}{2}-2}[(p-1)u^2+\delta].\end{aligned}$$

Since the inequality $b > 0$ holds on Σ_1 and, moreover, $a^{kj}u_{x_k}u_{x_j} \geqslant 0, p \geqslant 1$, $w \leqslant 0$, it follows that

$$\begin{aligned}\int_\Omega \{L^*(w)(u^2+\delta)^{p/2} &- w(u^2+\delta)^{\frac{p}{2}-1}c[(1-p)u^2+\delta]\}\,dx \\ &\leqslant \int_\Omega wp(u^2+\delta)^{\frac{p}{2}-1}uL(u)\,dx + \delta^{\frac{p}{2}}\int_{\Sigma_2 \cup \Sigma_3} bw\,d\sigma - \delta^{\frac{p}{2}}\int_{\Sigma_3}\frac{\partial w}{\partial \bar{\nu}}\,d\sigma.\end{aligned} \tag{1.2.5}$$

In (1.2.5) we now let δ approach zero. If $p > 1$, then $(u^2 + \delta)^{p/2-1} u \to |u|^{p-2}u$ as $\delta \to 0$. If $p = 1$, then $|(u^2 + \delta)^{-1/2} u| \leqslant 1$. For $p \geqslant 1$

$$\lim_{\delta\to 0} (u^2 + \delta)^{p/2} \{L^*(w) - cw(u^2 + \delta)^{-1} [(1-p) u^2 + \delta]\}$$
$$= |u|^p \{L^*(w) - (1-p) cw\}.$$

Therefore from (1.2.5) we obtain that

$$\int_\Omega |u|^p \, dx \leqslant \frac{p \max\limits_{\Omega\cup\Sigma} |w|}{\min\limits_{\Omega\cup\Sigma} [L^*(w) + (p-1) cw]} \int_\Omega |u|^{p-1} |L(u)| \, dx. \tag{1.2.6}$$

Applying Hölder's inequality to the integral on the right of (1.2.6), we obtain (1.2.3).

THEOREM 1.2.1. *If* $c < 0$ *in* $\Omega \cup \Sigma$, *then for all sufficiently large* p *such that* $-c^* + (1-p)c > 0$ *in* $\Omega \cup \Sigma$, *and for all functions* $u \in C^{(2)}(\Omega \cup \Sigma)$ *satisfying the condition* $u = 0$ *on* $\Sigma_2 \cup \Sigma_3$, *the estimate*

$$\|u\|_{\mathscr{L}_p(\Omega)} \leqslant \frac{p}{\min\limits_{\Omega\cup\Sigma} [-c^* + (1-p)c]} \|L(u)\|_{\mathscr{L}_p(\Omega)} \tag{1.2.7}$$

holds. If $c^* < 0$ *in* $\Omega \cup \Sigma$, *then estimate* (1.2.7) *holds for* p *sufficiently near to* 1 *such that* $-c^* + (1-p)c > 0$ *in* $\Omega \cup \Sigma$. *If* $c^* < 0$ *and* $c < 0$, *then* (1.2.7) *is satisfied for all* $p \geqslant 1$ *and for all* $u \in C^{(2)}(\Omega \cup \Sigma)$ *such that* $u = 0$ *on* $\Sigma_2 \cup \Sigma_3$.

The proof of Theorem 1.2.1 follows from Lemma 1.2.1, since in the cases considered we may set $w = -1$.

REMARK. If $c < 0, u \in C^{(2)}(\Omega \cup \Sigma)$, and $u = 0$ on $\Sigma_2 \cup \Sigma_3$, then in (1.2.7) one may pass to the limit as $p \to \infty$. In fact, suppose the number p_0 is such that $-c^* + (1-p_0)c > 0$. Then for $p > p_0$ we have $-c^* + (1-p)c > c(p_0 - p)$ and estimate (1.2.7) may be written in the form

$$\|u\|_{\mathscr{L}_p(\Omega)} \leqslant \frac{p}{(p-p_0) \min\limits_{\Omega\cup\Sigma} |c|} \|L(u)\|_{\mathscr{L}_p(\Omega)}. \tag{1.2.8}$$

Passing to the limit as $p \to \infty$ in (1.2.8), we obtain the following form for the maximum principle: for any function $u \in C^{(2)}(\Omega \cup \Sigma)$ with the condition $u = 0$ on $\Sigma_2 \cup \Sigma_3$, the inequality

$$\max_{\Omega\cup\Sigma} |u| \leqslant \frac{\max\limits_{\Omega\cup\Sigma} |L(u)|}{\min\limits_{\Omega\cup\Sigma} |c|} \tag{1.2.9}$$

holds.

THEOREM 1.2.2. *Suppose that for some* $p \geqslant 1$ *there exists a function* $w \in C^{(2)}(\Omega \cup \Sigma)$ *satisfying the following condition*:

$$w \leqslant 0, \quad L^*(w) + (p-1)cw > 0 \quad on \ \Omega \cup \Sigma, \quad w = 0 \quad on \ \Sigma_3.$$

Then if $u \in C^{(2)}(\Omega \cup \Sigma)$ *is any function such that* $L(u) = 0$, *the inequality*

$$\|u\|_{\mathscr{L}_p(\Omega)} \leqslant K_p \|u\|_{\mathscr{L}_p(\Sigma_2)} + K_p' \|u\|_{\mathscr{L}_p(\Sigma_1)} \tag{1.2.10}$$

holds, where

$$K_p = \left\{ \frac{\sup\limits_{\Sigma_2} |bw|}{\min\limits_{\Omega \cup \Sigma} [L^*(w) + (p-1)cw]} \right\}^{1/p}; \quad K_p' = \left\{ \frac{\sup\limits_{\Sigma_1} |a^{kj} w_{x_k} n_j|}{\min\limits_{\Omega \cup \Sigma} [L^*(w) + (p-1)cw]} \right\}^{1/p}.$$

PROOF. As in the proof of Lemma 1.2.1, we apply (1.1.14) to the functions w and $(u^2 + \delta)^{p/2}$. From (1.2.4) and the properties of the functions u and w, we deduce that

$$\int_\Omega \left\{ (u^2+\delta)^{p/2} L^*(w) - cw(u^2+\delta)^{\frac{p}{2}-1} [(1-p)u^2 + \delta] \right\} dx$$
$$\leqslant \int_{\Sigma_2} bw(u^2+\delta)^{p/2} d\sigma - \int_{\Sigma_1} (u^2+\delta)^{p/2} \frac{\partial w}{\partial \nu} d\sigma.$$

Letting δ tend to zero in the latter inequality, we find

$$\int_\Omega |u|^p [L^*(w) + (p-1)cw]\, dx \leqslant \int_{\Sigma_2} wb|u|^p dx + \int_{\Sigma_1} |u|^p \left| \frac{\partial w}{\partial \nu} \right| d\sigma. \tag{1.2.11}$$

This clearly implies (1.2.10). The theorem is proved.

§3. Existence of a solution of the first boundary value problem in the spaces $\mathfrak{L}_p(\Omega)$

Suppose the function u belongs to the class $C^{(2)}(\Omega \cup \Sigma)$ and $u = 0$ on $\Sigma_2 \cup \Sigma_3$. By V we denote the class of functions v belonging to $C^2(\Omega \cup \Sigma)$ such that $v = 0$ on $\Sigma_3 \cup \Sigma_1$. Then from (1.1.14) it follows that

$$\int_\Omega uL^*(v)\, dx = \int_\Omega vL(u)\, dx.$$

DEFINITION. The function $u \in \mathfrak{L}_p(\Omega)$ will be called a *weak solution* of

$$L(u) = f \text{ in } \Omega, \tag{1.3.1}$$

$$u = 0 \text{ on } \Sigma_2 \cup \Sigma_3, \tag{1.3.2}$$

if for any function v in class V the equation

$$\int_\Omega vf\,dx = \int_\Omega uL^*(v)\,dx \tag{1.3.3}$$

is satisfied.

THEOREM 1.3.1. *If* $c^* < 0$ *and* $c < 0$ *in* Ω, $p > 1$, *then for any* $f \in \mathfrak{L}_p(\Omega)$ *there exists a weak solution of the problem* (1.3.1), (1.3.2), *satisfying the inequality*

$$\inf_{u_0 \in Z} \| u + u_0 \|_{\mathscr{L}_p(\Omega)} \leqslant K \| f \|_{\mathscr{L}_p(\Omega)}, \quad K = \text{const}, \tag{1.3.4}$$

where Z *is the set of functions* u_0 *in the class* $\mathfrak{L}_p(\Omega)$ *satisfying the condition*

$$\int_\Omega u_0 L^*(v)\,dx = 0$$

for any function v *in* V.

PROOF. According to Theorem 1.2.1, any function v in class $C^{(2)}(\Omega \cup \Sigma)$ such that $v = 0$ on $\Sigma_3 \cup \Sigma_1$ satisfies

$$\| v \|_{\mathscr{L}_q(\Omega)} \leqslant \frac{q}{\min\limits_{\Omega \cup \Sigma} [-c + (1-q)c^*]} \| L^* v \|_{\mathscr{L}_q(\Omega)} \tag{1.3.5}$$

for any $q \geqslant 1$, since the set Σ_2 for the operator $L^*(v)$ is the same as the set Σ_1 for the operator $L(u)$.

We consider

$$\int_\Omega fv\,dx$$

as a functional on v in the space $\mathfrak{L}_q(\Omega)$, where $1/q + 1/p = 1$. Applying Hölder's inequality and estimate (1.3.5), we obtain

$$\begin{aligned} \left| \int_\Omega fv\,dx \right| &\leqslant \| f \|_{\mathscr{L}_p(\Omega)} \| v \|_{\mathscr{L}_q(\Omega)} \\ &\leqslant K_q \| L^*(v) \|_{\mathscr{L}_q(\Omega)} \| f \|_{\mathscr{L}_p(\Omega)}, \end{aligned} \tag{1.3.6}$$

where

$$K_q = \frac{q}{\min\limits_{\Omega \cup \Sigma} [-c + (1-q)c^*]}.$$

Let $\widetilde{\mathfrak{L}}_q(\Omega)$ be the subspace of the space $\mathfrak{L}_q(\Omega)$ obtained by closing, in $\mathfrak{L}_q(\Omega)$, the set of functions of the form $L^*(v)$, where v belongs to the class V. It is easy to see that the space of functionals on $\widetilde{\mathfrak{L}}_q(\Omega)$ coincides with the factor space $\mathfrak{L}_p(\Omega)/Z$, where Z is the subspace of $\mathfrak{L}_p(\Omega)$ consisting of functions z such that

$$\int_\Omega L^*(v)\, z\, dx = 0$$

for any v in V. In fact, by the Hahn-Banach Theorem [144] each functional on $\widetilde{\mathfrak{L}}_q(\Omega)$ may be extended onto $\mathfrak{L}_q(\Omega)$, and therefore each functional on $\widetilde{\mathfrak{L}}_q(\Omega)$ may be written in the form

$$\int_\Omega uL^*(v)\, dx, \tag{1.3.7}$$

where $u \in \mathfrak{L}_p(\Omega)$. Clearly each coset of the factor space $\mathfrak{L}_p(\Omega)/Z$ yields a unique functional of the form (1.3.7) in $\widetilde{\mathfrak{L}}_q(\Omega)$.

From (1.3.6) it follows that $\int_\Omega fv\, dx$ is a linear continuous functional of L^*v in $\widetilde{\mathfrak{L}}_q(\Omega)$ and therefore it may be represented in the form (1.3.7) with a function u in the space $\mathfrak{L}_p(\Omega)$. Consequently

$$\int_\Omega fv\, dx = \int_\Omega uL^*(v)\, dx,$$

where u is any element in the corresponding coset in $\mathfrak{L}_p(\Omega)/Z$. Clearly each such u satisfies the inequality

$$\left|\int_\Omega uL^*(v)\, dx\right| \leqslant \|f\|_{\mathscr{L}_p(\Omega)} \|v\|_{\mathscr{L}_q(\Omega)}$$
$$\leqslant K_q \|f\|_{\mathscr{L}_p(\Omega)} \|L^*(v)\|_{\mathscr{L}_q(\Omega)}.$$

From this it follows, according to the definition of the norm of functionals in $\widetilde{\mathfrak{L}}_q(\Omega)$, that

$$\inf_{u_0 \in Z} \|u + u_0\|_{\mathscr{L}_p(\Omega)} \leqslant K_q \|f\|_{\mathscr{L}_p(\Omega)}.$$

The theorem is proved.

THEOREM 1.3.2. *Let* $c < 0$ *in* $\Omega \cup \Sigma$, *let* $1/p + 1/q = 1$, *and let* q *be such that* $-c + (1-q)c^* > 0$ *in* $\Omega \cup \Sigma$. *Then, if* $f(x) \in \mathfrak{L}_p(\Omega)$, *there exists a generalized solution of the problem* (1.3.1), (1.3.2) *satisfying* (1.3.4).

THEOREM 1.3.3. *Let* $c^* < 0$ *in* $\Omega \cup \Sigma$, *let* q *be such that* $-c + (1-q)c^* > 0$ *in* $\Omega \cup \Sigma$, *and* $1/p + 1/q = 1$. *Then for any* $f(x) \in \mathfrak{L}_p(\Omega)$ *there exists a generalized solution of* (1.3.1), (1.3.2) *satisfying* (1.3.4).

Theorems 1.3.2 and 1.3.3 are proved exactly as was Theorem 1.3.1. From these theorems it follows that for $c < 0$ the problem (1.3.1), (1.3.2) is solvable for sufficiently large p, and for $c^* < 0$ this problem is solvable for p sufficiently close to 1.

Existence theorems for solutions of the problem (1.3.1), (1.3.2) in the spaces $\mathfrak{L}_p(\Omega)$ are proved by another method in Chapter I, §5.

§4. Existence of a weak solution of the first boundary value problem in Hilbert space

To prove the existence of a weak solution of the problem (1.3.1), (1.3.2), we use here a method, based on the Riesz theorem, which has been used previously in the study of boundary value problems for elliptic equations and systems (see, for example, [132, 39, 13]).

The definition of a weak solution of the problem (1.3.1), (1.3.2) in some Hilbert space $\mathcal{H}$ is based on the following considerations.

Let u be a function in $C^{(2)}(\Omega \cup \Sigma)$ equal to zero on Σ_2 and Σ_3, and let v be a function in $C^{(1)}(\Omega \cup \Sigma)$ equal to zero on Σ_3. Integrating by parts, we obtain the following identity:

$$\int_\Omega vL(u)\,dx = -\int_\Omega [a^{kj}v_{x_k}u_{x_j} + u(b^k - a^{kj}_{x_j})v_{x_k} + (b^k_{x_k} - a^{kj}_{x_k x_j} - c)uv]\,dx - \int_{\Sigma_1} uvb\,d\sigma.$$

The set of functions v in the class $C^{(1)}(\Omega \cup \Sigma)$ equal to zero on Σ_3 will be denoted by W. For functions in W we introduce the scalar product

$$(u, v)_{\mathcal{H}} \equiv \int_\Omega (a^{kj}u_{x_k}v_{x_j} + uv)\,dx + \int_{\Sigma_1 \cup \Sigma_2} uv\,|b|\,d\sigma \tag{1.4.1}$$

and denote by $\mathcal{H}$ the Hilbert space obtained by closing W with respect to the norm derived from this scalar product.

For u and v in W we consider the bilinear form

$$B(u, v) = -\int_\Omega (a^{kj}u_{x_k}v_{x_j} + u(b^k - a^{kj}_{x_j})v_{x_k} + (b^k_{x_k} - a^{kj}_{x_k x_j} - c)uv)\,dx - \int_{\Sigma_1} uvb\,d\sigma. \tag{1.4.2}$$

We show that the definition of $B(u, v)$ may be extended to apply to all functions u in $\mathcal{H}$ and v in W. In fact, for u and v in W we have

$$|B(u, v)| \leqslant M\left(\int_\Omega [v_{x_k}v_{x_k} + v^2]\,dx + \int_{\Sigma_1} v^2\,d\sigma\right)^{1/2} \|u\|_{\mathcal{H}}, \tag{1.4.3}$$

where M is a constant depending only on the coefficients of the equation (1.3.1). From (1.4.3) it follows that $B(u, v)$, for fixed v in W, may be considered as a linear bounded functional on u in $\mathcal{H}$.

Definition. Let $f \in \mathfrak{L}_2(\Omega)$. A function u in $\mathcal{H}$ will be called a *weak solution* of (1.3.1), (1.3.2), if for any function v in W the equation

$$\int_\Omega vf\,dx = B(u, v) \tag{1.4.4}$$

holds.

REMARK. A weak solution u in the space $\mathcal{H}$, defined above, is also a weak solution of the same problem in the space $\mathfrak{L}_2(\Omega)$ defined by (1.3.3) above.

In fact, if u in $\mathcal{H}$ satisfies (1.4.4), then there must exist a sequence u_n in W, approaching u in the norm of $\mathcal{H}$, and such that $B(u_n, v) \to B(u, v)$ as $n \to \infty$. If the function v belongs to class $C^{(2)}(\Omega \cup \Sigma)$ and is equal to zero on $\Sigma_1 \cup \Sigma_3$, then the bilinear form $B(u_n, v)$ may be transformed by integration by parts so that

$$B(u_n, v) = \int_\Omega L^*(v)\, u_n dx \quad \text{and} \quad \lim_{n\to\infty} B(u_n, v) = B(u, v) = \int_\Omega L^*(v)\, u\, dx.$$

Thus for such v the equation

$$B(u, v) = \int_\Omega L^*(v)\, u\, dx$$

is satisfied. This means that a weak solution u of the problem (1.3.1), (1.3.2) in the sense of (1.4.4) is also a weak solution of this problem in the sense of (1.3.3), since each function u in $\mathcal{H}$ is also an element of $\mathfrak{L}_2(\Omega)$.

THEOREM 1.4.1. *Suppose the inequality*

$$\frac{1}{2} b^k_{x_k} - \frac{1}{2} a^{kj}_{x_k x_j} - c \geqslant c_0 > 0$$

is satisfied in $\bar{\Omega}$, *and let* $f \in \mathfrak{L}_2(\Omega)$. *Then there exists a function* u *in* $\mathcal{H}$ *which is a weak solution of the problem* (1.3.1), (1.3.2) *in the sense of* (1.4.4).

PROOF. Since $B(u, v)$ is a linear continuous functional of u in $\mathcal{H}$ for each v in W, it follows by the Riesz theorem on the representation of linear continuous functionals in a Hilbert space that

$$B(u, v) = (u, T(v))_{\mathcal{H}},$$

where $T(v)$ is a linear operator on W with range in $\mathcal{H}$.

Let $v \in W$. Since by assumption $\tfrac{1}{2} b^k_{x_k} - \tfrac{1}{2} a^{kj}_{x_k x_j} - c > 0$ in Ω, it follows that

$$|B(v, v)| = \int_\Omega \left(a^{kj} v_{x_k} v_{x_j} + \left(\frac{1}{2} b^k_{x_k} - \frac{1}{2} a^{kj}_{x_k x_j} - c \right) v^2 \right) dx$$

$$+ \frac{1}{2} \int_{\Sigma_1} v^2 b\, d\sigma - \frac{1}{2} \int_{\Sigma_2} v^2 b\, d\sigma \geqslant \alpha \| v \|^2_{\mathcal{H}}; \quad \alpha = \text{const} > 0.$$

From this it follows in turn that

$$\| v \|_{\mathcal{H}}^2 \leqslant \frac{1}{\alpha} | B(v, v) | = \frac{1}{\alpha} | (v, T(v))_{\mathcal{H}} | \leqslant \frac{1}{\alpha} \| v \|_{\mathcal{H}} \| T(v) \|_{\mathcal{H}}$$

and, therefore,

$$\| v \|_{\mathcal{H}} \leqslant \frac{1}{\alpha} \| T(v) \|_{\mathcal{H}}. \tag{1.4.5}$$

From the estimate (1.4.5) it follows that the mapping of W into $\mathcal{H}$ defined by the operator T is one-to-one. We denote by $\mathcal{H}'$ the closure of the set $T(v)$ in the norm of $\mathcal{H}$, where $v \in W$.

Since

$$\left| \int_{\Omega} fv\,dx \right| \leqslant \| f \|_{\mathcal{L}_2(\Omega)} \| v \|_{\mathcal{L}_2(\Omega)} \leqslant \| f \|_{\mathcal{L}_2(\Omega)} \| v \|_{\mathcal{H}}$$

$$\leqslant \frac{1}{\alpha} \| f \|_{\mathcal{L}_2(\Omega)} \| Tv \|_{\mathcal{H}};$$

the integral $\int_{\Omega} fv\,dx$ may be considered as a linear continuous functional on $\mathcal{H}'$. Therefore by the Riesz theorem there exists a function u in $\mathcal{H}'$ such that

$$\int_{\Omega} fv\,dx = (u, Tv)_{\mathcal{H}} \equiv B(u, v)$$

for any v in W, i.e. u is a weak solution of (1.3.1), (1.3.2) in the sense of (1.4.4). The theorem is thereby proved.

§5. Solution of the first boundary value problem by the method of elliptic regularization

In the domain Ω with boundary Σ we consider the first boundary value problem for the equation

$$L(u) \equiv a^{kj} u_{x_k x_j} + b^k u_{x_k} + cu = f; \quad a^{kj}(x)\xi_k \xi_j \geqslant 0 \tag{1.5.1}$$

with boundary condition

$$u = g \text{ on } \Sigma_2 \cup \Sigma_3. \tag{1.5.2}$$

We shall assume that the coefficients of the operator $L(u)$, and also the coefficients of the adjoint operator $L^*(u)$, belong to some space $C^{(0,\alpha)}(\Omega \cup \Sigma)$, and we shall suppose that the domain Ω belongs to class $B^{(2,\alpha)}, \alpha > 0$. In this section we obtain a solution of the problem (1.5.1), (1.5.2) in various function spaces by the method of elliptic regularization.

DEFINITION. A bounded measurable function $u(x)$ will be called a *weak solution of the first boundary value problem* (1.5.1), (1.5.2) if for any function v in the class $C^{(2)}(\Omega \cup \Sigma)$, equal to zero on $\Sigma_1 \cup \Sigma_3$, the integral identity

$$\int_\Omega uL^*(v)\,dx = \int_\Omega vf\,dx - \int_{\Sigma_3} g\frac{\partial v}{\partial \bar\nu}\,d\sigma + \int_{\Sigma_2} bgv\,d\sigma \tag{1.5.3}$$

is fulfilled, where $\partial/\partial\bar\nu \equiv a^{kj}n_k\partial/\partial x_j$, $d\sigma$ is the surface area element on Σ, and f and g are bounded measurable functions given respectively on Ω and $\Sigma_2 \cup \Sigma_3$.

Below we shall give conditions under which such a solution exists in smooth and piecewise smooth domains, and also consider solutions in the spaces $\mathfrak{L}_p(\Omega)$ (see §3 above). It is clear that each classical solution of the problem (1.5.1), (1.5.2), if one exists, is also a weak solution of this problem in the sense of the integral identity (1.5.3). For smooth functions f and g, the weak solution of (1.5.1), (1.5.2) will be obtained as the limit as $\epsilon \to 0$ of solutions of the Dirichlet problem for elliptic equations

$$L_\varepsilon(u) \equiv \varepsilon\Delta u + L(u) = f, \quad \varepsilon = \text{const} > 0 \tag{1.5.4}$$

with suitable boundary conditions.

We denote by Γ the boundary of the set $\Sigma_0 \cup \Sigma_2$ on Σ. We assume that $a^{kj}_{x_l}$ are continuous in $\Omega \cup \Sigma$.

Suppose that in the neighborhood of some point P on the boundary Σ of the domain Ω the equation of the boundary Σ has the form

$$F(x_1, \ldots, x_m) = 0, \quad \operatorname{grad} F \neq 0, \quad F > 0 \text{ in } \Omega. \tag{1.5.5}$$

On Σ we consider the function

$$\beta \equiv L(F), \tag{1.5.6}$$

introduced in §1 above. At points internal to Σ^0 (or limits of internal points) and such that $(\partial a^{kj}/\partial\bar n)n_kn_j = 0$, the function β coincides with accuracy up to a positive factor with the function b (the Fichera function) defined by (1.1.3). In fact, since $a^{kj}F_{x_k} \equiv 0$ and $a^{kj}n_j \equiv 0$ on Σ^0, it follows that

$$\beta \equiv (a^{kj}F_{x_k})_{x_j} + (b^k - a^{kj}_{x_j})F_{x_k} = \frac{\partial}{\partial\bar n}(a^{kj}F_{x_k})n_j$$

$$+ (F_{x_k}F_{x_k})^{1/2}b = \frac{\partial a^{kj}}{\partial\bar n}F_{x_k}n_j + a^{kj}n_j\frac{\partial F_{x_k}}{\partial\bar n} + b(F_{x_k}F_{x_k})^{\frac12} = b(F_{x_k}F_{x_k})^{\frac12}.$$

The condition $(\partial a^{kj}/\partial\bar n)n_kn_j = 0$ at the point P on the boundary Σ^0 is easily seen to be satisfied, if in some full neighborhood of the point P the form $a^{kj}\xi_k\xi_j$ is defined and nonnegative, and $a^{kj}_{x_l}$ are continuous. In the general case $(\partial a^{kj}/\partial\bar n)n_kn_j \geqslant 0$ on Σ^0.

We denote by σ the set of points of Σ^0 where $\beta \leqslant 0$. We denote by M_j constants independent of ϵ. The following lemma will be important in future considerations.

LEMMA 1.5.1. *Let $u(x)$ satisfy in Ω the equation* (1.5.4), *and let $u \in C^{(2)}(\Omega \cup \Sigma)$, $u = 0$ on Σ, $|f| \leqslant M_0$, and $c(x) \leqslant -c_0 < 0$, where $c_0 = \text{const} > 0$. Suppose the set G on Σ is such that $\bar{G}$ lies inside $\Sigma_3 \cup \sigma$ and $\beta < 0$ at boundary points of Σ_3 belonging to $\bar{G}$. Then at points of G,*

$$|u_{x_j}| \leqslant M_1\varepsilon^{-1/2}, \quad j = 1, \ldots, m. \tag{1.5.7}$$

At all points of Σ,

$$|u_{x_j}| \leqslant M_2\varepsilon^{-1}, \quad j = 1, \ldots, m. \tag{1.5.8}$$

PROOF. Suppose $P_0 \subset G$. In a neighborhood of P_0 we pass to local coordinates $y_1, \cdots, y_m$ with origin at P_0, such that Σ in the neighborhood of P_0 lies in the plane $y_m = 0$; i.e. we set $y_k = F^k(x_1, \cdots, x_m)$, $k = 1, \cdots, m$, $F^m \equiv F(x_1, \cdots, x_m)$. Let $\delta > 0$ be so small that the set of points $Q^0\{\rho^2 = y_1^2 + \cdots + y_{m-1}^2 \leqslant 4\delta^2\}$ is contained in $\Sigma_3 \cup \sigma$, and $\beta < 0$ at boundary points of Σ_3 belonging to Q^0. We consider the function

$$\psi(\rho) = \begin{cases} \varepsilon^{1/2} & \text{for } \rho \leqslant \delta, \\ \varepsilon^{1/2}[1 - 27^{-1}(\rho^2 - \delta^2)^3\delta^{-6}) & \text{for } \delta \leqslant \rho \leqslant 2\delta. \end{cases}$$

In the domain $Q_\delta\{0 \leqslant \rho < 2\delta,\ 0 < y_m < \psi(\rho)\}$ we consider the function $w = K_0(e^{-z} - 1)$, where $z = K_1(y_m + \epsilon^{1/2} - \psi)\epsilon^{-1/2}$. Here K_0 and K_1 are constants, independent of ϵ, to be chosen later. Equation (1.5.4) in the variables $y_1, \cdots, y_m$ has the form

$$L_\varepsilon(u) \equiv \varepsilon\mu^{kj}u_{y_ky_j} + \varepsilon\nu^j u_{y_j} + \alpha^{kj}u_{y_ky_j} + \beta^j u_{y_j} + cu = f. \tag{1.5.9}$$

It is easy to see that at points of Σ the coefficient $\beta^m = \beta$. By definition, $\beta \leqslant 0$ on σ. We calculate $L_\epsilon(w)$. We have

$$\begin{aligned} L_\varepsilon(w) = K_0e^{-z}\Bigg\{ \varepsilon\Bigg[& \mu^{mm}K_1^2\varepsilon^{-1} - 2\sum_{j=1}^{m-1}\mu^{mj}K_1^2\psi_{y_j}\varepsilon^{-1} + \sum_{k,j=1}^{m-1}\mu^{kj}K_1^2\psi_{y_j}\psi_{y_k}\varepsilon^{-1} \\ & + \sum_{k,j=1}^{m-1}\mu^{kj}K_1\psi_{y_ky_j}\varepsilon^{-1/2} - \nu^m K_1\varepsilon^{-1/2} + \sum_{j=1}^{m-1}\nu^j K_1\psi_{y_j}\varepsilon^{-1/2}\Bigg] \\ & + \Bigg[\alpha^{mm}K_1^2\varepsilon^{-1} - 2\sum_{j=1}^{m-1}\alpha^{mj}K_1^2\psi_{y_j}\varepsilon^{-1} + \sum_{k,j=1}^{m-1}\alpha^{kj}K_1^2\psi_{y_k}\psi_{y_j}\varepsilon^{-1}\Bigg] \\ & + \sum_{k,j=1}^{m-1}\alpha^{kj}K_1\psi_{y_ky_j}\varepsilon^{-1/2} - \beta^m K_1\varepsilon^{-1/2} + \sum_{j=1}^{m-1}\beta^j K_1\psi_{y_j}\varepsilon^{-\frac{1}{2}} + c\Bigg\} - cK_0. \end{aligned} \tag{1.5.10}$$

Note that in the domain Q_δ we have

$$|\psi| \leqslant \varepsilon^{1/2}, \ |\psi_{y_j}| \leqslant M_3\varepsilon^{1/2}, \ |\psi_{y_k y_j}| \leqslant M_4 \varepsilon^{\frac{1}{2}}. \tag{1.5.11}$$

At points of $\sigma \cap Q^0$, the coefficient $\beta^m \leqslant 0$. Since by definition $\beta < 0$ at points of $\Sigma^0 \cap Q^0$ which are limit points of Σ_3, it follows that $\beta^m \leqslant 0$ in some neighborhood of the set $\sigma \cap Q^0$ on Σ, which we denote by G_0.

Therefore at points of a subdomain Q_δ^0 of Q_δ, whose projection on the plane $y_m = 0$ belongs to G_0, we have the inequality

$$-\beta^m K_1 \varepsilon^{-\frac{1}{2}} \geqslant -K_1 M_5. \tag{1.5.12}$$

At points of $Q^0 \backslash G_0$, clearly, $\alpha^{mm} \geqslant \alpha_0 > 0$, and so for small ϵ at points of the domain Q_δ, whose projections on $y_m = 0$ lie in $Q^0 \backslash G_0$, the inequality $\alpha^{mm} \geqslant \alpha_0/2 > 0$ holds. We call the set of such points Q_δ^1. At points of Q_δ^1 the terms in braces in (1.5.10) are bounded from below by a constant not depending on ϵ, since at these points (1.5.11) and (1.5.12) hold, the form $a^{kj}\xi_k\xi_j$ is nonnegative, and the terms in the first brackets are larger than $M_6 K_1^2 \epsilon^{-1}$, where $M_6 = \text{const} > 0$. Therefore, choosing K_1 sufficiently large, we obtain that $L_\epsilon(w) \geqslant c_0 K_0$ in Q_δ^0.

At points of Q_δ^1 the sign of $L_\epsilon(w)$ for sufficiently small ϵ is determined by the term $\alpha^{mm} K_1^2 \epsilon^{-1}$, and therefore $L_\epsilon(w) \geqslant c_0 K_0$ in Q_δ^1.

Choosing K_0 sufficiently large, we obtain that

$$L_\varepsilon(w \pm u) \geqslant c_0 K_0 \pm f > 0 \quad \text{in} \quad Q_\delta.$$

On the boundary Q_δ the inequality

$$w \pm u \leqslant 0,$$

is fulfilled if K_0 is sufficiently large, since $w \leqslant 0$ and $u = 0$ for $y_m = 0$, and $w \leqslant -\max_\Omega |u|$ for $y_m = \psi$ and for large K_0. Consequently, by virtue of the maximum principle for the elliptic equation (1.5.4), everywhere in Q_δ we have $w \pm u \leqslant 0$. Since $w = u = 0$ on Σ for $\rho \leqslant \delta$, at these points $(w \pm u)_{y_m} \leqslant 0$, and hence $|u_{y_m}| \leqslant K_0 K_1 \epsilon^{-1/2}$. This is sufficient for the proof of (1.5.7) at points of $\overline{G}$.

The estimate (1.5.8) on Σ is obtained by an analogous procedure. We consider the function $\widetilde{\psi}(\rho)$ defined by the equations

$$\widetilde{\psi}(\rho) = \begin{cases} \varepsilon & \text{for} \quad \rho < \delta, \\ \varepsilon\,[1 - 27^{-1}(\rho^2 - \delta^2)^3 \delta^{-6}] & \text{for} \quad \delta \leqslant \rho \leqslant 2\delta. \end{cases}$$

Suppose $\widetilde{w} = K_0(e^{-\widetilde{z}} - 1)$, where

$$\tilde{z} = K_1(y_m + \varepsilon - \tilde{\psi})\varepsilon^{-1}.$$

In the domain $\tilde{Q}_\delta\{0 \leqslant \rho < 2\delta, 0 < y_m < \tilde{\psi}(\rho)\}$ we compute $L_\epsilon(\tilde{w})$. We have

$$\begin{aligned} L_\varepsilon(\tilde{w}) = K_0 e^{-\tilde{z}} \Bigg\{ \varepsilon \Bigg[& \mu^{mm} K_1^2 \varepsilon^{-2} - 2\sum_{j=1}^{m-1} \mu^{mj} K_1^2 \tilde{\psi}_{y_j} \varepsilon^{-2} + \sum_{k,j=1}^{m-1} \mu^{kj} K_1 \tilde{\psi}_{y_k y_j} \varepsilon^{-1} \\ & + \sum_{k,j=1}^{m-1} \mu^{kj} K_1^2 \tilde{\psi}_{y_k} \tilde{\psi}_{y_j} \varepsilon^{-2} - \nu^m K_1 \varepsilon^{-1} + \sum_{j=1}^{m-1} \nu^j K_1 \tilde{\psi}_{y_j} \varepsilon^{-1} \Bigg] \\ & + \alpha^{mm} K_1^2 \varepsilon^{-2} + \sum_{k,j=1}^{m-1} \alpha^{kj} K_1^2 \tilde{\psi}_{y_k} \tilde{\psi}_{y_j} \varepsilon^{-2} - 2\sum_{j=1}^{m-1} \alpha^{mj} K_1^2 \tilde{\psi}_{y_j} \varepsilon^{-2} \\ & + \sum_{k,j=1}^{m-1} \alpha^{kj} K_1 \tilde{\psi}_{y_k y_j} \varepsilon^{-1} + \sum_{j=1}^{m-1} \beta^j K_1 \tilde{\psi}_{y_j} \varepsilon^{-1} - \beta^m K_1 \varepsilon^{-1} + c \Bigg\} - cK_0. \end{aligned} \tag{1.5.13}$$

The expression in brackets, multiplied by ϵ, is larger than $M_7 K_1^2 \epsilon^{-1}$ for small ϵ, where $M_7 = \text{const} > 0$; and the remaining terms on the right of (1.5.13) may be estimated from below by the quantity $M_8 K_1 \epsilon^{-1}$. Therefore for sufficiently large K_1 and small ϵ we have $L_\epsilon(w) \geqslant c_0 K_0$.

Furthermore, considering in $\tilde{Q}_\delta$ the function $\tilde{w} \pm u$, we obtain, as above, that $\tilde{w} \pm u \leqslant 0$ in $\tilde{Q}_\delta$ and $|u_{y_m}| \leqslant K_0 K_1 \epsilon^{-1}$ on Σ. Consequently inequality (1.5.8) holds. The lemma is proved.

REMARK. In the conditions of Lemma 1.5.1 the assumption that $\beta < 0$ at boundary points of Σ_3 may be replaced by the requirement that $\beta \leqslant 0$ in some neighborhood of the boundary of Σ_3.

THEOREM 1.5.1. *In equation* (1.5.1), *suppose that* $c(x) \leqslant -c_0 < 0$, *that* f *is a bounded measurable function in* Ω, g *a bounded measurable function on* $\Sigma_2 \cup \Sigma_3$, *and* $\beta \leqslant 0$ *at interior points of* $\Sigma_2 \cup \Sigma_0$. *Then in* Ω *there exists a weak solution of the boundary value problem* (1.5.1), (1.5.2) *which satisfies the inequality (maximum principle)*

$$|u| \leqslant \max\left\{\sup \frac{|f|}{c_0}, \sup|g|\right\}. \tag{1.5.14}$$

PROOF. Let $f_n \in C^{(1)}(\bar{\Omega})$, $f_n \to f$ as $n \to \infty$ in the norm of $\mathfrak{L}_2(\Omega)$ and $|f_n| \leqslant \sup |f|$; and let the functions g_n be defined in $\bar{\Omega}$, $g_n \in C^{(2,\alpha)}(\bar{\Omega})$, $g_n \to g$ as $n \to \infty$ in the norm of $\mathfrak{L}_2(\Sigma_2 \cup \Sigma_3)$ and $|g_n| \leqslant \sup_{\Sigma_2 \cup \Sigma_3} |g|$. Let $u_{\epsilon,n}$ be the solution of the problem

$$L_\epsilon(u) = f_n \text{ in } \Omega, \quad u = g_n \text{ on } \Sigma. \tag{1.5.15}$$

According to the smoothness assumptions on the domain Ω, the coefficients of the equation (1.5.1) and the functions g_n and f_n, such a solution $u_{\epsilon,n}$ of the elliptic problem (1.5.15) exists and belongs to the class $C^{(2,\alpha)}(\bar{\Omega})$ (see [81]). By virtue of the maximum principle for solutions $u_{\epsilon,n}$ of the elliptic equation (1.5.15) we have the inequality

$$|u_{\varepsilon,n}| \leqslant \max\left\{\sup \frac{|f|}{c_0},\ \sup |g|\right\}. \tag{1.5.16}$$

Since for $z_{\epsilon,n} = u_{\epsilon,n} - g_n$ we have

$$L_\epsilon(z_{\epsilon,n}) = f_n - L_\epsilon(g_n) \text{ in } \Omega;\ z_{\epsilon,n} = 0 \text{ on } \Sigma,$$

it follows that Lemma 1.5.1 is valid for $z_{\epsilon,n}$ for fixed n, and hence also for $u_{\epsilon,n}$. Suppose $v \in C^{(2)}(\Omega \cup \Sigma)$ and $v = 0$ on $\Sigma_1 \cup \Sigma_3$. Applying Green's formula, we obtain

$$\begin{aligned}\int_\Omega f_n v\,dx &= \int_\Omega \varepsilon \Delta v u_{\varepsilon,n}\,dx + \int_\Omega L^*(v)\,u_{\varepsilon,n}\,dx \\ &+ \int_\Sigma \varepsilon u_{\varepsilon,n} \frac{\partial v}{\partial \bar{n}}\,d\sigma + \int_{\Sigma_3} g_n \frac{\partial v}{\partial \bar{\nu}}\,d\sigma - \int_{\Sigma_2} b g_n v\,d\sigma \\ &- \int_{\Sigma_0 \cup \Sigma_2} \varepsilon v \frac{\partial u_{\varepsilon,n}}{\partial \bar{n}}\,d\sigma.\end{aligned} \tag{1.5.17}$$

From (1.5.16) it follows that $\{u_{\epsilon,n}\}$ is weakly compact in $\mathfrak{L}_2(\Omega)$. Suppose the subsequence $u_{\epsilon_k,n}$ converges weakly to u_n as $\epsilon_k \to 0$ in the space $\mathfrak{L}_2(\Omega)$. Passing to the limit in (1.5.17) as $\epsilon_k \to 0$, we obtain that u_n satisfies the integral identity

$$\int_\Omega u_n L^*(v)\,dx = \int_\Omega f_n v\,dx - \int_{\Sigma_3} g_n \frac{\partial v}{\partial \bar{\nu}}\,d\sigma + \int_{\Sigma_2} b g_n v\,d\sigma. \tag{1.5.18}$$

The last integral in (1.5.17) tends to zero as $\epsilon_k \to 0$. This is easy to see, since in the δ-neighborhood Γ^δ of the boundary Γ of the set $\Sigma_0 \cup \Sigma_2$, the derivative $\partial u_{\epsilon,n}/\partial n$ may be estimated for fixed n by using the inequality (1.5.8) because $v = 0$ on Γ and the area of Σ is finite; and because on $(\Sigma_0 \cup \Sigma_2)\setminus\Gamma^\delta$ the condition $\beta \leqslant 0$ is satisfied, so that (1.5.7) holds for $u_{\epsilon,n}$ on $(\Sigma_0 \cup \Sigma_2)\setminus\Gamma^\delta$. Hence this integral can be made as small as desired by choosing δ and ϵ small.

From the sequence $\{u_n\}$ we choose a subsequence weakly convergent in $\mathfrak{L}_2(\Omega)$ to a function u as $n_k \to \infty$. Then, passing to the limit in the integral identity (1.5.18) as $n_k \to \infty$, we obtain the assertion of the theorem. The estimate (1.5.14) for the function u follows from inequality (1.5.16) for the functions $u_{\epsilon,n}$.

THEOREM 1.5.2. *If the function* g *is continuous in a neighborhood of* $\Sigma_3 \cup \Sigma_2$ *on* Σ, *then the weak solution* $u(x)$ *constructed in Theorem* 1.5.1 *is continuous at points* Σ_3 *and points* Σ_2 *which are interior or limits of interior points of* Σ_2, *and it assumes the given values* g.

PROOF. Suppose P_0 is a point of $\Sigma_3 \cup \Sigma_2$. We transform equation (1.5.4) by using local coordinates $y_1, \cdots, y_m$ in a neighborhood of the point P_0; these coordinates are chosen so that Σ lies on the plane $y_m = 0$ and $y_m > 0$ in Ω. Suppose the equation (1.5.4) has the form (1.5.9) in the new coordinates. In a neighborhood of the point P_0 for $y_m \geqslant 0$ we consider the function

$$w = y_m^\gamma + \sum_{j=1}^{m-1} y_j^2, \quad \gamma = \text{const}, \quad 0 < \gamma < 1.$$

It is easy to see that if P_0 is a point of Σ_3, then

$$\begin{aligned} L_\varepsilon(w) &= \varepsilon\mu^{mm}\gamma(\gamma-1)y_m^{\gamma-2} + \varepsilon\sum_{j=1}^{m-1} 2\mu^{jj} \\ &+ \varepsilon\nu^m\gamma y_m^{\gamma-1} + \varepsilon\sum_{j=1}^{m-1}\nu^j 2y_j + \alpha^{mm}\gamma(\gamma-1)y_m^{\gamma-2} \\ &+ 2\sum_{j=1}^{m-1}\alpha^{jj} + \gamma\beta^m y_m^{\gamma-1} + \sum_{j=1}^{m-1}\beta^j 2y_j + cw < 0 \end{aligned}$$

in some neighborhood of P_0 for $y_m > 0$, since for small y_m the sign of $L_\epsilon(w)$ is determined by the term $\alpha^{mm}\gamma(\gamma-1)y_m^{\gamma-2}$. If P_0 is an interior point of Σ_2 or a limit of interior points of Σ_2, then according to the definition of Σ_2, at this point $\beta^m - \alpha^{mm}_{y_m} < 0$ and the sign of $L_\epsilon(w)$ for sufficiently small ϵ in some neighborhood of P_0 for $y_m > 0$ is determined by the terms $\alpha^{mm}\gamma(\gamma-1)y_m^{\gamma-2} + \gamma\beta^m y_m^{\gamma-1}$. Since $\alpha^{mm} = \alpha^{mm}_{y_m} y_m + O(y_m^2)$ and

$$\alpha^{mm}_{y_m}\gamma(\gamma-1)y_m^{\gamma-1} + \gamma\beta^m y_m^{\gamma-1} = \gamma\left(\beta^m - \alpha^{mm}_{y_m} + \gamma\alpha^{mm}_{y_m}\right)y_m^{\gamma-1},$$

these terms form a quantity less than zero if $\gamma > 0$ is sufficiently small. We may assume that the g_n converge to g as $n \to \infty$ uniformly on $\Sigma_2 \cup \Sigma_3$ in a neighborhood of P_0. Suppose $u_{\epsilon,n}$ is a solution of the problem (1.5.15). We consider the functions

$$V_\pm = \mp g_n(P_0) + \delta \pm u_{\varepsilon,n} + C_1 w; \; C_1, \delta = \text{const} > 0,$$

in the domain $G_\rho\{\sum_{j=1}^m y_j^2 < \rho, y_m > 0\}$. In this domain

$$L_\varepsilon(V_\pm) = c(\mp g_n(P_0) + \delta) \pm f_n + C_1 L_\varepsilon(w) < 0,$$

if the constant C_1 is sufficiently large. On the boundary of G_ρ for $y_m = 0$ we have $V_\pm \geqslant 0$, since $C_1 w \geqslant 0$ and, moreover, $\pm [u_{\epsilon,n} - g_n(P_0)] + \delta > 0$ for $y_m = 0$ and large n, if ρ is sufficiently small. Since $u_{\epsilon,n}$ are uniformly bounded with respect to ϵ and n by virtue of (1.5.16), it follows that for $\Sigma_{j=1}^m y_j^2 = \rho$ the inequality $V_\pm > 0$ holds, if C_1 is sufficiently large. According to the maximum principle, $V_\pm \geqslant 0$ in G_ρ; i.e. in G_ρ we have

$$-C_1 w - \delta \leqslant u_{\varepsilon,n} - g_n(P_0) \leqslant C_1 w + \delta, \tag{1.5.19}$$

where the constants C_1, δ and ρ do not depend on ϵ or n. These inequalities are clearly valid as well for $u(x)$, the weak limit of the $u_{\epsilon,n}$ in $\mathfrak{L}_2(\Omega)$. Therefore the continuity of the limit function $u(x)$ at P_0 follows from (1.5.19). The theorem is proved.

We now prove a theorem on the existence of a weak solution of (1.5.1), (1.5.2) in the spaces $\mathfrak{L}_p(\Omega)$.

THEOREM 1.5.3. *Suppose the coefficients in* (1.5.1) *satisfy* $c(x) < 0$ *in* $\Omega \cup \Sigma$, $f \in \mathfrak{L}_p(\Omega)$, $c^* \equiv a^{kj}_{x_k x_j} - b^j_{x_j} + c < 0$ *in* $\Omega \cup \Sigma$ *and* $\beta \leqslant 0$ *at interior points of* $\Sigma_2 \cup \Sigma_0$. *Then in* Ω *there exists a weak solution of the boundary problem* (1.5.1), (1.5.2) *with* $g = 0$, *in the space* $\mathfrak{L}_p(\Omega)$, *i.e. there exists a function* $u(x) \in \mathfrak{L}_p(\Omega)$, $1 \leqslant p < \infty$, *satisfying the integral identity*

$$\int_\Omega u L^*(v)\, dx = \int_\Omega v f\, dx \tag{1.5.20}$$

for every function v *in* $C^{(2)}(\Omega \cup \Sigma)$ *equal to zero on* $\Sigma_1 \cup \Sigma_3$. *For this weak solution we have the estimate*

$$\|u\|_{\mathfrak{L}_p(\Omega)} \leqslant \frac{p}{\min\limits_{\Omega\cup\Sigma} [-c^* + (1-p)c]} \|f\|_{\mathfrak{L}_p(\Omega)}. \tag{1.5.21}$$

If the condition $c^* < 0$ *in* $\Omega \cup \Sigma$ *is not satisfied, then such a weak solution of* (1.5.1), (1.5.2) *with* $g = 0$ *exists for sufficiently large* p *such that* $-c^* + (1-p)c \geqslant 0$ *in* $\Omega \cup \Sigma$.

PROOF. This theorem is proved in a manner analogous to the proof of Theorem 1.5.1. The function f is approximated in the norm of $\mathfrak{L}_p(\Omega)$ by functions f_n from $C^{(1)}(\Omega \cup \Sigma)$. Suppose $u_{\epsilon,n}$ is the solution of the equation

$$L_\epsilon(u) = f_n \text{ in } \Omega, \quad u_{\epsilon,n} = 0 \text{ on } \Sigma.$$

The functions $u_{\epsilon,n}$ are bounded in the norm of $\mathfrak{L}_p(\Omega)$ uniformly in ϵ and n.

This follows from (1.2.7), proved in Theorem 1.2.1. From the uniform boundedness of the norm of $u_{\epsilon,n}$ in $\mathcal{L}_p(\Omega)$ follows the weak compactness of this family of functions in $\mathcal{L}_p(\Omega)$. Choosing a weakly convergent subsequence $u_{\epsilon_k,n}$ as $\epsilon_k \to 0$, passing to the limit as $\epsilon_k \to 0$ in (1.5.17) and recalling that $g_n = 0$, as in the proof of Theorem 1.5.1, we obtain the fact that the limiting function $u_n(x)$ satisfies the integral identity (1.5.10). From (1.2.7) it follows that the estimate

$$\|u_n\|_{\mathcal{L}_p(\Omega)} \leqslant \frac{p}{\min\limits_{\Omega \cup \Sigma} [-c^* + (1-p)\,c]} \|f_n\|_{\mathcal{L}_p(\Omega)}$$

holds for $u_n(x)$. Further choosing from the family $\{u_n\}$ a weakly convergent subsequence in $\mathcal{L}_p(\Omega)$ as $n_k \to \infty$ and passing to the limit through this subsequence in the equation

$$\int_\Omega L^*(v)\, u_n dx = \int_\Omega v f_n dx,$$

we obtain that the limiting function u satisfies the integral identity (1.5.20). It is clear that $u(x)$ also satisfies (1.5.21).

The method of elliptic regularization may also be used to obtain a weak solution of the problem (1.5.1), (1.5.2) with $g = 0$ in the space $\mathcal{H}$, which was constructed in §4.

THEOREM 1.5.4. *Let $f \in \mathcal{L}_2(\Omega)$ and suppose the inequality*

$$\frac{1}{2} b^j_{x_j} - \frac{1}{2} a^{kj}_{x_k x_j} - c \geqslant \text{const} > 0$$

is satisfied in the domain Ω, with $\beta \leqslant 0$ at interior points of $\Sigma_2 \cup \Sigma_0$. Then there exists a function u in the space $\mathcal{H}$ constructed in §4 which satisfies (1.4.4) for any function $v \in C^{(1)}(\Omega \cup \Sigma)$ equal to zero on $\Sigma_3 \cup \Sigma_1$. This solution is the weak limit in $\mathcal{L}_2(\Omega)$ as $\epsilon \to 0$, $n \to \infty$ of the sequence of smooth solutions $u_{\epsilon,n}$ of the equation

$$L_\epsilon(u) = f_n \quad \text{in } \Omega, \quad u = 0 \quad \text{on } \Sigma,$$

where $f_n \in C^{(2)}(\Omega \cup \Sigma)$, $f_n \to f$ as $n \to \infty$ in the norm of $\mathcal{L}_2(\Omega)$.

PROOF. Suppose $v \in C^{(2)}(\Omega \cup \Sigma)$ and $v = 0$ on $\Sigma_3 \cup \Sigma_1$. Integrating by parts and taking into consideration the boundary conditions for $u_{\epsilon,n}$ on the boundary Σ of the region Ω, we obtain

$$\int_\Omega v L_\varepsilon(u_{\varepsilon,n})\, dx$$

$$= -\int_\Omega \left[a^{kj} v_{x_k} u_{\varepsilon,n x_j} + u_{\varepsilon,n}\left(b^k - a^{kj}_{x_j}\right) v_{x_k} + \left(b^k_{x_k} - a^{kj}_{x_k x_j} - c\right) u_{\varepsilon,n} v\right] dx$$

$$+\int_\Omega \varepsilon u_{\varepsilon,n}\Delta v\,dx-\int_{\Sigma_0\cup\Sigma_2}\varepsilon\frac{\partial u_{\varepsilon,n}}{\partial\bar n}v\,d\sigma=\int_\Omega vf_n\,dx, \tag{1.5.22}$$

where $\bar n$ is the direction of the interior normal to the boundary Σ. Thus we have

$$\int_\Omega vf_n\,dx=B(u_{\varepsilon,n},v)+\varepsilon\int_\Omega u_{\varepsilon,n}\Delta v\,dx-\int_{\Sigma_0\cup\Sigma_2}\varepsilon\frac{\partial u_{\varepsilon,n}}{\partial\bar n}v\,d\sigma. \tag{1.5.23}$$

Replacing v by $u_{\epsilon,n}$ in (1.5.22) and integrating by parts, we obtain

$$\int_\Omega u_{\varepsilon,n}f_n\,dx$$

$$=-\int_\Omega\left[a^{kj}u_{\varepsilon,nx_k}u_{\varepsilon,nx_j}+\left(\frac{1}{2}b^k_{x_k}-\frac{1}{2}a^{kj}_{x_kx_j}-c\right)u^2_{\varepsilon,n}\right]dx-\int_\Omega\varepsilon u_{\varepsilon,nx_k}u_{\varepsilon,nx_k}\,dx. \tag{1.5.24}$$

From (1.5.24) it follows that $u_{\epsilon,n}$ are bounded in the norm of $\mathcal{H}$ uniformly in ϵ and n, i.e.

$$\int_\Omega\left[a^{kj}u_{\varepsilon,nx_k}u_{\varepsilon,nx_j}+u^2_{\varepsilon,n}\right]dx+\varepsilon\int_\Omega u_{\varepsilon,nx_k}u_{\varepsilon,nx_k}\,dx\leqslant C\|f_n\|^2_{\mathcal{L}_2},$$

where C does not depend on ϵ or n. This means that there is a sequence $u_{\epsilon_k,n}$ weakly convergent as $\epsilon_k\to 0$ to a function $u_n(x)$ in the space $\mathcal{H}$. Passing to the limit in the equation (1.5.23) as $\epsilon_k\to 0$, we obtain that

$$\int_\Omega vf_n\,dx=B(u_n,v) \tag{1.5.25}$$

for any function $v\in C^{(1)}(\Omega\cup\Sigma)$ such that $v=0$ on $\Sigma_1\cup\Sigma_3$. The last integral in (1.5.23) approaches zero as $\epsilon\to 0$, as was shown in the proof of Theorem 1.5.1. Choosing now a weakly convergent (in $\mathcal{H}$) subsequence of the sequence $\{u_n\}$, and passing to the limit in (1.5.25) as $n_k\to\infty$, we obtain that the limit function $u(x)$ satisfies the required identity (1.4.4). The theorem is proved.

For equations of the type (1.1.1) there is some interest in considering the problem (1.1.4), (1.1.5) in domains with piecewise smooth boundary. For example, problem (1.1.4), (1.1.5) for parabolic equations is studied as a rule in a cylinder, which is an example of such a domain. For simplicity we shall examine the problem (1.1.4), (1.1.5) in a piecewise smooth domain Ω in class $B^{(2,\alpha)}$ with $g=0$ in condition (1.1.5).

The point P of the boundary Σ of the domain Ω will be called a point of smoothness of Σ if in some neighborhood of this point the surface Σ may be represented in the form

$$x_k=f_k(x_1,\ldots,x_{k-1},x_{k+1},\ldots,x_m)$$

for some k, where $f_k \in C^{(2,\alpha)}$, $0 < \alpha < 1$.

The set of points of Σ not satisfying this condition will be denoted by B.

A weak solution of (1.1.4), (1.1.5) with $g = 0$ in a domain $\Omega \in B^{(2,\alpha)}$ is defined to be a bounded measurable function $u(x)$ such that for any $v \in C^{(2)}(\Omega \cup \Sigma)$ equal to zero on $\Sigma_1 \cup \Sigma_3 \cup B$ the integral identity

$$\int_\Omega uL^*(v)\,dx = \int_\Omega vf\,dx \tag{1.5.26}$$

is fulfilled, where f is a bounded measurable function given in Ω.

THEOREM 1.5.5. *Suppose the boundary* Σ *of the domain* Ω *belongs to class* $B^{(2,\alpha)}$, f *is a bounded measurable function in* Ω, $g = 0$, $c(x) \leqslant -c_0 < 0$ *in* Ω, *and* $\beta \leqslant 0$ *at interior points of* $\Sigma_2 \cup \Sigma_0$. *Then in* Ω *there exists a weak solution* $u(x)$ *of the boundary value problem* (1.1.4), (1.1.5) *satisfying the inequality* (*maximum principle*)

$$|u| \leqslant \sup \frac{|f|}{c_0}. \tag{1.5.27}$$

PROOF. This theorem is proved in a manner similar to the proof of Theorem 1.5.1. We construct a sequence of domains Ω_δ such that outside a δ-neighborhood of the set B, all domains Ω_δ coincide with Ω, each Ω_δ belongs to the class $A^{(2,\alpha)}$, and $\Omega_\delta \subset \Omega$. Let $u^\delta_{\epsilon,n}$ be the solution of the Dirichlet problem for the elliptic equation

$$L_\epsilon(u) = f_n \quad \text{in } \Omega_\delta$$

with condition $u^\delta_{\epsilon,n} = 0$ on the boundary Σ^δ of the domain Ω_δ, where $f_n \in C^{(2)}(\Omega \cup \Sigma)$, $|f_n| \leqslant \sup|f|$, $f_n \to f$ as $n \to \infty$ in the norm of $\mathfrak{L}_2(\Omega)$. We apply Green's formula in the domain Ω to the functions $u^\delta_{\epsilon,n}$ and $v\phi_\delta$, where the infinitely differentiable function ϕ_δ is equal to zero in a δ-neighborhood of the set B and is equal to unity outside a 2δ-neighborhood of this set. Taking account of the fact that $u^\delta_{\epsilon,n} = 0$ on Σ^δ, we obtain

$$\begin{aligned} \int_\Omega f_n v\varphi_\delta\,dx = \int_\Omega \varepsilon\Delta(v\varphi_\delta)\,u^\delta_{\varepsilon,n}\,dx + \int_\Omega L^*(v\varphi_\delta)\,u^\delta_{\varepsilon,n}\,dx \\ -\int_{\Sigma_0 \cup \Sigma_2} \varepsilon v\varphi_\delta \frac{\partial u^\delta_{\varepsilon,n}}{\partial n}\,d\sigma. \end{aligned} \tag{1.5.28}$$

It is clear that the estimate

$$|u^\delta_{\varepsilon,n}| \leqslant \sup \frac{|f|}{c_0}$$

holds for the functions $u^\delta_{\epsilon,n}$. We extract a subsequence from the sequence $u^\delta_{\epsilon,n}$

which converges weakly to u_n^δ as $\epsilon_k \to 0$. Passing to the limit as $\epsilon_k \to 0$ in (1.5.28), we obtain

$$\int_\Omega f_n v\varphi_\delta dx = \int_\Omega L^*(v\varphi_\delta)\, u_n^\delta dx. \tag{1.5.29}$$

Suppose $u_{n_k}^\delta \to u^\delta$ as $n_k \to \infty$ weakly in the space $\mathfrak{L}_2(\Omega)$. It is clear from (1.5.29) that

$$\int_\Omega f v\varphi_\delta dx = \int_\Omega L^*(v\varphi_\delta)\, u^\delta dx. \tag{1.5.30}$$

We show that the last integral in (1.5.30) approaches $\int_\Omega L^*(v)u\,dx$, if $\delta \to 0$ through a sequence δ_k such that $u^{\delta_k} \to u$ weakly in the space $\mathfrak{L}_2(\Omega)$ as $\delta_k \to 0$. We have

$$\begin{aligned}\int_\Omega L^*(v\varphi_\delta)\, u^\delta dx = \int_\Omega L^*(v)\, u^\delta\varphi_\delta dx + 2\int_\Omega a^{kj} v_{x_k}\varphi_{\delta x_j} u^\delta dx \\ + \int_\Omega (L^*(\varphi_\delta) - c^*\varphi_\delta)\, u^\delta v\, dx.\end{aligned} \tag{1.5.31}$$

We note that the last two integrals in (1.5.31) are really taken over the 2δ-neighborhood of the set B, since the integrand in these integrals is equal to zero outside this neighborhood. This 2δ-neighborhood of B has measure no greater than $M\delta^2$. Since $v = 0$ at points of B, $|\phi_{\delta x_j}| = O(\delta^{-1})$, $|\phi_{\delta x_k x_j}| = O(\delta^{-2})$, and u^δ is bounded uniformly in δ, it follows that the last two integrals approach zero as $\delta \to 0$. Therefore passing to the limit in (1.5.31) as $\delta \to 0$, we obtain that the function $u(x)$ satisfies (1.5.26) and is therefore a weak solution of (1.1.4), (1.1.5). The theorem is proved.

One may also construct a weak solution of the problem (1.1.4), (1.1.5) in a domain Ω of class $B^{(2,\alpha)}$ in the space $\mathfrak{L}_p(\Omega)$, if $f \in \mathfrak{L}_p(\Omega)$.

§6. Uniqueness theorems for weak solutions of the first boundary value problem

In §§3–5 we have constructed weak solutions of the boundary value problem (1.1.4), (1.1.5) in various function spaces. In this section we shall present sufficient conditions for the uniqueness of weak solutions of (1.1.4), (1.1.5), and give examples wherein these conditions are violated and the weak solution of this problem is not unique.

The proof of the uniqueness theorem is based on the method of elliptic regularization of the adjoint equation.

We begin with some auxiliary propositions.

LEMMA 1.6.1 *Let $v(x)$ satisfy the equation*

$$\varepsilon\Delta v + L(v) = f, \quad \varepsilon = \text{const} > 0, \tag{1.6.1}$$

in a domain Ω *of class* $A^{(2,\alpha)}$, *as well as the condition*

$$v|_{\Sigma} = 0, \tag{1.6.2}$$

where $v \in C^{(2)}(\Omega \cup \Sigma)$, $a^{kj} \in C_{(2)}(\Omega)$, $b^j \in C_{(1)}(\Omega)$, $c \in C_0(\Omega)$, $f \in C_{(0)}(\Omega)$. *Suppose either*

$$c - \frac{1}{2} b^k_{x_k} + \frac{1}{2} a^{kj}_{x_k x_j} < \text{const} < 0, \quad \text{or} \quad c < \text{const} < 0 \text{ in } \Omega \cup \Sigma. \tag{1.6.3}$$

Assume that on the boundary Σ

$$\left|\frac{\partial v}{\partial x_j}\right| \leqslant C_1 \varepsilon^{-\frac{1}{2}}, \quad C_1 = \text{const}, \; j = 1, \ldots, m, \tag{1.6.4}$$

where C_1 *does not depend on* ϵ. *Then*

$$\int_\Omega \varepsilon^2 (\Delta v)^2 \, dx \leqslant C_2 \tag{1.6.5}$$

and the constant C_2 *is also independent of* ϵ.

PROOF. Multiply equation (1.6.1) by v and integrate over Ω. Integrating by parts, we obtain

$$\begin{gathered}\int_\Omega \varepsilon v_{x_k} v_{x_k} dx + \int_\Omega a^{kj} v_{x_k} v_{x_j} dx \\ - \int_\Omega \left[c - \frac{1}{2} b^j_{x_j} + \frac{1}{2} a^{kj}_{x_k x_j}\right] v^2 dx = - \int_\Omega f v dx.\end{gathered} \tag{1.6.6}$$

If $c < 0$, then the function v is bounded uniformly in ϵ by virtue of the maximum principle. Therefore if condition (1.6.3) is satisfied, then from (1.6.6) it follows that

$$\varepsilon \int_\Omega v_{x_k} v_{x_k} dx + \int_\Omega a^{kj} v_{x_k} v_{x_j} dx + \int_\Omega v^2 dx \leqslant C_3 \int_\Omega f^2 dx. \tag{1.6.7}$$

Here and in the following, we denote by C_j constants which are independent of ϵ.

We multiply (1.6.1) by $\epsilon \Delta v$ and integrate over Ω. Thus

$$\int_\Omega \varepsilon^2 (\Delta v)^2 dx + \int_\Omega \varepsilon \Delta v a^{kj} v_{x_k x_j} dx = \int_\Omega (f - cv - b^j v_{x_j}) \varepsilon \Delta v dx. \tag{1.6.8}$$

It is easy to see that

$$\left|\int_\Omega \varepsilon \Delta v (f - cv) \, dx\right| \leqslant \frac{1}{2} \int_\Omega \varepsilon^2 (\Delta v)^2 dx + \frac{1}{2} \int_\Omega (f - cv)^2 dx.$$

By assumption, at points of the boundary of Ω, (1.6.4) holds. Therefore, considering (1.6.7) and (1.6.4) and integrating by parts, we obtain

$$\left|\int_\Omega \varepsilon b^j v_{x_j} v_{x_k x_k} dx\right| \leqslant \left|\int_\Omega \varepsilon b^j v_{x_j x_k} v_{x_k} dx + \int_\Omega \varepsilon b^j_{x_k} v_{x_j} v_{x_k} dx + \int_\Sigma \varepsilon b^j v_{x_k} v_{x_j} n_k d\sigma\right|$$
$$\leqslant \left| -\int_\Omega \frac{1}{2} \varepsilon b^j_{x_j} v_{x_k} v_{x_k} dx + \int_\Omega \varepsilon b^j_{x_k} v_{x_j} v_{x_k} dx\right|$$
$$+ \left|\int_\Sigma \left[\varepsilon b^j v_{x_j} v_{x_k} n_k - \frac{\varepsilon}{2} b^j v_{x_k} v_{x_k} n_j\right] d\sigma\right| \leqslant C_4.$$

Here as always, $\bar{n} = (n_1, \cdots, n_m)$ is the unit vector internally normal to Σ.

Furthermore, by integration by parts we obtain

$$\int_\Omega \varepsilon v_{x_s x_s} a^{kj} v_{x_k x_j} dx = -\int_\Omega \varepsilon v_{x_s} a^{kj} v_{x_k x_j x_s} dx$$
$$-\int_\Omega \varepsilon v_{x_s} a^{kj}_{x_s} v_{x_k x_j} dx - \int_\Sigma \varepsilon a^{kj} v_{x_s} v_{x_k x_j} n_s d\sigma$$
$$= \int_\Omega \varepsilon a^{kj} v_{x_s x_k} v_{x_s x_j} dx + \int_\Omega \left[\varepsilon a^{kj}_{x_j} v_{x_s} v_{x_s x_k} - \varepsilon a^{kj}_{x_s} v_{x_s} v_{x_j x_k}\right] dx \tag{1.6.9}$$
$$+ \int_\Sigma \left[\varepsilon a^{kj} v_{x_s} v_{x_k x_s} n_j - \varepsilon a^{kj} v_{x_s} v_{x_k x_j} n_s\right] d\sigma \equiv I_1 + I_2 + I_3.$$

We note that the integral I_1 is nonnegative. We estimate I_2. We have

$$\left|\int_\Omega \varepsilon a^{kj}_{x_j} v_{x_s} v_{x_s x_k} dx\right| = \left|\int_\Omega \frac{1}{2} \varepsilon a^{kj}_{x_j x_k} v_{x_s} v_{x_s} dx + \int_\Sigma \frac{1}{2} \varepsilon a^{kj}_{x_j} v_{x_s} v_{x_s} n_k d\sigma\right| \leqslant C_5. \tag{1.6.10}$$

In exactly the same way we obtain

$$\left|\int_\Omega \varepsilon a^{kj}_{x_s} v_{x_s} v_{x_k x_j} dx\right|$$
$$= \left|\int_\Omega \varepsilon a^{kj}_{x_s} v_{x_s x_k} v_{x_j} dx + \int_\Omega \varepsilon a^{kj}_{x_s x_k} v_{x_s} v_{x_j} dx + \int_\Sigma \varepsilon a^{kj}_{x_s} v_{x_s} v_{x_j} n_k d\sigma\right|. \tag{1.6.11}$$

The last two integrals are bounded uniformly in ϵ by virtue of (1.6.4) and (1.6.7). We write the first integral on the right of (1.6.11) in the form

$$\int_\Omega \frac{1}{2} \varepsilon a^{kj}_{x_s} (v_{x_k} v_{x_j})_{x_s} dx = -\int_\Omega \frac{1}{2} \varepsilon a^{kj}_{x_s x_s} v_{x_k} v_{x_j} dx$$
$$-\int_\Sigma \frac{1}{2} \varepsilon a^{kj}_{x_s} v_{x_k} v_{x_j} n_s d\sigma.$$

From this and from (1.6.10) and (1.6.11) it follows that $|I_2| \leqslant C_6$.

To estimate I_3 we subdivide Σ into pieces Σ^t, $t = 1, \cdots, N$, and introduce in the neighborhood of Σ^t local coordinates

$$y_k = F^k(x_1, \ldots, x_m), \quad k = 1, \ldots, m,$$

so that the boundary Σ^t lies on the plane $y_m = 0$. Since $v = 0$ on Σ,

$$I_3 = \int_{\cup\Sigma^t} \varepsilon a^{kj} v_{y_m} F^m_{x_s} F^m_{x_j} \left(v_{y_m y_m} F^m_{x_k} F^m_{x_s} + v_{y_m y_l} F^m_{x_k} F^l_{x_s} + v_{y_m} F^m_{x_k x_s}\right) \varkappa\, dy_1 \ldots dy_{m-1}$$

$$- \int_{\cup\Sigma^t} \varepsilon a^{kj} v_{y_m} F^m_{x_s} F^m_{x_s} \left(v_{y_m y_m} F^m_{x_k} F^m_{x_j} + v_{y_m y_l} F^m_{x_k} F^l_{x_j} + v_{y_m} F^m_{x_k x_j}\right) \varkappa\, dy_1 \ldots dy_{m-1}$$

$$= - \int_{\cup\Sigma^t} \frac{1}{2} \varepsilon a^{kj} (v^2_{y_m})_{y_l} \left[F^m_{x_k} F^l_{x_j} F^m_{x_s} F^m_{x_s} - F^m_{x_k} F^l_{x_s} F^m_{x_s} F^m_{x_j}\right] \varkappa\, dy_1 \ldots dy_{m-1}$$

$$- \int_{\cup\Sigma^t} \varepsilon a^{kj} (v_{y_m})^2 F^m_{x_s} \left[F^m_{x_s} F^m_{x_k x_j} - F^m_{x_j} F^m_{x_k x_s}\right] \varkappa\, dy_1 \ldots dy_{m-1}. \tag{1.6.12}$$

Here $l \neq m$, κ is some bounded function, and it is assumed that the summation on l goes from 1 to $m - 1$, whereas the remaining repeated indices are summed from 1 to m. The last two integrals in (1.6.12) are bounded uniformly in ϵ. This is easily verified by transforming the first of them by integration by parts, and using (1.6.4). Thus Lemma 1.6.1 is proved.

Suppose that in a neighborhood of a point P the boundary Σ is given by the equation

$$F(x_1, \ldots, x_m) = 0, \quad F \in C^{(2)},$$

where grad F is different from zero on Σ and is directed along the internal normal $\bar{n}$. At points of Σ^0 we consider the function

$$\beta^* = L^*(F). \tag{1.6.13}$$

It was shown in §5 that at points interior to Σ^0 and also at points of Σ^0 which are limits of interior points, if there $(\partial/\partial\bar{n})(a^{kj}F_{x_k}F_{x_j}) = 0$, then the function β^* coincides in sign with the Fichera function (1.1.3) for the operator L^*.

THEOREM 1.6.1. *Suppose* $c^* < 0$ *in* $\Omega \cup \Sigma$, *suppose that* $\beta^* < 0$ *at points of* Σ_1, *and let* $\Omega \in A^{(2,\alpha)}$. *Assume that the coefficients of the equation* $L^*(v) = a^{kj}v_{x_k x_j} + b^{*k}v_{x_k} + c^*v = 0$ *may be continued to the* δ*-neighborhood* G_δ *of the set* $\overline{\Sigma_0 \cup \Sigma_2}$ *in such a way that* $a^{kj}\xi_k\xi_j \geqslant 0$ *in* $\Omega \cup G_\delta$, *while* $a^{kj} \in C_{(2)}(\Omega \cup G_\delta)$, $b^{*k} \in C_{(1)}(\Omega \cup G_\delta)$, $c^* \in C^{(0,\alpha)}(\overline{\Omega \cup G_\delta})$, $0 < \alpha < 1$. *Then the function* $u(x)$ *in the space* $\mathfrak{L}_2(\Omega)$ *is equal to zero almost everywhere in* Ω, *if for any* $v \in C^{(2)}(\bar{\Omega})$ *equal to zero on* $\Sigma_3 \cup \Sigma_1$,

$$\int_\Omega L^*(v)\, u\, dx = 0 \tag{1.6.14}$$

and if the set Γ *of boundary points for* $\Sigma_0 \cup \Sigma_2$ *on* Σ *may be divided into closed nonintersecting subsets* $\Gamma_1, \Gamma_2, \Gamma_3$ *such that the following conditions are fulfilled (in particular, any of the* Γ_i *may be empty):*

1) *In some neighborhood of each of its points the set* Γ_1 *lies on the intersection of the surface* $F(x_1, \cdots, x_m) = 0$ *(defining* Σ*) and a surface* $\Psi(x_1, \cdots, x_m) = 0$, *the functions* F *and* Ψ *belong to class* $C^{(2)}$, *and the vector* $\bar{n}$ *normal to* Σ *on* Γ_1 *is not orthogonal to the surface* $\Psi = 0$. *The function* $u(x)$ *belongs to class* $\mathfrak{L}_3(Q)$ *in some neighborhood* Q *of the set* Γ_1.

2) *In some neighborhood of each of its points the set* Γ_2 *lies on the intersection of the surfaces* $F = 0$ *(defining* Σ*) and* $\Psi = 0$, *where* $F, \Psi \in C^{(2)}$, *and the vector* $\bar{n}$ *is not orthogonal to the surface* $\Psi = 0$. *At the points of* Γ_2,

$$a^{kj}\Psi_{x_k}\Psi_{x_j} = 0.$$

3) *The area of the* δ*-neighborhood of* Γ_3 *on* Σ *is of the order* δ^q, *where* $q \geqslant 2$.

Proof. We construct the domain Ω_δ such that $\Omega_\delta \subset \Omega \cup G_\delta$ and $\Omega_\delta \supset \Omega$. We shall assume that the domain Ω_δ with boundary Σ^δ belongs to class $A^{(2,\alpha)}$, $0 < \alpha < 1$, and the set $\Sigma_2 \cup \Sigma_0$ lies inside Ω_δ. By Γ^δ we shall denote a δ-neighborhood of the set Γ. Suppose $a(x) \in C^{(2,\alpha)}(\bar{\Omega}_\delta)$, $a = 0$ in Ω, and $a > 0$ in $\bar{\Omega}_\delta \setminus \bar{\Omega}$. Let v^1 satisfy the equation

$$\varepsilon\Delta v^1 + L^*(v^1) + a\Delta v^1 = \Phi, \quad \varepsilon > 0, \tag{1.6.15}$$

in Ω_δ, together with the condition

$$v^1 = 0 \text{ on } \Sigma^\delta, \tag{1.6.16}$$

where $\Phi \in C_0^\infty(\Omega)$. We note that on the part of the boundary Σ^δ which does not belong to Σ we have $a^{kj}n_kn_j + an_jn_j \neq 0$. At points of Σ belonging to Σ^δ, either $a^{kj}n_kn_j \neq 0$ or $\beta^* < 0$. Therefore, applying Lemma 1.5.1 to the operator $L^*(v) + a\Delta v$, we obtain that for v^1 the inequality

$$\left|\frac{\partial v^1}{\partial x_j}\right| \leqslant C_1\varepsilon^{-\frac{1}{2}} \text{ on } \Sigma^\delta$$

is fulfilled, where the constant C_1 does not depend on ϵ. From this and Lemma 1.6.1 it follows that estimate (1.6.5) is valid for v^1. Let $\phi^\delta(x)$ be a function such that

$$\varphi^\delta \in C^{(2)}(\overline{\Omega \cup G_\delta}), \quad \varphi^\delta = 0 \text{ in } \Gamma^\delta, \quad \varphi^\delta = 1 \text{ outside } \gamma^\delta,$$

where γ^δ is a neighborhood of Γ containing Γ^δ, so that $\Omega \setminus \gamma^\delta$ contains any given closed region in Ω if δ is chosen sufficiently small, and where $0 \leqslant \phi^\delta \leqslant 1$.

The function $v = v^1\phi^\delta$ may be substituted into the identity (1.6.14), since $v = 0$ on $\Sigma_3 \cup \Sigma_1$. We have

$$0 = \int_\Omega L^*(v^1\varphi^\delta)\,u\,dx = \int_\Omega \varphi^\delta L^*(v^1)\,u\,dx + \int_\Omega v^1(L^*(\varphi^\delta) - c^*\varphi^\delta)\,u\,dx + \int_\Omega 2a^{kj}\varphi^\delta_{x_k} v^1_{x_j} u\,dx.$$

From this, using (1.6.15), we obtain that

$$\int_\Omega \varphi^\delta \Phi u\,dx = \int_\Omega \varepsilon\varphi^\delta \Delta v^1 u\,dx - \int_\Omega v^1(L^*(\varphi^\delta) - c^*\varphi^\delta)\,u\,dx - \int_\Omega 2a^{kj}\varphi^\delta_{x_k} v^1_{x_j} u\,dx. \tag{1.6.17}$$

Since $\Phi \in C_0^\infty(\Omega)$ it follows that $\Phi\phi^\delta = \Phi$ for sufficiently small δ, and the integral on the left of (1.6.17) coincides with the integral $\int_\Omega \Phi u\,dx$. We show that the right side of (1.6.17) is as small as desired for sufficiently small δ and $\epsilon(\delta)$. This means that the left side of this equation is equal to zero for any smooth function Φ of compact support, so that $u = 0$ almost everywhere in Ω.

Suppose $u_n \in C_0^\infty(\Omega)$ and $u_n \to u$ in the norm of $\mathfrak{L}_2(\Omega)$ as $n \to \infty$. Since (1.6.5) holds for the function v^1 and v^1 is bounded uniformly in ϵ and δ because of the maximum principle, we have

$$\begin{aligned}\left|\int_\Omega \varepsilon\Delta v^1\varphi^\delta u\,dx\right| &= \left|\int_\Omega \varepsilon\Delta v^1\varphi^\delta((u - u_n) + u_n)\,dx\right| \\ &\leqslant \left|\int_\Omega \varepsilon\Delta v^1\varphi^\delta(u - u_n)\,dx\right| + \left|\int_\Omega \varepsilon\Delta(\varphi^\delta u_n)\,v^1 dx\right| \\ &\leqslant \left(\int_\Omega \varepsilon^2(\Delta v^1)^2\,dx\right)^{1/2}\left(\int_\Omega (u - u_n)^2\,dx\right)^{1/2} + \left|\int_\Omega \varepsilon v^1\Delta(\varphi^\delta u_n)\,dx\right|.\end{aligned} \tag{1.6.18}$$

It is easy to see that the right side of this last inequality approaches zero as $n \to \infty$ and $\epsilon_n \to 0$ for fixed δ. We note that the two last integrals in (1.6.17) are really taken only over the intersection of the γ^δ-neighborhood of Γ with the domain Ω, since $\phi^\delta \equiv 1$ outside γ^δ. Suppose $\gamma^\delta = \gamma_1^\delta \cup \gamma_2^\delta \cup \gamma_3^\delta$, where γ_k^δ is the intersection of γ^δ with some neighborhood of Γ_k $(k = 1, 2, 3)$.

In order to estimate those integrals taken over $\gamma_1^\delta \cap \Omega$, we decompose the neighborhood γ_1^δ of the set Γ_1 into pieces $Q_{1,\delta}^t$ $(t = 1, \cdots, N_1)$ so that in $Q_{1,\delta}^t$ we may introduce local coordinates $y_1^t, \cdots, y_m^t$ such that

$$y_m^t = F(x_1, \ldots, x_m), \; y_{m-1}^t = \Psi(x_1, \ldots, x_m),$$

and in $Q_{1,\delta}^l \cap Q_{1,\delta}^k$ we have $y_m^l = y_m^k$, $y_{m-1}^l = y_{m-1}^k$.

In the new coordinates $y_1, \cdots, y_m$ the operator $\epsilon\Delta v^1 + a\Delta v^1 + L^*(v^1)$ will have some form

$$(\varepsilon + a)\left(\mu^{kj} v^1_{y_k y_j} + \nu^k v^1_{y_k}\right) + \alpha^{kj} v^1_{y_k y_j} + \beta^{*j} v^1_{y_j} + c^* v^1.$$

We denote $\Omega \cap Q_{1,\delta}^t$ by R_1^t. In the neighborhood of Γ_1 for the function ϕ^δ we take

$$\varphi^\delta = \varphi\left(\frac{\delta y_{m-1}^2 + y_m^2}{\delta^2}\right),$$

where $\phi(s) = 0$ for $s \leqslant 1$, $\phi(s) = 1$ for $s \geqslant 2$, $0 < \phi < 1$ for $1 < s < 2$, and $\phi(s)$ is a smooth function of s. We have

$$\int_{\Omega \cap \gamma_1^\delta} (L^*(\varphi^\delta) - c^*\varphi^\delta) v^1 u\,dx = \int_{\bigcup R_1^t} \left(\alpha^{kj}\varphi^\delta_{y_k y_j} + \beta^{*j}\varphi^\delta_{y_j}\right) u v^1 \varkappa\,dy, \quad (1.6.19)$$

where κ is a smooth positive function. We note that the measure of the domain γ_1^δ, where $\phi^\delta_{y_j}$ and $\phi^\delta_{y_j y_k}$ may differ from zero, is of order $\delta^{3/2}$, and in this domain

$$\begin{gathered}\varphi^\delta_{y_m} = O(\delta^{-1}), \; \varphi^\delta_{y_{m-1}} = O\left(\delta^{-\frac{1}{2}}\right), \; \varphi^\delta_{y_m y_m} = O(\delta^{-2}), \\ \varphi^\delta_{y_{m-1} y_{m-1}} = O(\delta^{-1}), \; \varphi^\delta_{y_{m-1} y_m} = O(\delta^{-3/2}).\end{gathered} \quad (1.6.20)$$

Since $\alpha^{kj} n_k n_j = 0$, $\alpha^{kj} n_k = 0$, $\alpha^{mm}_{y_j} = 0$ on Γ_1, in the domain γ_1^δ we have

$$\alpha^{mm} = O(\delta), \quad \alpha^{m-1,\, m-1} = O(1), \quad \alpha^{m,\, m-1} = O\left(\delta^{\frac{1}{2}}\right). \quad (1.6.21)$$

Using relations (1.6.20) and (1.6.21), we estimate integral (1.6.19). Using Hölder's inequality, we obtain

$$\left|\int_{R_1^t} \left(\alpha^{kj}\varphi^\delta_{y_k y_j} + \beta^{*j}\varphi^\delta_{y_j}\right) v^1 u \varkappa\,dy\right| \leqslant \int_{R_1^t} C_2 \delta^{-1} |u|\,dy$$

$$\leqslant C_2\delta^{-1}\left(\int_{R_1^t} dy\right)^{2/3}\left(\int_{R_1^t} |u|^3\,dy\right)^{1/3} \leqslant C_3\left(\int_{R_1^t} |u|^3\,dy\right)^{1/3}.$$

The last integral tends to zero as $\delta \to 0$, since by assumption $u \in \mathfrak{L}_3(Q_1)$, where Q_1 is some neighborhood of Γ_1 and $\operatorname{mes} \gamma_1^\delta = O(\delta^{3/2})$.

We now estimate the last integral in (1.6.17), taken over $\Omega \cap \gamma_1^\delta$:

$$\left|\int_{\Omega \cap \gamma_1^\delta} a^{kj} v^1_{x_j} \phi^\delta_{x_k} u\,dx\right| \leqslant \left(\int_{\Omega \cap \gamma_1^\delta} a^{kj} v^1_{x_k} v^1_{x_j}\,dx\right)^{½} \cdot \left(\int_{\Omega \cap \gamma_1^\delta} a^{kj}\phi^\delta_{x_k}\phi^\delta_{x_j} u^2\,dx\right)^{½}. \quad (1.6.22)$$

In order to estimate the first integral on the right side of (1.6.22) we multiply the equation (1.6.15) by $v^1(1-\phi^{2\delta})\eta$ and integrate over $\Omega_\delta \cap Q^t_{1,2\delta}$. Here the function $\eta(y_1, \cdots, y_{m-2}) \in C^{(2)}(Q^t_{1,2\delta})$ is chosen so that $\eta > 0$ inside $Q^t_{1,2\delta}$, and $(1-\phi^{2\delta})\eta = 0$ together with its first derivatives on the boundary of $Q^t_{1,2\delta}$. Integrating by parts and considering that $v^1 = 0$ on Σ^δ and $a = O(\delta)$ in Ω_δ, we obtain

$$\int_{\Omega_\delta \cap Q^t_{1,2\delta}} \left[-(\epsilon + a)v^1_{x_k}v^1_{x_j}(1-\phi^{2\delta})\eta - a^{kj}v^1_{x_k}v^1_{x_j}(1-\phi^{2\delta})\eta\right] dx$$

$$+ \int_{\Omega_\delta \cap Q^t_{1,2\delta}} \sum_{k=1}^{m} \frac{1}{2}\left[(\epsilon + a)(1-\phi^{2\delta})\eta\right]_{x_k x_k}(v^1)^2\, dx$$

$$+ \int_{\Omega_\delta \cap Q^t_{1,2\delta}} \frac{1}{2}\left[a^{kj}(1-\phi^{2\delta})\eta\right]_{x_k x_j}(v^1)^2\, dx$$

$$+ \int_{\Omega_\delta \cap Q^t_{1,2\delta}} \left[c^*(1-\phi^{2\delta})\eta - \frac{1}{2}(b^{*j}(1-\phi^{2\delta})\eta)_{x_j}\right](v^1)^2\, dx$$

$$= \int_{\Omega_\delta \cap Q^t_{1,2\delta}} \Phi(1-\phi^{2\delta})\eta v^1\, dx. \tag{1.6.23}$$

From this, using relations (1.6.20) and considering that measure $\gamma_1^{2\delta}$ is of order $\delta^{3/2}$, we obtain

$$\int_{\Omega_\delta \cap Q^t_{1,2\delta}} a^{kj}v^1_{x_k}v^1_{x_j}(1-\phi^{2\delta})\eta\, dx \leqslant C_4(\delta^{1/2} + \epsilon\delta^{-1/2}) + \int_{\Omega_\delta \cap Q^t_{1,2\delta}} \left|a^{kj}\phi^{2\delta}_{x_k x_j}\right|\eta(v^1)^2\, dx. \tag{1.6.24}$$

In order to estimate the last integral, we introduce the local coordinates $y_1, \cdots, y_m$ in the domain $Q^t_{1,2\delta}$. Using (1.6.20) and (1.6.21), we obtain

$$\int_{\Omega_\delta \cap Q^t_{1,2\delta}} \left|a^{kj}\phi^{2\delta}_{x_k x_j}\right|\eta(v^1)^2\, dx \leqslant \int_{\Omega_\delta \cap Q^t_{1,2\delta}} \left|\alpha^{kj}\phi^{2\delta}_{y_k y_j}\right|\eta(v^1)^2\kappa\, dy$$

$$+ \int_{\Omega_\delta \cap Q^t_{1,2\delta}} \left|a^{kj}\phi_{y_l}(y_l)_{x_k x_j}\right|\eta(v^1)^2\kappa\, dy \leqslant C_5\delta^{1/2},$$

where the constants C_4 and C_5 do not depend on ϵ and δ. It is evident that

$$\int_{\Omega \cap \gamma_1^\delta} a^{kj}v^1_{x_k}v^1_{x_j}\, dx \leqslant C \int_{\underset{t}{\cup}(\Omega_\delta \cap Q^t_{1,2\delta})} a^{kj}v^1_{x_k}v^1_{x_j}(1-\phi^{2\delta})\eta\, dy.$$

Furthermore we obtain

$$\int_{\Omega\cap\gamma_1^\delta} a^{kj}\phi^\delta_{x_k}\phi^\delta_{x_j}u^2\,dx = \int_{\cup R_1^t} \alpha^{kj}\phi^\delta_{y_k}\phi^\delta_{y_j}u^2\kappa\,dy$$

$$\leqslant C_6\delta^{-1}\Big(\int_{\cup R_1^t} dy\Big)^{1/3}\Big(\int_{\cup R_1^t}|u|^3\,dy\Big)^{2/3} \leqslant C_7\delta^{-1/2}\Big(\int_{\cup R_1^t}|u|^3\,dy\Big)^{2/3}.$$

Hence for $\epsilon < \delta$ it follows from (1.6.22) that

$$\Big|\int_{\Omega\cap\gamma_1^\delta} a^{kj}v^1_{x_k}\varphi^\delta_{x_j}u\,dx\Big| \leqslant C_8\Big(\int_{\Omega\cap\gamma_1^\delta}|u|^3dx\Big)^{1/3},$$

and by virtue of the assumptions on u the right side of the latter inequality tends to zero as $\delta\to 0$ and $\epsilon<\delta$.

We now estimate the two last integrals in (1.6.17), taken over $\Omega\cap\gamma_2^\delta$, where γ_2^δ is a neighborhood of Γ_2 such that $\phi^\delta\equiv 1$ outside γ_2^δ. As the function ϕ^δ in the neighborhood of Γ_2 we choose

$$\varphi^\delta = \varphi\left(\frac{y_{m-1}^2+y_m^2}{\delta^2}\right),$$

where the function $\phi(s)$ was defined above. The estimate of these integrals, taken over $\Omega\cap\gamma_2^\delta$, is carried out exactly as it was for the domain $\Omega\cap\gamma_1^\delta$. For this we divide the neighborhood γ_2^δ of the set Γ_2 into domains $Q^t_{2,\delta}$, $t=1,\cdots,N_2$, and in $Q^t_{2,\delta}$ pass to new coordinates $y_1,\cdots,y_m$ such that

$$y_m = F(x_1\ldots,x_m),\quad y_{m-1}=\Psi(x_1,\ldots,x_m).$$

It is easy to see that in this process $\alpha^{m-1,m-1}=a^{kj}\Psi_{x_k}\Psi_{x_j}$. Therefore in the δ-neighborhood of Γ_2,

$$\begin{gathered}\varphi^\delta_{y_j}=O(\delta^{-1}),\quad \varphi^\delta_{y_ky_j}=O(\delta^{-2}),\\ \alpha^{m,m}=O(\delta^2),\quad \alpha^{m-1,m-1}=O(\delta),\quad \alpha^{m-1,m}=O(\delta).\end{gathered}\tag{1.6.25}$$

It is clear that the measure of the domain γ_2^δ is of order δ^2. Considering (1.6.25) and setting $\Omega\cap Q^t_{2,\delta}=R^t_2$, we obtain

$$\Big|\int_{\Omega\cap\gamma_2^\delta}(L^*(\varphi^\delta)-c^*\varphi^\delta)v^1u\,dx\Big| = \Big|\int_{\cup R_2^t}(\alpha^{kj}\varphi^\delta_{y_ky_j}$$

$$+\beta^{*k}\varphi^\delta_{y_k})v^1u\varkappa\,dy\Big| \leqslant C_9\delta^{-1}\int_{\cup R_2^t}|u|\,dy \leqslant C_{10}\Big(\int_{\cup R_2^t}|u|^2dy\Big)^{1/2}.$$

The last integral tends to zero as $\delta\to 0$ since $u\in\mathfrak{L}_2(\Omega)$.

An inequality of the type (1.6.22) is clearly valid also for the domain $\Omega\cap\gamma_2^\delta$. Taking account of (1.6.25) in a neighborhood of Γ_2, we obtain

$$\left| \int_{R_2^t} a^{kj} \varphi^\delta_{y_k} \varphi^\delta_{y_j} u^2 \varkappa dy \right| \leqslant C_{11} \delta^{-1} \int_{R_2^t} | u |^2 dy.$$

From (1.6.23) taken for $\Omega_\delta \cap Q^t_{2,2\delta}$ we deduce that

$$\int_{\Omega \cap \lambda_2^\delta} a^{kj} v^1_{x_k} v^1_{x_j} dx \leqslant \int_{\bigcup_t (\Omega_\delta \cap Q^t_{2,2\delta})} a^{kj} v^1_{x_k} v^1_{x_j} (1 - \phi^{2\delta}) \eta \, dx \leqslant C_{12} \epsilon + C_{13} \delta.$$

Thus for $\epsilon < \delta$ we obtain

$$\left| \int_{\Omega \cap \gamma_2^\delta} a^{kj} v^1_{x_k} \varphi^\delta_{x_j} u dx \right| \leqslant C_{14} \left(\int_{\Omega \cap \gamma_2^\delta} | u |^2 dx \right)^{\frac{1}{2}}$$

which means that the last integral in (1.6.17), taken over $\Omega \cap \gamma_2^\delta$, approaches zero as $\delta \to 0$, $\epsilon < \delta$.

We now consider the neighborhood γ_3^δ of Γ_3 and show that the two last integrals in (1.6.17), taken over $\Omega \cap \gamma_3^\delta$, tend to zero as $\delta \to 0$ and $\epsilon < \delta$. By assumption the δ-neighborhood of the set Γ_3 on Σ has area not surpassing $C_{15}\delta^2$. We divide the neighborhood γ_3^δ into sufficiently small regions $Q^t_{3,\delta}$, $t = 1, \cdots, N_3$. In $Q^t_{3,\delta}$ we introduce local coordinates $y_1, \cdots, y_m$ such that Σ lies in the plane $y_m = 0$, and $y_m > 0$ in Ω. Let the smooth function $\phi_1^\delta (y_1, \cdots, y_{m-1})$ be defined at points of $\Sigma \cap Q^t_{3,\delta}$ so that $\phi_1^\delta = 0$ in the δ-neighborhood of Γ_3, $\phi_1^\delta > 0$ outside the closure of this neighborhood, and $\phi_1^\delta \equiv 1$ outside the 2δ-neighborhood of Γ_3, whereas everywhere on Σ we have $0 \leqslant \phi_1^\delta \leqslant 1$. Let the smooth function $\phi_2^\delta(y_m) = 0$ for $y_m \leqslant \delta$, $\phi_2^\delta(y_m) = 1$ for $y_m \geqslant 2\delta$, and $0 < \phi_2^\delta < 1$ for $\delta < y_m < 2\delta$. The function ϕ^δ in a neighborhood of Γ_3 is defined by the equation

$$\varphi^\delta = \varphi_1^{\sqrt{\delta}} \cdot \varphi_2^\delta.$$

It is easy to see that the measure of the subset of γ_3^δ, where $\phi^\delta_{y_j}$ and $\phi^\delta_{y_k y_j}$ may be different from zero, does not surpass $C_{16}\delta^2$. We may also assume that

$$\varphi^\delta_{1y_j} = O(\delta^{-1}), \quad \varphi^\delta_{1y_k y_j} = O(\delta^{-2}), \quad \varphi^\delta_{2y_m} = O(\delta^{-1}), \quad \varphi^\delta_{2y_m y_m} = O(\delta^{-2}).$$

Therefore we have the relations

$$\begin{aligned} &\varphi^\delta_{y_m} = O(\delta^{-1}); \; \varphi^\delta_{y_j} = O\left(\delta^{-\frac{1}{2}}\right), \quad \text{if} \quad j \neq m; \\ &\varphi^\delta_{y_m y_m} = O(\delta^{-2}); \; \varphi^\delta_{y_m y_j} = O\left(\delta^{-\frac{3}{2}}\right), \; j \neq m; \\ &\varphi^\delta_{y_k y_j} = O(\delta^{-1}), \quad \text{if} \quad k \neq m, \; j \neq m. \end{aligned} \tag{1.6.26}$$

In the region γ_3^δ we have

$$\alpha^{mm}=O(\delta),\ \alpha^{mj}=O\left(\delta^{\frac{1}{2}}\right),\qquad (1.6.27)$$
$$\alpha^{kj}=O(1)\ \text{ for }\ k\neq m,\ j\neq m.$$

Exactly as in the preceding case, we obtain that the second integral on the right of (1.6.17), taken over $\Omega\cap\gamma_3^\delta$, does not surpass

$$C_{17}\left(\int_{\Omega\cap\gamma_3^\delta}|u|^2\,dx\right)^{\frac{1}{2}},$$

and the last integral in (1.6.17), taken over $\Omega\cap\gamma_3^\delta$, is estimated exactly as for γ_2^δ, with use of (1.6.23), (1.6.26), (1.6.27).

The theorem is thereby proved.

We note that the condition

$$\beta^*<0 \text{ on } \Sigma_1$$

of Theorem 1.6.1 is always fulfilled if all points of Σ_1 are limits of interior points of Σ^0, and, moreover, either the coefficients of the operator L^* may be continued to a neighborhood of Σ_1 so that $a^{kj}\xi_k\xi_j\geqslant 0$, $a^{kj}\in C^{(1)}$, or $\partial a^{kj}F_{x_k}F_{x_j}/\partial\bar{n}$ is sufficiently small on Σ_1.

Examples show that the assumptions of Theorem 1.6.1 are essential. In particular, if the δ-neighborhood of the set Γ on Σ has area of the order δ, then in this case the weak solution of (1.1.4), (1.1.5) may be nonunique in the class $\mathfrak{L}_p(\Omega)$ for any $p<3$, i.e. Condition 1) of Theorem 1.6.1 may not be weakened.

We consider the heat equation

$$L(u)\equiv u_{xx}-u_t=0 \qquad (1.6.28)$$

in a domain Ω on the (x, t) plane bounded by a smooth closed curve Σ which in a neighborhood of the point $(0, 0)$ coincides with the curve $t=|x|^{2+\epsilon}$, $\epsilon=$ const >0. We assume that the curve Σ has no horizontal tangent except at the points $(0, 0)$ and $(0, 1)$. In this case the point $(0, 0)$ belongs to Σ_2, the point $(0, 1)$ belongs to Σ_1, and all remaining points of Σ belong to Σ_3. Performing the substitution $u=e^{\alpha t}v$, $\alpha=$ const >0, in (1.6.28), we obtain an equation for which $c^*<0$ in $\Omega\cup\Sigma$. We may suppose that the boundary Σ in the neighborhood of $(0, 1)$ is such that $\beta^*<0$ at the point $(0, 1)$. The function

$$w=t^{-1/2}e^{-\frac{x^2}{4t}}$$

satisfies (1.6.28) in Ω. On the boundary Σ of the domain Ω it determines a continuous function, which in a neighborhood of the point $(0, 0)$ is given by

$$w|_\Sigma = |x|^{-\left(1+\frac{\varepsilon}{2}\right)} e^{-\frac{1}{4}|x|^{-\varepsilon}}.$$

It is easy to show that there exists a function $W(x, t)$, continuous in $\Omega \cup \Sigma$, coinciding with w at points of Σ and satisfying (1.6.28) in Ω. The existence of such a function $W(x, t)$ is proved, for example, in [61] and [100]. Since the boundary Σ by assumption is smooth, it is known [37] that the function $W(x, t)$ has continuous derivatives W_x, W_{xx}, W_t everywhere in $\Omega \cup \Sigma$ except perhaps at the points (0, 0) and (0, 1).

We show that the function $u = W(x, t) - w(x, t)$ is a weak solution of (1.1.4), (1.1.5) in the domain Ω with $f \equiv 0$ and $g \equiv 0$, and lies in the space $\mathfrak{L}_p(\Omega)$ for $p < 3$. Clearly $u \neq 0$ on a set of positive measure in Ω since $W(x, t)$ is continuous in $\Omega \cup \Sigma$, and $w(x, t) \to \infty$ for $x = 0$ and $t \to 0$. Thus the problem (1.1.4), (1.1.5) in the case considered has at least two solutions: the function u and the identically zero solution.

We shall show that $u(x, t) \in \mathfrak{L}_p(\Omega)$ for $p < 3$. Since $W(x, t)$ is bounded in Ω, it is sufficient to show that $w(x, t) \in \mathfrak{L}_p(\Omega)$ for $p < 3$. We have

$$\int_\Omega w^p\,dx\,dt = \int_\Omega t^{-\frac{p}{2}} e^{-\frac{px^2}{4t}}\,dx\,dt \leqslant \int_0^{t_0} \left(t^{-\frac{p}{2}} \int_{-t^{1/2+\varepsilon}}^{t^{1/2+\varepsilon}} dx \right) dt + C_1$$

$$= 2\int_0^{t_0} t^{-\frac{p}{2}+\frac{1}{2+\varepsilon}}\,dt + C_1 < C_2; \quad t_0,\ C_1,\ C_2 = \text{const},$$

if $-p/2 + 1/(2+\epsilon) > -1$. When $p < 3$, the inequality $-p/2 + 1/(2+\epsilon) > -1$ is satisifed if the constant $\epsilon > 0$ is chosen sufficiently small. We show now that $u(x, t)$ satisfies integral identity (1.6.14). Suppose $v \in C^{(2)}(\Omega \cup \Sigma)$ and $v = 0$ on $\Sigma_3 \cup \Sigma_1$. By continuity $v = 0$ at all points of Σ. Since $u(x, t)$ is a smooth function in the region $\Omega \cap \{\delta \leqslant t \leqslant 1-\delta\}$, $\delta = \text{const} > 0$, according to Green's formula (1.1.14) the identity

$$\int_{\Omega\cap\{\delta\leqslant t\leqslant 1-\delta\}} L^*(v)\,u\,dx\,dt = \int_{\Omega\cap\{t=1-\delta\}} uv\,dx - \int_{\Omega\cap\{t=\delta\}} uv\,dx \tag{1.6.29}$$

holds. It is easy to see that

$$|u(x, t)| \leqslant C_3 t^{-\frac{1}{2}}, \quad \operatorname{mes}(\Omega\cap\{t=\delta\}) = 2\delta^{\frac{1}{2+\varepsilon}}$$

for small δ. Since v is a smooth function equal to zero on Σ, it follows that $|v| \leqslant C_4 t^{1/(2+\epsilon)}$ in Ω. Hence the integrals on the right side of (1.6.29) approach zero as $\delta \to 0$, if ϵ is sufficiently small. Consequently the function $u(x, t)$

satisfies (1.6.14) and the problem (1.1.4), (1.1.5) may have a nonunique solution in class $\mathfrak{L}_p(\Omega)$, $p < 3$, if the area of a δ-neighborhood of Γ on Σ is of order δ.

One may show by an analogous argument that if the set Γ_3 in Theorem 1.6.1 is not empty, problem (1.1.4), (1.1.5) may have a nonunique solution in the class $\mathfrak{L}_p(\Omega)$ for $p < 2$. For this we consider the heat equation

$$L(u) \equiv \Delta u - u_t = 0 \tag{1.6.30}$$

in a smooth domain Ω of the space $R^m(x_1, \cdots, x_{m-1}, t)$, bounded in a neighborhood of the origin by the surface $t = r^{2+\epsilon}$, where $r^2 = \Sigma_{j=1}^{m-1} x_j^2$, $\epsilon = \text{const} > 0$. We shall assume that the boundary Σ of the domain Ω has normal vector parallel to the t axis only at two points P_1 and P_2, where P_1 is the origin. The function

$$w(x, t) = t^{-\frac{m-1}{2}} e^{-\frac{r^2}{4t}}$$

is continuous on the surface Σ. Therefore there exists a function $W(x, t)$, continuous in $\Omega \cup \Sigma$, satisfying (1.6.30) in $\Omega \cup \Sigma$, and smooth everywhere in $\Omega \cup \Sigma$, with the exception of possibly the two points P_1 and P_2 (see [37, 61]). The functions $u(x, t) = W(x, t) - w(x, t)$, as is easily verified, belong to $\mathfrak{L}_p(\Omega)$ for $p < 1 + 2/(m-1)$ and satisfy (1.6.14) if ϵ is sufficiently small.

Thus in the space R^3 the solution of (1.1.4), (1.1.5) may be nonunique in any class $\mathfrak{L}_p(\Omega)$ for $p < 2$, if Γ_3 is not empty. In this sense the condition $u \in \mathfrak{L}_2(\Omega)$ of Theorem 1.6.1 is exact.

The condition $c^* < 0$ is also essential in Theorem 1.6.1. The function r^α, where $r^2 = \Sigma_{j=1}^2 x_j^2$, $-1 < \alpha < 0$, satisfies the equation

$$r^2 \Delta u - \alpha^2 u = 0 \tag{1.6.31}$$

in the sense of (1.5.3) in any ball $\Omega = \{r < \beta\}$, $\beta > 0$. Since $c = -\alpha^2 < 0$, according to Theorem 1.5.1 there exists a bounded weak solution $u(x)$ of (1.6.31) in Ω with the condition $u|_{r=\beta} = \beta^\alpha$. The function $u - r^\alpha$ belongs to class $\mathfrak{L}_p(\Omega)$ for $|\alpha p| < 2, p \geqslant 1$, and satisfies the identity (1.6.14). Clearly the boundary of the domain Ω belongs to Σ_3. It is easy to see that $u - r^\alpha \neq 0$ on a set of positive measure in Ω, and that $c^* = -\alpha^2 + 4 > 0$.

One may prove the following theorem exactly as Theorem 1.6.1 was proved, except that certain simplifications are introduced. This theorem has to do with uniqueness, in the class of bounded measurable functions, of weak solutions of (1.1.4), (1.1.5), in the sense of identity (1.5.3).

THEOREM 1.6.2. *Suppose* $c^* < 0$ *in* $\Omega \cup \Sigma$, *suppose that* $\beta^* < 0$ *at* *points of* Σ_1, *and suppose* $\Omega \in A^{(2,\alpha)}$. *Assume that the coefficients of* $L^*(v)$ *may be continued into the* δ*-neighborhood* G_δ *of the set* $\Sigma_0 \cup \Sigma_2$ *so that*

$$a^{kj}\xi_k\xi_j > 0 \quad in \quad \overline{\Omega \cup G_\delta}, \quad a^{kj} \in C_{(2)}(\overline{\Omega \cup G_\delta}),$$

$$b^{*k} \in C_{(1)}(\overline{\Omega \cup G_\delta}), \quad c^* \in C^{(0,\alpha)}(\overline{\Omega \cup G_\delta}), \quad 0 < \alpha < 1.$$

Then a measurable bounded function $u(x)$ *is equal to zero almost everywhere in* Ω *if, for any function* v *in* $C^2(\Omega \cup \Sigma)$ *and equal to zero on* $\Sigma_1 \cup \Sigma_3$,

$$\int_\Omega L^*(v)\, u\, dx = 0$$

and the area of the δ*-neighborhood of the set* Γ *on* Σ *has order* δ^q, $q > 0$.

In the proof of Theorem 1.6.2 the function ϕ^δ is constructed in a neighborhood of the set Γ exactly as the function ϕ^δ in the proof of Theorem 1.6.1 was constructed in a neighborhood of Γ_3.

The following theorem on the maximum principle follows from Theorems 1.6.2 and 1.5.1.

THEOREM 1.6.3 *Let the hypotheses of Theorems* 1.6.2 *and* 1.5.1 *be fulfilled. Then any bounded weak solution* $u(x)$ *of* (1.1.4), (1.1.5), *in the sense of* (1.5.3), *satisfies the maximum principle*:

$$|u| \leqslant \max\left\{\sup_\Omega \frac{|f|}{c_0}, \sup_{\Sigma_2 \cup \Sigma_3} |g|\right\}. \tag{1.6.32}$$

An example of a problem (1.1.4), (1.1.5) constructed in [102] shows that the hypothesis of Theorem 1.6.2 regarding the set Γ is essential, since if the area of the δ-neighborhood of the set Γ on Σ does not tend to zero as $\delta \to 0$, then the solution of (1.1.4), (1.1.5) may be nonunique in the class of measurable bounded functions.

We examine a smooth domain Ω on the (x, t) plane bounded by the segment $0 \leqslant x \leqslant 1$ of the line $t = 1$, the curve $t = f(x)$ for $0 \leqslant x \leqslant 1$, and by two smooth curves K_1 and K_2 joining the points (0, 0), (0, 1) and (1, 0), (1, 1) respectively, and such that the boundary of the domain Ω is intersected by straight lines parallel to the t axis at no more than two points. The function $f(x)$ is constructed so that $f \in C^{(2,\alpha)}$, $f(x) = 0$ on a Cantor set E of positive measure lying on the segment $0 \leqslant x \leqslant 1$, and $0 < f(x) < 1$ at points of this segment which do not belong to E. Let

$$L(u) \equiv t^2 u_{tt} - u_t - 2u = 0. \tag{1.6.33}$$

This ordinary differential equation has a general solution of the form

$$u = C_1(1 - 2t + 2t^2) + C_2 t^2 e^{-\frac{1}{t}}, \tag{1.6.34}$$

where C_1 and C_2 are arbitrary constants. Performing the substitution $u = u^1 e^{\alpha t}$ in (1.6.33) and choosing the constant $\alpha > 0$ such that $-\alpha + t^2\alpha^2 - 4\alpha t < 0$, $t^2\alpha^2 - \alpha - 2 < 0$ for $t \geqslant 0$, we obtain the equation

$$L_1(u^1) \equiv t^2 u^1_{tt} - u^1_t + 2\alpha t^2 u^1_t + (t^2\alpha^2 - \alpha - 2)\, u^1 = 0, \tag{1.6.35}$$

for which $c^* = (t^2\alpha^2 - \alpha - 2) - 4\alpha t + 2 < 0$ and $c < 0$.

It is easy to see that for the domain Ω constructed above and for equations (1.6.33) and (1.6.35), the segment $0 \leqslant x \leqslant 1$ of the line $t = 1$ belongs to Σ_3, the set E on the line $t = 0$ belongs to Σ_2, the points of the curve $t = f(x)$ where $f(x) \neq 0$ belong to Σ_3, and the points of the curves K_1 and K_2 belong to Σ_3 and Σ_0. Therefore all hypotheses of Theorem 1.6.2 are fulfilled for equation (1.6.35) with the exception of the condition regarding Γ. For this equation, the set Γ contains the set E and has positive measure on the segment $0 \leqslant x \leqslant 1$.

The function v in (1.6.14) should be zero on Σ_3, but since v is smooth in $\Omega \cup \Sigma$, it also vanishes on E. Let $w(t)$ be a solution of (1.6.35) such that $w(1) = 0$ and $w(t) \not\equiv 0$. From (1.6.34) it easily follows that such a solution $w(t)$ exists. It is easily verified that the function $u_1(x, t)$ defined by the equation $u_1(x, t) = w(t)$ if $x \in E$, and equal to zero at the remaining points of $\Omega \cup \Sigma$, is a weak solution of the homogeneous problem $L_1(u) = 0$ in Ω, $u = 0$ on $\Sigma_3 \cup \Sigma_2$, where $L_1(u)$ is the operator (1.6.35). It is clear that $u(x, t)$ is a bounded measurable function and that $u(x, t)$ differs from zero on a set of positive measure in Ω.

REMARK 1. The condition $\beta^* < 0$ at the points Σ_1 of the boundary Σ of the region Ω in Theorem 1.6.1 may be replaced by the following conditions: 1) $\beta^* \leqslant 0$ at interior points of Σ_1. 2) The coefficients of the operator $L^*(v)$ may be continued to some neighborhood of the set Γ^1 of boundary points of Σ_1 in Σ so that the conditions indicated in Theorem 1.6.1 for the neighborhood G_δ will be satisfied. 3) The set Γ^1 may be divided into subsets, each of which satisfies one of conditions 1), 2) or 3) formulated in Theorem 1.6.1 for Γ_1, Γ_2 and Γ_3 respectively.

The proof of Theorem 1.6.1 under this hypothesis is carried out in the manner of the previous proof which used the condition $\beta^* < 0$ on Σ_1. It is only necessary to include a δ-neighborhood of the set Γ^1 in the domain Ω_δ.

In exactly the same way we may replace the condition $\beta^* < 0$ on Σ_1 in Theorem 1.6.2 by the above formulated conditions 1) and 2) together with the additional assumption that the area of the δ-neighborhood of Γ^1 on Σ is of order δ^q for some $q > 0$.

REMARK 2. The condition $\beta^* < 0$ on Σ_1 in Theorems 1.6.1 and 1.6.2 may be replaced by the condition $\beta^* \leqslant 0$ on Σ_1 and also in some neighborhood of the set $\bar{\Sigma}_3 \cap \Sigma_1$.

THEOREM 1.6.4. *Suppose* $c^* < 0$ *in* $\Omega \cup \Sigma$ *and suppose* $\beta^* \leqslant 0$ *at points of* $\Sigma_1 \cup \Sigma_0$ *and on* Σ_3 *in some neighborhood of the set* $\bar{\Sigma}_3 \cap (\Sigma_1 \cup \Sigma_0)$. *Assume that* $\Omega \in A^{(2,\alpha)}$ *and that the coefficients of the operator* $L^*(v)$ *may be continued to the* δ*-neighborhood* G^δ *of the set* Σ_2 *so that* $a^{kj} \in C_{(2)}(\overline{\Omega \cup G^\delta})$, $b^{*k} \in C_{(1)}(\overline{\Omega \cup G^\delta})$ *and* $c^* \in C^{(0,\alpha)}(\overline{\Omega \cup G^\delta})$, $0 < \alpha < 1$, *and* $a^{kj}\xi_k\xi_j \geqslant 0$ *in* $\overline{\Omega \cup G^\delta}$. *Then the function* $u(x) \in \mathfrak{L}_p(\Omega)$, $p \geqslant 2$, *is equal to zero almost everywhere in* Ω *if for any function* $v \in C^{(2)}(\bar{\Omega})$ *equal to zero on* $\Sigma_3 \cup \Sigma_1$,

$$\int_\Omega L^*(v)\, u\, dx = 0$$

and the set Γ^2 *of boundary points for* Σ_2 *on* Σ *may be divided into closed nonintersecting subsets* $\Gamma_1, \Gamma_2, \Gamma_3$ *such that conditions* 1), 2) *and* 3) *in Theorem* 1.6.1 *hold for these subsets.*

Theorem 1.6.4 is also proved in a manner analogous to the proof of Theorem 1.6.1. In this case the domain $\Omega\delta$ contains only the δ-neighborhood of the set Σ_2. We should also take into consideration the fact that in Lemma 1.5.1 the condition $\beta < 0$ at points on the boundary of Σ_3 and belonging to $\bar{G}$ may be replaced by the condition $\beta \leqslant 0$ on Σ_3 in some neighborhood of the boundary points of Σ_3 (see the remark to Lemma 1.5.1).

In exactly the same way we obtain the following theorem:

THEOREM 1.6.5. *Suppose the coefficients of* (1.5.1) *and the boundary of* Ω *satisfy the assumptions of Theorem* 1.6.4, *but the condition on* Γ^2 *is replaced by the condition that the* δ*-neighborhood of the set* Γ^2 *on* Σ *has area of the order* δ^q *for* $q > 0$. *Then a weak solution of* (1.5.1), (1.5.2) *in the sense of* (1.5.3) *is unique in the class of bounded measurable functions in* Ω.

The following uniqueness theorem for (1.5.1), (1.5.2) does not assume the possibility of a smooth continuation of the coefficients of (1.5.1) through any portion of the boundary, as was assumed in Theorems 1.6.1–1.6.5.

THEOREM 1.6.6. *Suppose* $c^* < 0$ *in* $\Omega \cup \Sigma$ *and suppose* $\beta^* \leqslant 0$ *at points of* $\Sigma_1 \cup \Sigma_0$, *as well as on* Σ_3 *in some neighborhood of* $\bar{\Sigma}_3 \cap (\Sigma_1 \cup \Sigma_0)$. *Assume that* $\Omega \in A^{(2,\alpha)}$ *and that there exists a neighborhood* σ_2 *of the set* Σ_2 *on* Σ, *such that*

$$\frac{1}{\delta}\int_{\sigma_2^\delta} u^2(x)\,dx \to 0 \quad \text{as} \quad \delta \to 0, \tag{1.6.36}$$

where σ_2^δ is the set of points of Ω whose distance from σ_2 does not surpass δ. Then the function $u(x)$ is equal to zero almost everywhere in Ω if $u(x) \in \mathfrak{L}_2(\Omega)$ and for any function $v \in C^{(2)}(\bar{\Omega})$ equal to zero on $\Sigma_3 \cup \Sigma_1$,

$$\int_\Omega L^*(v)\,u\,dx = 0. \tag{1.6.37}$$

PROOF. We consider the equation

$$\varepsilon\Delta v + \sqrt{\varepsilon}\,a(x,\varepsilon)\,\Delta v + L^*(v) = \Phi \tag{1.6.38}$$

in Ω, with the condition

$$v = 0 \text{ on } \Sigma, \tag{1.6.39}$$

where $\Phi \in C_0^\infty(\Omega)$, $\epsilon = \text{const} > 0$, and $a(x, \epsilon)$ is a smooth function equal to unity in the $\sqrt{\epsilon}$-neighborhood of σ_2 and equal to zero outside the $2\sqrt{\epsilon}$-neighborhood of σ_2, and such that $0 \leqslant a(x, \epsilon) \leqslant 1$ for $0 < \epsilon \leqslant 1$. According to Lemma 1.5.1 and the remarks attached thereto, the estimate

$$\left|\frac{\partial v}{\partial x_j}\right| \leqslant C_1 \varepsilon^{-\frac{1}{2}} \text{ on } \Sigma,\ j = 1, \ldots, m,$$

is valid for the solution $v(x)$ of (1.6.38), (1.6.39), where C_1 does not depend on ϵ.

Multiplying equation (1.6.38) by v, integrating over Ω and transforming the individual terms of this equation by integration by parts, we obtain

$$\begin{aligned}\int_\Omega \Phi v\,dx = \int_\Omega \Big[&-\varepsilon v_{x_j} v_{x_j} - \sqrt{\varepsilon}\,a v_{x_j} v_{x_j} - a^{kj} v_{x_k} v_{x_j} \\ &+ \Big(-\frac{1}{2} b^{*j}_{x_j} + c^* + \frac{1}{2} a^{kj}_{x_k x_j} + \frac{1}{2}\sqrt{\varepsilon}\,a_{x_j x_j}\Big) v^2\Big]\,dx.\end{aligned} \tag{1.6.40}$$

Since $c^* < 0$, according to the maximum principle the function v is bounded in Ω uniformly in ϵ. We may assume that

$$|a_{x_j}| \leqslant C_2 \varepsilon^{-\frac{1}{2}}, \quad |a_{x_j x_j}| \leqslant C_3 \varepsilon^{-1}; \quad C_2, C_3 = \text{const}.$$

Clearly the measure of the set where $a_{x_j x_j} \neq 0$ has order $\sqrt{\epsilon}$. Therefore

$$\sqrt{\varepsilon}\int_\Omega a_{x_j x_j} v^2\,dx < C_4,$$

and from (1.6.40) it follows that

$$\varepsilon \int_\Omega v_{x_j} v_{x_j}\,dx < C_5.$$

We denote by C_j constants independent of ϵ. Furthermore, exactly as was proved in Lemma 1.6.1, we find

$$\varepsilon^2 \int_\Omega (\Delta v)^2 \, dx + \varepsilon^{3/2} \int_\Omega a (\Delta v)^2 \, dx \leqslant C_6. \tag{1.6.41}$$

For the proof of Theorem 1.6.6 we substitute the function $v(x)$, a solution of (1.6.38), (1.6.39), into (1.6.37). According to (1.6.38) we have

$$0 = \int_\Omega L^*(v) \, u \, dx = \int_\Omega \Phi u dx - \int_\Omega \varepsilon \Delta v u dx - \int_\Omega \sqrt{\varepsilon} \, a \Delta v u dx. \tag{1.6.42}$$

We show that the two last integrals in (1.6.42) tend to zero as $\epsilon \to 0$. The integral

$$\varepsilon \int_\Omega \Delta v u dx$$

is estimated with the help of (1.6.41) exactly as the integral

$$\varepsilon \int \Delta v \varphi^8 u dx$$

was estimated in the proof of Theorem 1.6.1. The last integral in (1.6.42) is taken over the region $\sigma_2^{2\sqrt{\epsilon}}$, since according to the definition of $a(x, \epsilon)$ the integrand is equal to zero outside this domain. We have

$$\begin{aligned} \left| \sqrt{\varepsilon} \int_\Omega a \Delta v u dx \right| &\leqslant \sqrt{\varepsilon} \left(\int_\Omega a (\Delta v)^2 \, dx \right)^{1/2} \left(\int_\Omega a u^2 dx \right)^{1/2} \\ &\leqslant \left(\varepsilon^{3/2} \int_\Omega a (\Delta v)^2 \, dx \right)^{1/2} \left(\frac{1}{\sqrt{\varepsilon}} \int_{\sigma_2^{2\sqrt{\varepsilon}}} u^2 dx \right)^{1/2}. \end{aligned} \tag{1.6.43}$$

Bearing in mind the inequality (1.6.41) and the assumption (1.6.36), we deduce from (1.6.43) that

$$\sqrt{\varepsilon} \int_\Omega a \Delta v u dx \to 0 \quad \text{as} \quad \varepsilon \to 0.$$

Consequently

$$\int_\Omega u \Phi dx = 0,$$

which means that $u \equiv 0$ almost everywhere in Ω. The theorem is proved.

We note that in Theorem 1.6.6 we do not explicitly impose any conditions on the set Γ or Γ^2. It is easy to verify that condition (1.6.36) is not fulfilled for the equation (1.6.28) or (1.6.33) for solutions of the homogeneous boundary value problems.

Theorems 1.5.1 and 1.5.2 yield conditions under which the weak solution of (1.5.1), (1.5.2) has the property that

$$\frac{1}{\delta}\int_{\sigma_2^\delta} (u-\tilde{g})^2\,dx \to 0 \quad \text{as} \quad \delta\to 0,$$

where $\tilde{g}$ is a continuous prolongation of the function g, given on $\Sigma_2 \cup \Sigma_3$, to the interior of the domain Ω. The function g here is assumed continuous on $\Sigma_2 \cup \Sigma_3$.

REMARK 3. In Theorem 1.6.6, condition (1.6.36) may be replaced by the condition that for any closed set G_2 consisting of interior points of Σ_2

$$\frac{1}{\delta}\int_{G_2^\delta} u^2(x)\,dx \to 0 \quad \text{as} \quad \delta \to 0, \tag{1.6.44}$$

where G_2^δ denotes the intersection of the δ-neighborhood of G_2 with Ω. In addition, it is necessary to assume that the set Γ^2 of boundary points of Σ_2 on Σ may be divided into closed disjoint subsets $\Gamma_1, \Gamma_2, \Gamma_3$ for which conditions 1), 2) and 3) of Theorem 1.6.1 are satisfied, and in the δ-neighborhoods of Γ_1, Γ_2 and Γ_3 the coefficients a^{kj} satisfy the conditions (1.6.21), (1.6.25) and (1.6.27) respectively.

This assertion may be proved in the same way as Theorem 1.6.6. For this purpose one considers the solution v^1 of (1.6.39), (1.6.38), where in the construction of the function $a(x, \epsilon)$ in place of the set σ_2 one takes a $\delta/2$-neighborhood of the set Σ_2 on Σ. In equation (1.6.37) we replace v by the function $v^1\phi^\delta$, where ϕ^δ is the function constructed in the proof of Theorem 1.6.1. We have

$$0=\int_\Omega L^*(v^1\varphi^\delta)\,u\,dx=\int_\Omega \varphi^\delta L^*(v^1)\,u\,dx+\int_\Omega v^1(L^*(\varphi^\delta)-c^*\varphi^\delta)\,u\,dx$$
$$+\int_\Omega 2a^{kj}\varphi^\delta_{x_k}v^1_{x_j}u\,dx.$$

From this it follows that

$$\int_\Omega \varphi^\delta\Phi u\,dx=\int_\Omega \varphi^\delta u(\varepsilon\Delta v^1+\sqrt{\varepsilon}\,a\Delta v^1)\,dx-\int_\Omega v^1(L^*(\varphi^\delta)$$
$$-c^*\varphi^\delta)\,u\,dx-\int_\Omega 2a^{kj}\varphi^\delta_{x_k}v^1_{x_j}u\,dx. \tag{1.6.45}$$

Just as in the proof of Theorems 1.6.1 and 1.6.6, we show that the first integral on the right of (1.6.45) can be made as small as desired for fixed δ by choosing ϵ small; and the two last integrals in (1.6.45) are arbitrarily small for $\epsilon < \delta$, if δ is sufficiently small. It is necessary to take into consideration that the integrand in the integral on the left of (1.6.22) is equal to zero in the δ-neighborhood of Γ^2. Therefore in deducing the inequality analogous to (1.6.24), we may consider the function $(1-\phi^{2\delta})\phi^{3\delta/4}\eta$ in place of the function $(1-\phi^{2\delta})\eta$. As a result, the integral

$$\int_{\Omega \cap Q^t_{1,2\delta}} \sqrt{\epsilon}\, a(x,\epsilon)\Delta v^1 v^1 (1-\phi^{2\delta})\phi^{\frac{3}{4}\delta}\eta\, dx$$

in the equation analogous to (1.6.23) will vanish for sufficiently small ϵ, and all estimates may be carried out as in the proof of Theorem 1.6.1.

We note that the condition on Γ^2 mentioned in Remark 3 may not be omitted, i.e. condition (1.6.36) of Theorem 1.6.6 may not be replaced by condition (1.6.44). In fact, in the boundary value problems constructed for equations (1.6.28) and (1.6.33), the condition (1.6.44) is fulfilled, since the set of interior points of Σ_2 in these cases is empty. But these problems have nonunique solutions in the class $\mathfrak{L}_p$, $p = 2$.

Weak solutions of the problem (1.3.1), (1.3.2) in the Hilbert space $\mathcal{H}$ were constructed in §§4 and 5. (See Theorems 1.4.1 and 1.5.4). A uniqueness theorem for such solutions was proved in [102]. The proof of this theorem was based on the theory of symmetric first order systems.

We denote by Γ' the union of the boundaries of the sets Σ_0, Σ_1, Σ_2 and Σ_3 on Σ.

THEOREM 1.6.7. (R. PHILLIPS AND L. SARASON). *Suppose that* Γ' *consists of two closed nonintersecting sets* Γ_0 *and* Γ_l *such that the area of the* δ*-neighborhood of* Γ_l *on* Σ *is of order* δ^2, *and* Γ_0 *consists of a finite number of* $(m-2)$*-dimensional smooth manifolds. Suppose that* $2c - b^j_{x_j} + a^{kj}_{x_k x_j} \leqslant -c_1 < 0$ *in* Ω *and suppose that at points of each individual manifold of* Γ_0, *either* $a^{kj}\nu_k\nu_j \equiv 0$ *or* $a^{kj}\nu_k\nu_j \neq 0$, *where* $\bar{\nu}$ *is the vector normal to* Γ_0 *on* Σ. *Then the function* u *in the space* $\mathcal{H}$ *is equal to zero almost everywhere in* Ω *if*

$$\int_\Omega uL^*(v)\,dx = 0$$

for all v *in* $C^{(2)}(\Omega \cup \Sigma)$ *equal to zero on* $\Sigma_3 \cup \Sigma_1$.

We shall not give the proof of Theorem 1.6.7 found in [102] because of its complexity; rather we shall prove a closely related theorem.

THEOREM 1.6.8. *Suppose that every closed set* G_2 *of interior points of* Σ_2 *has a neighborhood in which all coefficients of the operator* $L^*(v)$ *are defined, and in which* $a^{kj}\xi_k\xi_j \geqslant 0$, $a^{kj} \in C_{(2)}$, $b^{*j} \in C_{(1)}$, $c^* \in C^{(0,\alpha)}$, $0 < \alpha < 1$. *Suppose* $\beta^* \leqslant 0$ *on* $\Sigma_1 \cup \Sigma_0$, $c^* < 0$ *in* $\Omega \cup \Sigma$, *and the set of boundary points of* Σ_3 *and of* Σ_2 *consists of two closed nonintersecting sets* Γ_0 *and* Γ_l *such that the area of the* δ*-neighborhood of* Γ_l *on* Σ *is of order* δ^2, *and* Γ_0 *consists of a finite number of* $(m-2)$*-dimensional smooth manifolds, on each of which either* $a^{kj}\nu_k\nu_j \equiv 0$ *or* $a^{kj}\nu_k\nu_j \neq 0$. *Then the function* u *in* $\mathfrak{L}_2(\Omega)$ *is equal to zero almost everywhere in* Ω *if*

$$\int_{\Omega} uL^*(v)\,dx = 0 \tag{1.6.46}$$

for all v *in* $C^{(2)}(\Omega \cup \Sigma)$ *equal to zero on* $\Sigma_3 \cup \Sigma_1$, *and the function* u *in a neighborhood of the points of* Γ_0 *where* $a^{kj}\nu_k\nu_j \neq 0$ *coincides with some function in the space* $\mathcal{H}$. *(The condition of smooth continuability of the coefficients of the equation* $L^*(v) = 0$ *to a neighborhood of* Σ_2 *may be replaced by condition* (1.6.44).)

PROOF. Let G^δ be the set of points of Σ_2 whose distance from the boundary of Σ_2 is greater than δ. We consider the domain Ω_δ, which contains the $\delta/2$-neighborhood of G^δ and outside this neighborhood coincides with Ω. Suppose $\Omega_\delta \in A^{(2,\alpha)}$ and suppose $a_1(x)$ is a function in $C_{(2)}(\Omega_\delta)$ such that $a_1 = 0$ in Ω and $a_1 > 0$ in $\bar{\Omega}_\delta \setminus \bar{\Omega}$ We construct the function $a(x, \delta)$ such that $a(x, \delta) \in C_{(2)}(\Omega_\delta)$, $a(x, \delta)$ is equal to unity in the $\delta/2$-neighborhood of the set $\Gamma_0 \cup \Gamma_l$, is zero outside the δ-neighborhood of this set, and $0 \leqslant a \leqslant 1$. In the domain Ω_δ we consider the equation

$$\varepsilon\Delta v + a_1(x)\Delta v + a(x, \delta)\Delta v + L^*(v) = \Phi, \tag{1.6.47}$$

where $\Phi \in C_0^\infty(\Omega)$, $\epsilon = \text{const} > 0$, with the condition $v = 0$ on the boundary of Ω_δ.

If, in place of the condition of continuability of the coefficients of $L^*(v)$ in a neighborhood of Σ_2, condition (1.6.44) is fulfilled, then we examine the equation

$$\varepsilon\Delta v + a(x, \delta)\Delta v + L^*(v) = \Phi \text{ in } \Omega;\ v = 0 \text{ on } \Sigma, \tag{1.6.48}$$

in which $a(x, \delta) \in C^\infty(\bar{\Omega})$, $a(x, \delta) = 1$ in the $\delta/2$-neighborhood of $\Sigma_2 \cup \Gamma_0 \cup \Gamma_l$, $a(x, \delta) = 0$ outside the ¾ δ-neighborhood of this set, and $0 \leqslant a(x, \delta) \leqslant 1$.

By virtue of our assumptions and of Lemma 1.5.1 and 1.6.1, the estimate

$$\varepsilon^2 \int_{\Omega_\delta} (\Delta v)^2\,dx \leqslant C_1, \tag{1.6.49}$$

is valid for a solution of (1.6.47) vanishing on the boundary of Ω_δ; here the constant C_1 does not depend on ϵ, but depends on δ. (This same inequality is valid for solutions of (1.6.48).) Suppose the smooth function $\phi^\delta(x)$ is equal to zero in the δ-neighborhood of $\Gamma_0 \cup \Gamma_l$, $0 \leqslant \phi^\delta(x) \leqslant 1$, and $\phi^\delta(x) = 1$ outside some neighborhood γ^δ of the set $\Gamma_0 \cup \Gamma_l$. (In the case of (1.6.48), in the construction of $\phi^\delta(x)$ we use the set $\Sigma_2 \cup \Gamma_0 \cup \Gamma_l$ in place of the set $\Gamma_0 \cup \Gamma_l$.)

In (1.6.46) we use the function $v^1\phi^\delta$ is place of the function v, where v^1 satisfies (1.6.47) in Ω_δ and vanishes on the boundary of this domain. We have

$$\int_{\Omega} L^*(v^1\varphi^\delta)u\,dx = 0.$$

From this it follows that

$$\int_\Omega \varphi^\delta \Phi u\,dx = \int_\Omega \varepsilon \Delta v^1 \varphi^\delta u\,dx - \int_\Omega v^1 [L^*(\varphi^\delta) - c^* \varphi^\delta]\, u\,dx - \int_\Omega 2a^{kj} \varphi^\delta_{x_k} v^1_{x_j} u\,dx. \tag{1.6.50}$$

We have used the fact that $\phi^\delta a(x, \delta) = 0$ and $a_1(x) = 0$ in Ω. The first integral on the right of (1.6.50) is as small as desired, with fixed δ, for sufficiently small ϵ. This is proved with the aid of (1.6.49) exactly as we estimated the analogous integral in the proof of Theorem 1.6.1. The last two integrals on the right of (1.6.50) (we denote them by I_1 and I_2 respectively) are, by the special choice of ϕ^δ and for $\epsilon < \delta$, as small as desired for sufficiently small δ. We note that in these integrals, the integrand can differ from zero only in γ^δ.

The construction of the function ϕ^δ in a neighborhood of the set Γ_l and the estimation of the integrals I_1 and I_2 taken over the corresponding neighborhood of this set are carried through exactly as for the set Γ_3 in the proof of Theorem 1.6.1. We note only that in the derivation of the inequality analogous to (1.6.24), we should multiply (1.6.47) (or (1.6.48)) by $\eta\phi^{3\delta/4}(1 - \phi^{2\delta})v^1$ and integrate over Ω_δ.

The construction of the function ϕ^δ and the estimate of the integrals I_1 and I_2 taken over that part of Γ_0 where $a^{kj}\nu_k\nu_j \equiv 0$ are carried out exactly as for the set Γ_2 in the proof of Theorem 1.6.1.

In the neighborhood of the set of points of Γ_0 where $a^{kj}\nu_k\nu_j \neq 0$ we construct the function ϕ^δ as was done for the neighborhood of Γ_1 in the proof of Theorem 1.6.1. Since, by assumption, in a neighborhood of Γ_0 the function $u(x)$ coincides with a function in $\mathfrak{H}$ and $a^{kj}\nu_k\nu_j \neq 0$, it is easy to prove that

$$\frac{1}{\delta^{1/2}} \int_{\gamma_1^\delta} u^2 dx \to 0 \quad \text{as} \quad \delta \to 0, \tag{1.6.51}$$

where γ_1^δ denotes a neighborhood of that part of Γ_0 where $a^{kj}\nu_k\nu_j \neq 0$, in which $1 - \phi_\delta \neq 0$. For this purpose we use local coordinates $y_1, \cdots, y_m$ in a neighborhood of a point P belonging to Γ_0, so that the y_m axis is directed along the interior normal to Σ, and the direction of the y_{m-1} axis coincides with the direction of ν. We may suppose that in the operator $L^*(v)$, after transformation, the coefficients of derivatives of the form $v_{y_{m-1}y_j}$, $j \neq m-1$, are equal to zero. From the fact that in a neighborhood of Γ_0 the function $u(x)$ coincides with a function in $\mathfrak{H}$, it follows that

$$\int_{\gamma_1^\delta} u^2_{y_{m-1}} dy < \infty$$

and

$$\int_{\gamma_1^\delta} u^2 dx \leqslant C_2 \sqrt{\delta} \int_{\Gamma_0^\delta} (u_{y_{m-1}}^2 + u^2)\, dy,$$

for sufficiently small δ, where the constant C_2 does not depend on δ, and where Γ_0^δ is some domain containing γ_1^δ with mes $\Gamma_0^\delta \to 0$ as $\delta \to 0$. The further estimation of the integrals I_1 and I_2 proceeds as for the corresponding integrals for Γ_1 in the proof of Theorem 1.6.1; one need only replace the condition $u \in \mathfrak{L}_3(\Omega)$ by the relation (1.6.51).

In the case when condition (1.6.44) is satisfied, the function ϕ^δ is constructed in a neighborhood of the boundary of Σ_2 analogously, depending on whether this boundary belongs to Γ_0 or to Γ_1. Besides, ϕ^δ is assumed to vanish at points no further than $\delta/4$ away from Σ_2, and ϕ^δ is equal to unity at points whose distance from Σ_2 along normals to Σ is greater than δ. We thus obtain the proof of Theorem 1.6.8.

It is easy to see that a weak solution of (1.3.1), (1.3.2) is unique in any class $\mathfrak{L}_p(\Omega)$ for $p \geqslant 1$, if the problem

$$L^*(v) = \Phi \text{ in } \Omega; \quad v|_{\Sigma_1 \cup \Sigma_3} = 0$$

has a solution $v(x)$ belonging to the class $C^{(2)}(\Omega \cup \Sigma)$ for any function $\Phi \in C_0^\infty(\Omega)$. Such conditions will be given in §8.

We now prove a uniqueness theorem for piecewise smooth domains Ω. For such domains the existence Theorem 1.5.5 holds.

THEOREM 1.6.9. *Suppose* $\Omega \in B^{(2,\alpha)}$, *let the set* Γ *of boundary points of* $\Sigma_2 \cup \Sigma_0$ *be such that the area of the* δ*-neighborhood of the set* Γ *on* Σ *is of order* δ^q, $q > 0$, *and let the hypotheses of Theorem* 1.6.2 *hold with respect to the coefficients of the operator* $L^*(v)$ *and the function* β^*. *Then a generalized solution of* (1.1.4), (1.1.5) *defined by the integral identity* (1.5.26) *is unique in the class of bounded measurable functions in* Ω.

PROOF. This theorem may be proved in the same way as Theorem 1.6.1. First we construct a domain $\Omega_\delta \in B^{(2,\alpha)}$ exactly as in the proof of that theorem, and then we construct a domain $\widetilde{\Omega}_\delta$ such that $\widetilde{\Omega}_\delta \in A^{(2,\alpha)}$, $\widetilde{\Omega}_\delta \subset \Omega_\delta$ and $\widetilde{\Omega}_\delta$ coincides with Ω_δ outside the δ-neighborhood of the set of points of B which lie on the boundary of Ω_δ. We denote them by B'. (The set B is defined preceding Theorem 1.5.5.) In the domain $\widetilde{\Omega}_\delta$ we consider the equation

$$\varepsilon \Delta v + a(x) \Delta v + a_1(x, \delta) \Delta v + L^*(v) = \Phi, \tag{1.6.52}$$

where $a(x)$ is a smooth function such that $a(x) = 0$ in Ω and $a(x) > 0$ in

$\overline{\Omega}_\delta \backslash \overline{\Omega}$, the smooth function $a_1(x, \delta) = 1$ in the δ-neighborhood of the set B', and $a_1(x, \delta) = 0$ outside the 2δ-neighborhood of this set. In the integral identity (1.6.46) we replace the function v by the function $v^1 \phi^{2\delta}$, where v^1 is a solution of (1.6.52) equal to zero on the boundary of the domain $\widetilde{\Omega}_\delta$, and ϕ^δ is a smooth function equal to zero in the δ-neighborhood of the set $B' \cup \Gamma$ and equal to unity outside some neighborhood γ^δ of this set. In a neighborhood of Γ the function ϕ^δ is constructed exactly as it was constructed in a neighborhood of Γ_3 in the proof of Theorem 1.6.1. For the points of B' which are limiting points of Σ_1 we construct the function ϕ^δ as was done for the set Γ_1 in Theorem 1.6.1. For the remaining points of B', which we denote by B_1, in place of the function ϕ^δ we choose a smooth function equal to zero in a δ-neighborhood of B_1, equal to unity outside the 2δ-neighborhood of this set, and such that $0 \leqslant \phi^\delta \leqslant 1$,

$$\varphi^{\delta}_{x_k x_j} = O(\delta^{-2}); \quad \varphi^{\delta}_{x_j} = O(\delta^{-1}). \tag{1.6.53}$$

Using the standard technique of barriers, one may show that v^1 is small in a neighborhood of B_1 uniformly in ϵ and δ. Considering this estimate for v^1 and equation (1.6.23), we prove that

$$\int_{B_1^\delta} a^{kj} v^1_{x_k} v^1_{x_j} dx \to 0 \quad \text{as} \quad \delta \to 0, \tag{1.6.54}$$

where B_1^δ denotes the δ-neighborhood of the set B_1. Using relations (1.6.53) and (1.6.54), we estimate the integrals I_1 and I_2 on the right of (1.6.50).

In a smiilar manner, uniqueness theorems for solutions of (1.1.4), (1.1.5) may be proved in a domain Ω of class $B^{(2,\alpha)}$ in spaces $\mathfrak{L}_p(\Omega)$, analogous to Theorems 1.6.1, 1.6.4, 1.6.6 and 1.6.8 which were proved for smooth domains.

§7. A lemma on nonnegative quadratic forms

We prove here an inequality having to do with nonnegative quadratic forms, which lemma will be used in an essential manner in the next section for the study of the smoothness of weak solutions of the boundary value problem (1.1.4), (1.1.5), in §6 of Chapter II for the study of hypoellipticity of general second order equations, and in §2 of Chapter III in connection with degenerating hyperbolic equations.

LEMMA 1.7.1. *Let $a^{kj}(x)\xi_k\xi_j \geqslant 0$ for all x in R^m and all $\xi = (\xi_1, \cdots, \xi_m)$, and suppose $a^{kj}(x) \in C_{(2)}(R^m)$. Then for any $v \in C_{(2)}(R^m)$*

$$\left(a^{kj}_{x_p} v_{x_k x_j}\right)^2 \leqslant M a^{kj} v_{x_s x_k} v_{x_s x_j}, \tag{1.7.1}$$

where M depends only on the second derivatives of the functions a^{kj}. (In (1.7.1), as everywhere, repeated indexes are assumed to be summed from 1 to m.)

PROOF. Any nonnegative function defined for all values of x and belonging to class $C_{(2)}(R^1)$ satisfies the inequality

$$f_x^2 \leqslant 2\{\sup|f_{xx}|\} f. \tag{1.7.2}$$

In fact, suppose that this inequality is not satisfied at some point x_0, so that

$$f_x^2(x_0) > 2\{\sup|f_{xx}|\} f(x_0), \quad f(x_0) \neq 0. \tag{1.7.3}$$

Consider the point $\widetilde{x}_0 = x_0 - 2f(x_0)/f_x(x_0)$. By Taylor's formula,

$$(\widetilde{x}_0) = f(x_0) - 2\frac{f(x_0)}{f_x(x_0)} f_x(x_0) + 2\frac{f^2(x_0)}{f_x^2(x_0)} f_{xx}(\bar{x})$$

or

$$f(\widetilde{x}_0) = -f(x_0)\left[1 - 2\frac{f(x_0)}{f_x^2(x_0)} f_{xx}(\bar{x})\right].$$

Since by assumption inequality (1.7.3) is fulfilled, we have

$$1 - 2\frac{f(x_0)}{f_x^2(x_0)} f_{xx}(\bar{x}) > 0$$

and consequently $f(\widetilde{x}_0) < 0$, which contradicts the assumption that $f(x) \geqslant 0$ for all $x \in R^1$.

Suppose x' is any point of R^m. We perform the change of independent variables $y_s = \alpha_{ks} x_k$, where α_{ks} are constants, and $\|\alpha_{ks}\|$ is an orthogonal matrix such that the matrix $\|a^{kj}\alpha_{ks}\alpha_{jp}\|$ is diagonal at the point x'.

It is easy to verify that at the point x' for $v \in C^{(2)}(R^m)$ the equations

$$a^{kj} v_{x_k x_s} v_{x_j x_s} = \widetilde{a}^{ss} v_{y_s y_p} v_{y_s y_p}, \tag{1.7.4}$$

$$\left(a^{kj}_{x_p} v_{x_k x_j}\right)^2 = \left(\widetilde{a}^{sl}_{x_p} v_{y_s y_l}\right)^2 \tag{1.7.5}$$

hold. Clearly $\widetilde{a}^{sl}\xi_s\xi_l \geqslant 0$ at all points of R^m. Therefore by virtue of (1.7.2),

$$\left(\widetilde{a}^{ss}_{x_p}\right)^2 \leqslant C_1 \widetilde{a}^{ss} \text{ in } R^m, \tag{1.7.6}$$

where the constant C_1 depends on the second order derivatives of a^{kj}. Since $\widetilde{a}^{ss} + 2\widetilde{a}^{sl} + \widetilde{a}^{ll} \geqslant 0$ in R^m, we may employ (1.7.2) for this function, obtaining

$$(\widetilde{a}^{sl}_{x_p})^2 \leqslant C_2(\widetilde{a}^{ss} + 2\widetilde{a}^{sl} + \widetilde{a}^{ll}) + C_3(\widetilde{a}^{ss} + \widetilde{a}^{ll}) \tag{1.7.7}$$

at all points of R^m; here C_2 and C_3 are constants.

At the point x' we have $\widetilde{a}^{sl} = 0$ for $s \neq l$. Therefore from (1.7.7) it follows that at x'

$$(\widetilde{a}^{sl}_{x_p})^2 \leqslant C_4(\widetilde{a}^{ss} + \widetilde{a}^{ll}), \tag{1.7.8}$$

where the constants C_j depend only on the second derivatives of the a^{kj}.

Relations (1.7.4), (1.7.5) and estimates (1.7.6), (1.7.8) imply the assertion of the lemma for v in class $C^{(2)}(R^m)$. A passage to a limit yields (1.7.1) for v in $C_{(2)}(R^m)$.

Several corollaries of inequality (1.7.2) and of Lemma 1.7.1 will be used in the following.

COROLLARY 1. *The quadratic form* $a^{kj}(x)\xi_k\xi_j \geqslant 0$ *for all* ξ *and* $x \in R^m$. *Hence, considering it a function of* ξ_j *and applying* (1.7.2), *one has*

$$|a^{kj}(x)\xi_k|^2 \leqslant 2a^{jj}(x)a^{kl}(x)\xi_k\xi_l \tag{1.7.9}$$

for all ξ *and* $x \in R^m$.

COROLLARY 2. *Let* $\eta = (\eta_1, \cdots, \eta_m)$ *where the* η_j *are complex numbers. Then*

$$|a^{kj}(x)\eta_k|^2 \leqslant 2a^{jj}(x)a^{kl}(x)\eta_k\overline{\eta_l} \tag{1.7.10}$$

for all $x \in R^m$ *and all complex values of* $\eta_1, \cdots, \eta_m$.

Inequality (1.7.10) follows immediately from (1.7.9). In fact, suppose $\eta_k = \xi_k + i\tilde{\xi}_k$ where ξ_k and $\tilde{\xi}_k$ are real numbers. Then (1.7.10) may be written in the form

$$|a^{kj}(x)\xi_k|^2 + |a^{kj}(x)\tilde{\xi}_k|^2 \leqslant 2a^{jj}(x)[a^{kl}(x)\xi_k\xi_l + a^{kl}(x)\tilde{\xi}_k\tilde{\xi}_l].$$

This latter inequality is a direct consequence of (1.7.9).

COROLLARY 3. *Suppose* v *is a complex valued function, and* $v \in C_{(2)}(R^m)$, $D_j = -i\partial/\partial x_j$. *Then*

$$|a^{kj}_{x_p}D_kD_jv|^2 \leqslant Ma^{kj}D_kD_lv\overline{D_jD_lv}. \tag{1.7.11}$$

The constant M *depends only on the second derivatives of the function* a^{kj}.

Inequality (1.7.11) also immediately follows from (1.7.1), if the functions D_kD_jv are written in the form

$$D_kD_jv = D_kD_jv_1 + iD_kD_jv_2,$$

where v_1 and v_2 are real functions, and these expressions are substituted into (1.7.11).

§8. On smoothness of weak solutions of the first boundary value problem. Conditions for existence of solutions with bounded derivatives

In this section we establish a series of theorems on the smoothness of the weak solution of (1.1.4), (1.1.5) constructed in §5.

We denote by E the set of points of $\Omega \cup \Sigma$ where the determinant of the matrix $\|a^{kj}\|$ is equal to zero. We note that in some neighborhood of each point P of the set

$\Omega \backslash E$ the weak solution of the problem (1.1.4), (1.1.5) is a weak solution of an elliptic equation of second order, and by virtue of known results (see [81, 118]) it belongs to class $C^{(k+2,\alpha)}$ in some neighborhood of the point P if the coefficients of the equation and the function f in a neighborhood of this point belong to class $C^{(k,\alpha)}$. An analogous assertion is valid also for those points of Ω in a neighborhood of which equation (1.14) is parabolic and one of the variables x_j plays the role of time (see [37, 61]).

In §§5 and 6 of Chapter II, far-reaching sufficient conditions on the coefficients of (1.1.4) will be indicated, under which the weak solution of (1.1.4), (1.1.5) will possess the analogous property of local smoothness.

In the present section we derive conditions on the coefficients of the equation, the boundary of Ω, and the functions f and g, under which the weak solution belongs to some class $C_{(k)}(\Omega \cup \Sigma)$. We use the fact that the weak solution, as shown in §5, may be characterized as a limit of solutions of elliptic equations (1.5.4) as $\epsilon \to 0$. Therefore for the proof of the smoothness of weak solutions we first establish estimates for solutions of (1.5.4) which are uniform with respect to ϵ.

We introduce the notation

$$B_1(x) = c(x) + \frac{1}{4} Mm + \max_s \left\{ b^s_{x_s} + \frac{1}{2} \sum_{\substack{j=1 \\ j \neq s}}^{m} | b^j_{x_s} | + \frac{1}{2} \sum_{\substack{j=1 \\ j \neq s}}^{m} | b^s_{x_j} | \right\}, \tag{1.8.1}$$

where M is the constant in inequality (1.7.1), which depends on the second derivatives of the coefficients a^{kj}.

LEMMA 1.8.1. *Suppose $u_\epsilon(x)$ is the solution of the equation*

$$L_\varepsilon(u) = \varepsilon \Delta u + a^{kj} u_{x_k x_j} + b^k u_{x_k} + cu = f \text{ in } \Omega \tag{1.8.2}$$

with the condition

$$u = g \text{ on } \Sigma, \tag{1.8.3}$$

where $c < 0$ in $\Omega \cup \Sigma$ and $a^{kj} \xi_k \xi_j \geqslant 0$ in $\Omega \cup \Sigma$ for any ξ. Moreover, let the following conditions hold.

1) *In some domain ω contained in Ω the coefficients a^{kj}, b^k, c, and f belong to the class $C_{(1)}(\omega)$; g is a bounded function, $u_\epsilon \in C^{(0)}(\Omega \cup \Sigma)$ and $u_\epsilon \in C^{(1)}(\bar{\omega})$, $u_\epsilon \in C^{(3)}(\omega)$.*

2) *In $\bar{\omega}$ inequality* (1.7.1) *is fulfilled (this inequality, in particular, is always satisfied if $\bar{\omega} \subset \Omega$ and if in the domain Ω, the coefficients $a^{kj} \in C_{(2)}(\Omega)$ and $a^{kj} \xi_k \xi_j \geqslant 0$ for all ξ).*

3) *In the* δ*-neighborhood* E^δ *of the set* E *for some* $\delta > 0$, *the inequality*

$$\sup_{E^\delta \cap \bar{\omega}} B_1 < 0$$

is fulfilled.

Then the solution u_ϵ *of the problem* (1.8.2), (1.8.3) *satisfies the estimate*

$$\max_{\bar{\omega}} |\operatorname{grad} u_\varepsilon|^2 \leqslant M_1 + \max_{\sigma} |\operatorname{grad} u_\varepsilon|^2, \tag{1.8.4}$$

where the constant M_1 *does not depend on* ϵ *and* σ *is the boundary of* ω.

PROOF. Since $c < 0$ in $\Omega \cup \Sigma$, it follows that u_ϵ is bounded uniformly with respect to ϵ, by virtue of the maximum principle. In the region $\omega \backslash G_\delta$, where $G_\delta = E^\delta \cap \omega$, equation (1.8.2) is uniformly elliptic for $\epsilon \geqslant 0$, i.e. at the points of $\omega \backslash G_\delta$ the inequality

$$a^{kj}\xi_k\xi_j \geqslant \alpha_0 \sum_{k=1}^m \xi_k^2 \tag{1.8.5}$$

is fulfilled with a positive constant α_0 independent of ϵ. In the domain G_δ the condition

$$B_1 \leqslant \alpha_1 < 0, \quad \alpha_1 = \text{const} \tag{1.8.6}$$

is fulfilled. We shall construct an equation which $v = p^1 + C_1 u_\epsilon^2$ satisfies, where $p^1 = |\operatorname{grad} u_\epsilon|^2$ and C_1 is a positive constant which we shall choose below.

We differentiate (1.8.2) with respect to x_s, multiply it by u_{x_s} and sum with respect to s from 1 to m, obtaining an equation for p^1 of the form

$$\frac{1}{2}\varepsilon\Delta p^1 - \varepsilon u_{x_k x_s} u_{x_k x_s} + \frac{1}{2} a^{kj} p^1_{x_k x_j} + \frac{1}{2} b^k p^1_{x_k}$$
$$+ \left[-a^{kj} u_{x_s x_k} u_{x_s x_j} + a^{kj}_{x_s} u_{x_k x_j} u_{x_s}\right] + b^k_{x_s} u_{x_k} u_{x_s} + c p^1 + c_{x_s} u u_{x_s} = f_{x_s} u_{x_s}. \tag{1.8.7}$$

Clearly u_ϵ satisfies

$$\frac{1}{2}\varepsilon\Delta(u^2) + \frac{1}{2} a^{kj} (u^2)_{x_k x_j} + \frac{1}{2} b^k (u^2)_{x_k} + cu^2 - \varepsilon u_{x_k} u_{x_k} - a^{kj} u_{x_k} u_{x_j} = fu. \tag{1.8.8}$$

From (1.87) and (1.88) we obtain an equation for v of the form

$$\begin{aligned} &\frac{1}{2}\varepsilon\Delta v + \frac{1}{2} a^{kj} v_{x_k x_j} + \frac{1}{2} b^k v_{x_k} + cv \\ &- \varepsilon u_{x_k x_s} u_{x_k x_s} + \left[-a^{kj} u_{x_s x_k} u_{x_s x_j} + a^{kj}_{x_s} u_{x_k x_j} u_{x_s}\right] \\ &+ b^k_{x_s} u_{x_k} u_{x_s} + c_{x_s} u u_{x_s} - \varepsilon C_1 u_{x_k} u_{x_k} - C_1 a^{kj} u_{x_k} u_{x_j} \\ &\qquad = f_{x_s} u_{x_s} + C_1 f \cdot u. \end{aligned} \tag{1.8.9}$$

If the maximum of v is attained at a point Q_1 inside the domain ω, then at this point $\Delta v \leqslant 0$, $a^{kj} v_{x_k x_j} \leqslant 0$ and $v_{x_j} = 0$. It is clear that for fixed s the inequality

$$|a^{kj}_{x_s} u_{x_k x_j} u_{x_s}| \leqslant \frac{1}{Mm} (a^{kj}_{x_s} u_{x_k x_j})^2 + \frac{Mm}{4} u^2_{x_s}$$

holds, where M is the constant of (1.7.1). By virtue of Lemma 1.7.1,

$$-a^{kj} u_{x_s x_k} u_{x_s x_j} + \frac{1}{Mm} \sum_{s=1}^{m} (a^{kj}_{x_s} u_{x_k x_j})^2 \leqslant 0.$$

Therefore at the point Q_1 where v attains its greatest value we have

$$\frac{Mm}{4} v + cv + b^s_{x_s} u^2_{x_s} + \frac{1}{2} \sum_{\substack{j=1 \\ j \neq s}}^{m} |b^j_{x_s}| u^2_{x_s} + \frac{1}{2} \sum_{\substack{s=1 \\ s \neq j}}^{m} |b^j_{x_s}| u^2_{x_j} + 2\tau v + \frac{1}{4\tau} \sum_{s=1}^{m} |c_{x_s}|^2 u^2$$
$$+ \frac{1}{4\tau} \sum_{j=1}^{m} |f_{x_j}|^2 - C_1 a^{kj} u_{x_k} u_{x_j} + |C_1 f u| \geqslant 0, \tag{1.8.10}$$

where $\tau = \text{const} > 0$. We choose the constant C_1 so large that at points of $\omega \backslash G_\delta$

$$-C_1 \alpha_0 + c + 2\tau + \max \sum_{s,j=1}^{m} |b^j_{x_s}| + \frac{Mm}{4} < 0.$$

If $Q_1 \in \Omega \backslash G_\delta$, then with this choice of C_1 it follows from (1.8.10) that

$$v < M_1,$$

where the constant M_1 does not depend on ϵ. If $Q_1 \in G_\delta$, then by choosing $\tau > 0$ sufficiently small and considering condition (1.8.6), we also deduce from (1.8.10) that $v \leqslant M_1$. The estimate (1.8.4) is thereby proved.

LEMMA 1.8.2. *Assume that for some domain $\omega \subset \Omega$ inequality* (1.7.1) *is fulfilled, and that the coefficients a^{kj}, b^k, c and f belong to the class $C_{(\mu)}(\bar{\omega})$, $c < 0$ in $\bar{\Omega}$. Let $\bar{S}_l = (s_1, \cdots, s_l)$, where the s_j run through the values* $1, \cdots, m$. *Assume that for any $l \leqslant \mu$ the inequality*

$$\sup_{E^\delta \cap \bar{\omega}} B_l < 0 \tag{1.8.11}$$

is fulfilled, where E^δ is the δ-neighborhood of the set E, and

$$B_l = c + \frac{1}{4} Mml$$
$$+ \max_{\bar{S}_l} \left\{ \sum_{\rho=1}^{l} b^{s_\rho}_{x_{s_\rho}} + \sum_{\substack{\rho,\mu=1 \\ \rho \neq \mu}}^{l} a^{s_\rho s_\mu}_{x_{s_\rho} x_{s_\mu}} + \frac{1}{2} \sum_{\rho=1}^{l} \sum_{\substack{k=1, \\ k \neq s_\rho}}^{m} (|b^k_{x_{s_\rho}}| + |b^{s_\rho}_{x_k}|) + \frac{1}{2} \sum_{\rho,\varkappa=1}^{l} \sum_{k,j=1}^{m} (|a^{kj}_{x_{s_\rho} x_{s_\varkappa}}| + |a^{s_\rho s_\varkappa}_{x_k x_j}|) \right\} \tag{1.8.12}$$

the last summation being subject to the conditions

$$\rho<\varkappa,\ (k,j)\neq(s_\rho,s_\varkappa),\ (j,k)\neq(s_\rho,s_\varkappa).$$

Then for solutions u_ϵ *of the problem* (1.8.2), (1.8.3) *such that* $u_\epsilon\in C^{(\mu+2)}(\omega)$, $u_\epsilon\in C^{(\mu)}(\bar\omega)$, *the estimate*

$$\max_{\bar\omega} p^l\leqslant M_l+\sum_{\nu=1}^{l}\max_{\sigma} p^\nu,\ l\leqslant\mu, \tag{1.8.13}$$

holds, where $p^l\equiv\Sigma_{\bar S_l}(D_{\bar S_l}u_\epsilon)^2$, $D_{\bar S_l}=(\partial/\partial x_{s_1})\cdots(\partial/\partial x_{s_l})$, M_l *are constants independent of* ϵ, *and* σ *is the boundary of* ω.

PROOF. We prove (1.8.13) by induction. For $\mu=1$ this inequality is proved in Lemma 1.8.1. We assume that (1.8.13) is fulfilled for p^l for $l\leqslant\mu_0$, where $1\leqslant\mu_0<\mu$, and prove it for μ_0+1. We apply the operator $D_{\bar S_l}$ to the right and left sides of the equation (1.8.2), multiply the equation obtained by $D_{\bar S_l}u$, and sum over all possible $\bar S_l=(s_1,\cdots,s_l)$, $s_j=1,\cdots,m$. In this manner we obtain the following equation for the sum of squares of all derivatives of u_ϵ of order l, which sum we denote by p^l:

$$\begin{aligned}&\frac{1}{2}\varepsilon\Delta p^l-\varepsilon\sum_{\bar S_l}D_{\bar S_l}u_{x_k}D_{\bar S_l}u_{x_k}+\frac{1}{2}a^{kj}p^l_{x_kx_j}\\&+\frac{1}{2}b^kp^l_{x_k}+cp^l-\sum_{\bar S_l}a^{kj}D_{\bar S_l}u_{x_k}D_{\bar S_l}u_{x_j}\\&+\sum_{\bar S_l}\sum_{\rho=1}^{l}a^{kj}_{x_{s_\rho}}D_{\bar S_{l-1}}u_{x_kx_j}D_{\bar S_l}u+\sum_{\bar S_l}\sum_{\substack{\rho,\mu=1\\ \rho<\mu}}^{l}a^{kj}_{x_{s_\rho}x_{s_\mu}}D_{\bar S_{l-2}}u_{x_kx_j}D_{\bar S_l}u\\&+\sum_{\bar S_l}\sum_{\rho=1}^{l}b^k_{x_{s_\rho}}D_{\bar S_{l-1}}u_{x_k}D_{\bar S_l}u+\sum_{\bar S_l}D_{\bar S_l}uR_{\bar S_l}u=0,\end{aligned}\tag{1.8.14}$$

where we denote by $R_{\bar S_l}u$ terms admitting an estimate

$$|R_{\bar S_l}u|^2\leqslant K_1\sum_{\nu=1}^{l-1}p^\nu+K_2,$$

where the constants K_1 and K_2 do not depend on ϵ. Suppose $v=p^{\mu_0+1}+C_{\mu_0}p^{\mu_0}$, where C_{μ_0} is a constant to be chosen later. With the help of (1.8.14) for $l=\mu_0$ and $l=\mu_0+1$ we obtain for v the equation

$$
\begin{aligned}
&\frac{1}{2}\varepsilon\Delta v+\frac{1}{2}a^{kj}v_{x_kx_j}+\frac{1}{2}b^k v_{x_k}+cv\\
&\quad -C_{\mu_0}\varepsilon\sum_{\bar{S}_{\mu_0}}D_{\bar{S}_{\mu_0}}u_{x_k}D_{\bar{S}_{\mu_0}}u_{x_k}-\varepsilon\sum_{\bar{S}_{\mu_0+1}}D_{\bar{S}_{\mu_0+1}}u_{x_k}D_{\bar{S}_{\mu_0+1}}u_{x_k}\\
&\quad -C_{\mu_0}\sum_{\bar{S}_{\mu_0}}a^{kj}D_{\bar{S}_{\mu_0}}u_{x_k}D_{\bar{S}_{\mu_0}}u_{x_j}-\sum_{\bar{S}_{\mu_0+1}}a^{kj}D_{\bar{S}_{\mu_0+1}}u_{x_k}D_{\bar{S}_{\mu_0+1}}u_{x_j}\\
&\qquad +\sum_{\bar{S}_{\mu_0+1}}D_{\bar{S}_{\mu_0+1}}uR_{\bar{S}_{\mu_0+1}}u+Q_1^{\mu_0}\\
&+\left[\sum_{\bar{S}_{\mu_0+1}}\sum_{\rho=1}^{\mu_0+1}a^{kj}_{x_{s_\rho}}D_{\bar{S}_{\mu_0}}u_{x_kx_j}D_{\bar{S}_{\mu_0+1}}u+C_{\mu_0}\sum_{\bar{S}_{\mu_0}}\sum_{\rho=1}^{\mu_0}a^{kj}_{x_{s_\rho}}D_{\bar{S}_{\mu_0-1}}u_{x_kx_j}D_{\bar{S}_{\mu_0}}u\right]\\
&+\sum_{\bar{S}_{\mu_0+1}}\left\{\sum_{\rho,\varkappa=1,\ \rho<\varkappa}^{\mu_0+1}a^{kj}_{x_{s_\rho}x_{s_\varkappa}}D_{\bar{S}_{\mu_0-1}}u_{x_kx_j}D_{\bar{S}_{\mu_0+1}}u+\sum_{\rho=1}^{\mu_0+1}b^k_{x_{s_\rho}}D_{\bar{S}_{\mu_0}}u_{x_k}D_{\bar{S}_{\mu_0+1}}u\right\}=0,
\end{aligned}
\tag{1.8.15}
$$

where the terms Q_1^l admit an estimate

$$|Q_1^l|\leqslant K_3\sum_{\nu=1}^{l}p^\nu+K_4,$$

with constants K_3 and K_4 independent of ϵ. It is clear that the inequality

$$\left|a^{kj}_{x_{s_\rho}}D_{\bar{S}_{l-1}}u_{x_kx_j}D_{\bar{S}_l}u\right|\leqslant\frac{\gamma}{2Mml}\left(a^{kj}_{x_{s_\rho}}D_{\bar{S}_{l-1}}u_{x_kx_j}\right)^2+\frac{Mml}{2\gamma}\left(D_{\bar{S}_l}u\right)^2 \tag{1.8.16}$$

holds, where $\gamma=\text{const}>0$. We use Lemma 1.7.1 to estimate the first term on the right side of (1.8.16). Clearly

$$\sum_{\bar{S}_l}a^{kj}D_{\bar{S}_l}u_{x_k}D_{\bar{S}_l}u_{x_j}=\sum_{\bar{S}_{l-1}}a^{kj}D_{\bar{S}_{l-1}}u_{x_kx_\rho}D_{\bar{S}_{l-1}}u_{x_jx_\rho}.$$

The sum

$$\sum_{\bar{S}_l}\sum_{\rho=1}^{l}a^{kj}_{x_{s_\rho}}D_{\bar{S}_{l-1}}u_{x_kx_j}D_{\bar{S}_l}u$$

may contain no more than ml terms with the same factor $D_{\bar{S}_{l-1}}u_{x_kx_j}$. Therefore according to (1.7.1) we have

$$\frac{1}{Mml}\sum_{\bar{S}_l}\sum_{\rho=1}^{l}\left(a^{kj}_{x_{s_\rho}}D_{\bar{S}_{l-1}}u_{x_kx_j}\right)^2\leqslant\sum_{\bar{S}_{l-1}}a^{kj}D_{\bar{S}_{l-1}}u_{x_kx_\rho}D_{\bar{S}_{l-1}}u_{x_jx_\rho}. \tag{1.8.17}$$

We use inequalities (1.8.16), (1.8.17) with $l=\mu_0$ and $l=\mu_0+1$ and γ equal respectively to 1 and 2, to estimate the terms in square brackets on the left of (1.8.15).

In the following summations we assume $(k,j)\neq(s_\rho,s_\varkappa)$, $(j,k)\neq(s_\rho,s_\varkappa)$, $\rho<\varkappa$. Then

$$\left|\sum_{\bar{S}_{\mu_0+1}}\sum_{\rho,\varkappa=1}^{\mu_0+1} a^{kj}_{x_{s_\rho}x_{s_\varkappa}} D_{\bar{S}_{\mu_0-1}}u_{x_kx_j}D_{\bar{S}_{\mu_0+1}}u\right|$$

$$\leqslant\frac{1}{2}\sum_{\bar{S}_{\mu_0+1}}\left\{\sum_{\rho,\varkappa=1}^{\mu_0+1}\sum_{k,j=1}^{m}\left|a^{kj}_{x_{s_\rho}x_{s_\varkappa}}\right|\left|D_{\bar{S}_{\mu_0+1}}u\right|^2+\sum_{\rho,\varkappa=1}^{\mu_0+1}\left|a^{kj}_{x_{s_\rho}x_{s_\varkappa}}\right|\left|D_{\bar{S}_{\mu_0-1}}u_{x_kx_j}\right|^2\right\}.$$

This latter summation may be written in the form

$$\frac{1}{2}\sum_{\bar{S}_{\mu_0+1}}\sum_{\rho,\varkappa=1}^{\mu_0+1}\sum_{k,j=1}^{m}\left|a^{s_\rho s_\varkappa}_{x_kx_j}\right|(D_{\bar{S}_{\mu_0+1}}u)^2.$$

Furthermore, we clearly have the estimate

$$\left|\sum_{\bar{S}_{\mu_0+1}}\sum_{\substack{\rho=1,\\ s_\rho\neq k}}^{\mu_0+1} b^k_{x_{s_\rho}} D_{\bar{S}_{\mu_0}}u_{x_k}D_{\bar{S}_{\mu_0+1}}u\right|$$

$$\leqslant\frac{1}{2}\sum_{\bar{S}_{\mu_0+1}}\sum_{k=1}^{m}\sum_{\substack{\rho=1,\\ s_\rho\neq k}}^{\mu_0+1}\left|b^k_{x_{s_\rho}}\right|(D_{\bar{S}_{\mu_0+1}}u)^2+\frac{1}{2}\sum_{\bar{S}_{\mu_0}}\sum_{k=1}^{m}\sum_{\substack{\rho=1,\\ s_\rho\neq k}}^{\mu_0+1}\left|b^k_{x_{s_\rho}}\right|(D_{\bar{S}_{\mu_0}}u_{x_k})^2.$$

We write the latter summation in the form

$$\frac{1}{2}\sum_{\bar{S}_{\mu_0+1}}\sum_{\substack{\rho=1,\\ k\neq s_\rho}}^{\mu_0+1}\sum_{k=1}^{m}\left|b^{s_\rho}_{x_k}\right|(D_{\bar{S}_{\mu_0+1}}u)^2.$$

We now assume that v takes on its greatest value at a point Q_1 lying inside ω. Then from (1.8.15) and the estimates obtained above it follows that at the point Q_1

$$\frac{Mm(\mu_0+1)}{4}v+cv+\tau v-\frac{1}{2}C_{\mu_0}\sum_{\bar{S}_{\mu_0}}a^{kj}D_{\bar{S}_{\mu_0}}u_{x_k}D_{\bar{S}_{\mu_0}}u_{x_j}$$

$$+\sum_{\bar{S}_{\mu_0+1}}\sum_{\substack{\rho,\varkappa=1,\\ \rho\neq\varkappa}}^{\mu_0+1}a^{s_\rho s_\varkappa}_{x_{s_\rho}x_{s_\varkappa}}(D_{\bar{S}_{\mu_0+1}}u)^2+\sum_{\bar{S}_{\mu_0+1}}\sum_{\rho=1}^{\mu_0+1}b^{s_\rho}_{x_{s_\rho}}(D_{\bar{S}_{\mu_0+1}}u)^2+$$

$$+\frac{1}{2}\sum_{\substack{\bar{S}_{\mu_0+1} \\ \rho<\varkappa,\ (k,j)\neq(s_\rho,\, s_\varkappa),\ (j,k)\neq(s_\rho,s_\varkappa)}}\ \sum_{\rho,\varkappa=1}^{\mu_0+1}\ \sum_{k,j=1}^{m}\left\{\left|a^{kj}_{x_{s_\rho}x_{s_\varkappa}}\right|(D_{\bar{S}_{\mu_0+1}}u)^2+\left|a^{s_\rho s_\varkappa}_{x_k x_j}\right|(D_{\bar{S}_{\mu_0+1}}u)^2\right\}$$

$$+\frac{1}{2}\sum_{\bar{S}_{\mu_0+1}}\ \sum_{\rho=1}^{\mu_0+1}\ \sum_{\substack{k=1\\ k\neq s_\rho}}^{m}\left\{\left|b^{k}_{x_{s_\rho}}\right|(D_{\bar{S}_{\mu_0+1}}u)^2+\left|b^{s_\rho}_{x_k}\right|(D_{\bar{S}_{\mu_0+1}}u)^2\right\}+Q_2^{\mu_0}>0, \tag{1.8.18}$$

where τ is an arbitrary small positive number. If Q_1 belongs to $\omega\setminus E^\delta$, then for sufficiently large C_{μ_0} it follows from (1.8.18) that at the point Q_1

$$v<M_{\mu_0+1}, \tag{1.8.19}$$

where the constant M_{μ_0+1} does not depend on ϵ. However if $Q_1\in E^\delta$, then, bearing condition (1.8.11) in mind, we see from (1.8.18) also that (1.8.19) is satisfied at the point Q_1. As a consequence, (1.8.13) holds for $l=\mu_0+1$. The lemma is proved.

LEMMA 1.8.3. *Suppose* σ_1 *is a closed set of points on* Σ *such that* $g=0$ *in a neighborhood of* σ_1 *and such that at each point of* σ_1 *either* $a^{kj}n_kn_j\neq 0$ *or* $\beta<0$, *where the function* β *is defined by* (1.5.6). *Suppose that* $\Omega\in A_{(2)}$, *that the coefficients of equation* (1.8.2) *and the functions* f *and* g *are bounded in* $\bar{\Omega}$, *and that the solution* u_ϵ *of* (1.8.2), (1.8.3) *belongs to* $C^{(0)}(\bar{\Omega})$ *and has continuous derivatives of first order at the points of* σ_1, *with* $c<0$ *in* $\bar{\Omega}$. *Then*

$$\max_{\sigma_1}|\operatorname{grad} u_\varepsilon|\leqslant C_1, \tag{1.8.20}$$

where the constant C_1 *does not depend on* ϵ.

PROOF. Suppose the point P_1 belongs to σ_1. In a neighborhood of the point P_1 we introduce local coordinates $y_1,\cdots,y_m$ with origin at the point P_1 and such that the surface Σ near P_1 given by (1.5.5) lies in the plane $y_m=0$. Also suppose that $y_m>0$ at points of Ω. In the new coordinates the equation (1.8.2) takes on the form

$$L_\varepsilon(u)=\varepsilon\mu^{kj}u_{y_ky_j}+\varepsilon\nu^ku_{y_k}+\alpha^{kj}u_{y_ky_j}+\beta^ku_{y_k}+cu=f. \tag{1.8.21}$$

According to the hypotheses, at points of σ_1 either $\alpha^{mm}>0$ or $\beta^m<0$. Suppose $\psi(y_1,\cdots,y_{m-1})$ is a smooth function defined on Σ in the 2δ-neighborhood of P_1 for sufficiently small δ, with $\psi=1$ in the $\delta/2$-neighborhood of P_1 and $\psi=0$ outside the δ-neighborhood of P_1, with $0\leqslant\psi\leqslant 1$. In the closed domain $Q_{\delta,\tau}\{0\leqslant y_m\leqslant\tau\psi\}$ we consider the function $w=e^{\gamma(\tau\psi-y_m)}$. We have

$$
\begin{aligned}
L_\varepsilon(w)=e^{\gamma(\tau\psi-y_m)}\Bigg[&\sum_{k,j=1}^{m-1}(\varepsilon\mu^{kj}\gamma^2\tau^2\psi_{y_k}\psi_{y_j}+\varepsilon\mu^{kj}\gamma\tau\psi_{y_ky_j})+\varepsilon\mu^{mm}\gamma^2-2\sum_{k=1}^{m-1}\varepsilon\mu^{km}\gamma^2\tau\psi_{y_k}\\
&+\sum_{k,j=1}^{m-1}(\alpha^{kj}\gamma^2\tau^2\psi_{y_k}\psi_{y_j}+\alpha^{kj}\gamma\tau\psi_{y_ky_j})+\alpha^{mm}\gamma^2 \qquad (1.8.22)\\
&-2\sum_{k=1}^{m-1}\alpha^{km}\gamma^2\tau\psi_{y_k}+\varepsilon\sum_{k=1}^{m-1}\nu^k\gamma\tau\psi_{y_k}-\varepsilon\nu^m\gamma+\sum_{k=1}^{m-1}\beta^k\gamma\tau\psi_{y_k}-\beta^m\gamma+c\Bigg].
\end{aligned}
$$

Since at each point of $Q_{\delta,\tau}$ for sufficiently small δ and τ either $\alpha^{mm}>0$ or $\beta^m<0$, it is easy to see that $L_\epsilon(w)>\max|f|$ in $Q_{\delta,\tau}$ when τ is chosen sufficiently small and γ sufficiently large. In this process τ and γ may be chosen independently of ϵ. Hence the functions $v_\pm = w \pm u_\epsilon$ may take on positive maxima only on the boundary of $Q_{\delta,\tau}$. At points of Σ we have $v_\pm = e^{\gamma\tau\psi}$, and on the remaining portion of the boundary of $Q_{\delta,\tau}$ we have $v_\pm = 1 \pm u_\epsilon$. Since u_ϵ is bounded uniformly with respect to ϵ, it follows that we may choose the constant γ so large that $e^{\gamma\tau} > 1 + \max_{\bar{\Omega}} |u_\epsilon|$. This means that the functions $v_\pm$ take their greatest positive value for $y_m = 0$ at points in the $\delta/2$-neighborhood of P_1, where $\psi = 1$. Thus at these points

$$(v_\pm)_{y_m}\leqslant 0 \quad\text{and}\quad |u_{\epsilon y_m}|\leqslant\gamma e^{\gamma\tau}.$$

Since $u_{\epsilon y_j} = 0$ for $j \neq m$, the assertion of the lemma follows from the estimate obtained for $u_{\epsilon y_m}$.

LEMMA 1.8.4. *Let σ_1 be a closed set of points on Σ with $g = 0$ in a neighborhood of σ_1 and $a^{kj}n_kn_j \neq 0$ at each point of σ_1. Suppose that $\Omega \in A^{(\mu+2)}$, that the coefficients of* (1.8.2) *and the function f belong to the class $C_{(\mu)}(\Omega)$, $\mu \geqslant 1$, that g is a bounded function, and that the solution $u_\epsilon(x)$ of the problem* (1.8.2), (1.8.3) *belongs to class $C^{(\mu+2)}(\bar{\Omega})$. Assume that at points of $\bar{\Omega} \cap E$ the functions B_l defined by* (1.8.12) *are negative for $l \leqslant \mu$. Finally, assume that* (1.7.1) *holds for the domain Ω, and that $c < 0$ in $\bar{\Omega}$. Then, at each point of the set σ_1 for $l \leqslant \mu$,*

$$p^l\leqslant M_l+C_l\sqrt{\max_{\Sigma} p^l}, \qquad (1.8.23)$$

where the constants M_l and C_l depend only on $\max_{\bar{\Omega}} |u_\epsilon|$ and $\max_\Sigma p^\nu$ for $\nu \leqslant l-1$; $p^\nu = \Sigma_{\bar{S}_\nu}(D_{\bar{S}_\nu}u_\epsilon)^2$.

PROOF. Since $c<0$ in $\Omega \cup \Sigma$, we know that u_ϵ is bounded uniformly in ϵ. For $\mu = 1$ the inequality (1.8.23) was proved in Lemma 1.8.3 with the constant $C_1 = 0$. We now prove (1.8.23) for $\mu = 2$. For $\mu > 2$ the proof proceeds in an analogous manner. Suppose the point P_1 belongs to σ_1. In a neighborhood of the point P_1 we introduce local coordinates $y_1, \cdots, y_m$ so that σ_1 lies in the plane $y_m = 0$, and $y_m > 0$ at points of Ω. In the domain $Q_\delta\{0 < y_m < \kappa\psi/\gamma\}$ we consider the function $w = e^{-\gamma y_m + \kappa\psi}$, where $\psi(y_1, \cdots, y_{m-1})$ is the function defined in the proof of Lemma 1.8.3, and γ and κ are constants. Let $K_0 = \max_{\bar{Q}_\delta, \rho} |u_{\epsilon y_\rho}|$. Choose κ large enough so that the inequality $e^\kappa > 1 + K_0$ is satisfied. In the coordinates $y_1, \cdots, y_m$ we have

$$L_\varepsilon(w) = e^{-\gamma y_m + \kappa\psi}\Bigg[\varepsilon\mu^{mm}\gamma^2 - 2\varepsilon\sum_{k=1}^{m-1}\mu^{mk}\gamma\kappa\psi_{y_k} + \varepsilon\sum_{k,j=1}^{m-1}(\mu^{kj}\kappa^2\psi_{y_k}\psi_{y_j} + \mu^{kj}\kappa\psi_{y_k y_j})$$

$$-\varepsilon\nu^m\gamma + \varepsilon\sum_{k=1}^{m-1}\kappa\nu^k\psi_{y_k} + \alpha^{mm}\gamma^2 - 2\sum_{k=1}^{m-1}\alpha^{mk}\gamma\kappa\psi_{y_k} + \sum_{k,j=1}^{m-1}\alpha^{kj}\kappa^2\psi_{y_k}\psi_{y_j}$$

$$+\sum_{k,j=1}^{m-1}\alpha^{kj}\kappa\psi_{y_k y_j} - \beta^m\gamma + \sum_{k=1}^{m-1}\beta^k\kappa\psi_{y_k} + c\Bigg] \geqslant a_0\gamma^2$$

for $\gamma \geqslant \gamma_0$, where γ_0 is a sufficiently large number $a_0 > 0$, and a_0 and γ_0 do not depend on ϵ.

Suppose $\rho \neq m$. Differentiating equation (1.8.21) with respect to y_ρ, we obtain

$$L_\varepsilon(u_{y_\rho}) = f_{y_\rho} - \alpha^{kj}_{y_\rho}u_{y_k y_j} - \beta^k_{y_\rho}u_{y_k} - \varepsilon\mu^{kj}_{y_\rho}u_{y_k y_j} - \varepsilon\nu^k_{y_\rho}u_{y_k} - c_{y_\rho}u. \tag{1.8.24}$$

According to inequality (1.8.13) for $l = 2$, which was proved in Lemma 1.8.2 under the condition $B_t < 0$ on $\bar{\Omega} \cap E^\delta$ for $t = 1, 2$, it follows from (1.8.24) that

$$|L_\varepsilon(u_{\varepsilon y_\rho})| \leqslant K_1 + K_2\sqrt{\max_\Sigma p^2},$$

where the positive constants K_1 and K_2 depend on $\max_\Sigma p^1$. Let

$$\gamma = \gamma_0 + \left(a_0^{-1}\left(K_1 + K_2\sqrt{\max_\Sigma p^2}\right)\right)^{1/2}.$$

We consider the functions $v_\pm = w \pm u_{\epsilon y_\rho}$, $\rho \neq m$. It is clear that

$$L_\varepsilon(w \pm u_{\varepsilon y_\rho}) \geqslant a_0\gamma^2 - \left(K_1 + K_2\sqrt{\max_\Sigma p^2}\right) > 0$$

in Q_δ. Since $c < 0$ in $\Omega \cup \Sigma$, the maximum principle says that the functions $v_\pm$ take on their greatest positive value on the boundary Q_δ. On that portion of the boundary where $\psi = 1$ we have $v_\pm = e^\kappa$, and on the remaining part of the boundary lying in the plane $y_m = 0$ we have $v_\pm \leqslant e^\kappa$. The constant κ was so chosen that for $y_m = \kappa\psi/\gamma$ the inequality

$$v_\pm \leqslant 1 + \max_{\bar{Q}_\delta}|u_{\varepsilon y_\rho}| \leqslant 1 + K_0 < e^\kappa,$$

is satisfied; hence $v_\pm$ takes on its greatest positive value for $y_m = 0$, where $\psi = 1$. Consequently at the point P_1

$$(v_\pm)_{y_m} \leqslant 0 \text{ and } |u_{\epsilon y_\rho y_m}| \leqslant \gamma\epsilon^\kappa.$$

Since $\alpha^{mm} \neq 0$ at P_1, it follows from (1.8.21) that at P_1

$$|u_{\epsilon y_m y_m}| \leqslant K_3 + K_4 \sum_{\rho=1}^{m-1} |u_{\epsilon y_\rho y_m}| \leqslant K_3 + K_5 \gamma,$$

where the constants K_3, K_4 and K_5 depend on $\max_\Sigma p^1$. Therefore by the choice of γ we have

$$p^2 \leqslant M_2 + C_2 \sqrt{\max_\Sigma p^2},$$

at the point P_1, which is what was required to be proved.

For $l > 2$ the estimate (1.8.23) is obtained in the same way. First we estimate at the point P_1 those derivatives of order l containing only one differentiation with respect to y_m; the remaining derivatives of order l are then found by using equation (1.8.2) and equations obtained from it by differentiating with respect to $y_1, \cdots, y_m$. The lemma is thereby proved.

A simple corollary of Lemmas 1.8.1 and 1.8.3 is a theorem on the existence of bounded derivatives of first order for the generalized solution of (1.1.4), (1.1.5).

REMARK 1. Under the conditions of Lemma 1.8.4 and for $l \leqslant \mu$, the inequality

$$p^l \leqslant M_l + C_l \sqrt{\max_{Q_1} p^l}$$

is valid at each point of the set σ_1, where Q_1 is some neighborhood of σ_1 and M_l and C_l depend on $\max_\Omega |u_\epsilon|$ and $\max_{Q_1} p^\nu$ with $\nu \leqslant l-1$. We shall use this remark later in the proof of Theorems 1.8.3–1.8.8.

THEOREM 1.8.1. *Assume that the boundary* Σ *of the domain* Ω *is such that* $\Sigma_2 \cup \Sigma_3$ *and* $\Sigma_0 \cup \Sigma_1$ *do not have points in common, and that* $\Omega \in A_{(2)}$. *Suppose* $g = 0$ *on* $\Sigma_2 \cup \Sigma_3$. *Furthermore, let the following conditions hold.*

1) *The coefficients of the equation* (1.1.4) *and the function* f *belong to the class* $C_{(1)}(\Omega \cup G_{\delta_1})$ *for some* $\delta_1 \neq 0$, *where* G_{δ_1} *is the* δ_1*-neighborhood of the set* $\Sigma_0 \cup \Sigma_1$, *and* $a^{kj}\xi_k\xi_j \geqslant 0$ *in* $\Omega \cup G_{\delta_1}$ *for all* ξ.

2) *For some* $\delta > 0$ *the inequality*

$$\sup_{E^\delta \cap (\Omega \cup G_{\delta_1})} B_1 < 0$$

is fulfilled, where E^δ *is the* δ*-neighborhood of the set of points at which* $\det \|a^{kj}\| = 0$.

3) *At the points of* Σ_2 *the function* β, *defined by* (1.5.6), *is negative*, $c < 0$ *in* $\Omega \cup G_{\delta_1}$ *and the assumptions of Theorem* 1.6.2 *about uniqueness of the weak solution are fulfilled.*

4) *Inequality* (1.7.1) *is satisfied at the points of* $\Omega \cup G_{\delta_1}$. (*Condition* (4) *may be replaced by the requirement that in some neighborhood of* $\bar{\Omega}$ *the functions* $a^{kj} \in C_{(2)}$ *and* $a^{kj}\xi_k\xi_j \geqslant 0$ *for all* ξ).

Then a weak solution of the problem (1.1.4), (1.1.5) *satisfies a Lipschitz condition in* $\bar{\Omega}$.

PROOF. We consider a domain Ω_1 such that $\Omega_1 \subset (\Omega \cup G_{\delta_1})$, such that Ω_1 contains $\Sigma_0 \cup \Sigma_1 \cup \Omega$, and such that the boundary of Ω_1, which we denote by S_1, contains $\Sigma_2 \cup \Sigma_3$. Suppose $\Omega_1 \in A_{(2)}$. In Ω_1 we consider the equation

$$\varepsilon\Delta u + a\Delta u + L(u) = f, \tag{1.8.25}$$

where $a(x)$ is a smooth function such that $a = 0$ in Ω and $a > 0$ in $\bar{\Omega}_1 \backslash \bar{\Omega}$. Suppose $u_\epsilon(x)$ is the solution of (1.8.25) satisfying

$$u|_{S_1} = 0. \tag{1.8.26}$$

Since on S_1 either $a^{kj}n_kn_j + an_kn_k \neq 0$ or $\beta < 0$, it follows from Lemmas 1.8.3 and 1.8.1 that the derivatives of first order of the function $u_\epsilon(x)$ are bounded in Ω_1 uniformly with respect to ϵ. As is shown in the proof of Theorem 1.5.1, as $\epsilon \to 0$ the function u_ϵ converges to the weak solution $u(x)$, which satisfies the integral identity

$$\int_{\Omega_1} \Delta(av)u\,dx + \int_{\Omega_1} L^*(v)u\,dx = \int_{\Omega_1} fv\,dx \tag{1.8.27}$$

for any function $v \in C^{(2)}(\bar{\Omega}_1)$ such that $v = 0$ on $S_1 \backslash \Sigma_2$. It is easy to show that $u(x)$ is also a weak solution of the problem (1.1.4), (1.1.5) in the domain Ω with $g = 0$ on $\Sigma_2 \cup \Sigma_3$, i.e. the integral identity

$$\int_\Omega L^*(v)\,u\,dx = \int_\Omega fv\,dx \tag{1.8.28}$$

holds for any function v in $C^{(2)}(\bar{\Omega})$ equal to zero on $\Sigma_1 \cup \Sigma_3$. From (1.8.27) it follows that (1.8.28) is fulfilled for functions v in $C^{(2)}(\bar{\Omega})$ equal to zero on Σ_3 and in $\bar{\Omega}_1 \backslash \Omega$. Let v_1 be any function in $C^{(2)}(\bar{\Omega})$ equal to zero on $\Sigma_3 \cup \Sigma_1$. Then in (1.8.28) we replace $v(x)$ by the function $v_1\phi^\delta$, where the function ϕ^δ is infinitely differentiable, $\phi^\delta = 0$ in the δ-neighborhood of $\Sigma_0 \cup \Sigma_1$ and in $\Omega_1 \backslash \Omega$, $\phi^\delta = 1$ in Ω outside the 2δ-neighborhood of the set $\Sigma_0 \cup \Sigma_1$, and finally $0 \leqslant \phi^\delta \leqslant 1$. We shall suppose that in the local coordinates $y_1, \cdots, y_m$ the function ϕ^δ depends only on y_m, and that

$$|\varphi^\delta_{y_m}| = O(\delta^{-1}), \quad \varphi^\delta_{y_m y_m} = O(\delta^{-2}).$$

We have

$$\int_\Omega L^*(v_1)\,\varphi^\delta u\,dx + \int_\Omega (L^*(\varphi^\delta) - c^*\varphi^\delta)\,v_1 u\,dx + \int_\Omega a^{kj}\varphi^\delta_{x_k} v_{1x_j} u\,dx = \int_\Omega f v_1 \varphi^\delta dx. \tag{1.8.29}$$

We show that as $\delta \to 0$ this last inequality becomes (1.8.28) for the function $v = v_1$. For this it is sufficient to prove that the two last integrals on the left of (1.8.29) tend to zero as $\delta \to 0$. These integrals are clearly taken over the 2δ-neighborhood of $\Sigma_0 \cup \Sigma_1$, where $\phi^\delta_{x_k}$ and $\phi^\delta_{x_k x_j}$ may be different from zero. We denote this neighborhood by $h_{2\delta}$. In these integrals we pass to local coordinates $y_1, \cdots, y_m$ such that the boundary $\Sigma_0 \cup \Sigma_1$ lies in the plane $y_m = 0$. Since ϕ^δ depends only on y_m, these integrals have the form

$$\int_{h_{2\delta}} (\alpha^{mm}\varphi^\delta_{y_m y_m} + \beta^{*m}\varphi^\delta_{y_m})\, v_1 u \varkappa\, dy + \int_{h_{2\delta}} \alpha^{mm}\varphi^\delta_{y_m} v_{1y_m} u \varkappa\, dy, \tag{1.8.30}$$

where $\alpha^{mm} = O(\delta^2)$, $\beta^{*m} = 0$ at interior points of Σ_0 and κ is some bounded function. We recall that $v_1 = 0$ at points of Σ_1, by assumption. Since the measure of the domain $h_{2\delta}$ is of order δ, it clearly follows that the integral (1.8.30) tends to zero as $\delta \to 0$. The theorem is proved.

THEOREM 1.8.2 *Assume the following hypotheses:*

1) *The boundary Σ of the domain Ω is such that no two of the sets $\bar\Sigma_2$, $\bar\Sigma_3$, $\Sigma_0 \cup \Sigma_1$ have points in common; also $\Omega \in A^{(\mu+2)}$; $g = 0$ on $\Sigma_2 \cup \Sigma_3$.*

2) *For some $\delta_1 > 0$ the coefficients of the equation* (1.1.4) *and the function f belong to the class $C_{(\mu)}$ in $\Omega \cup G^1_{\delta_1}$ and to the class $C_{(2\mu+1)}$ in $G^2_{\delta_1}$; here $G^1_{\delta_1}$ is the δ_1-neighborhood of $\Sigma_0 \cup \Sigma_1$, and $G^2_{\delta_1}$ is the δ_1-neighborhood of the set Σ_2; also $a^{kj}\xi_k\xi_j \geqslant 0$ in $\Omega \cup G^1_{\delta_1} \cup G^2_{\delta_1}$ for any ξ and $\mu \geqslant 2$.*

3) *For some $\delta > 0$ the inequality*

$$\sup_{E^\delta \cap (\Omega \cup G^1_{\delta_1} \cup G^2_{\delta_1})} B_l < 0 \tag{1.8.31}$$

is satisfied for $l \leqslant \mu$.

4) *In the domain $\Omega \cup G^1_{\delta_1} \cup G^2_{\delta_1}$ the inequality* (1.7.1) *is fulfilled, or else $a^{kj} \in C_{(2)}$ and $a^{kj}\xi_k\xi_j \geqslant 0$ in a neighborhood of Ω; also $c < 0$ in $\Omega \cup G^1_{\delta_1} \cup G^2_{\delta_1}$.*

5) *The assumptions of Theorem* 1.6.2 *are fulfilled.*

Then the weak solution $u(x)$ of (1.1.4), (1.1.5) *belongs to the class $C_{(\mu)}(\bar\Omega)$.*

PROOF. We consider a domain Ω_1 such that $\Omega_1 \subset (\Omega \cup G^1_{\delta_1} \cup G^2_{\delta_1})$, such that Ω_1 contains $\Sigma_2 \cup \Sigma_0 \cup \Sigma_1 \cup \Omega$, and such that the boundary of Ω_1, which we denote by S_1, contains Σ_3. Suppose $\Omega_1 \in A^{(\mu+2)}$.

In the domain Ω_1 we construct a function w such that $L(w) - f$ vanishes on Σ_2 together with its derivatives to order $\mu - 1$ inclusive. We first construct the function w locally in a neighborhood of an arbitrary point P_1 of the boundary of Σ_2. For this purpose we introduce local coordinates $y_1, \cdots, y_m$ in a neighborhood of P_1 such that the boundary of Σ_2 lies in the plane $y_m = 0$. In this neighborhood equation (1.1.4) in new coordinates takes the form

$$L(u) \equiv \alpha^{kj} u_{y_k y_j} + \beta^k u_{y_k} + cu = f. \tag{1.8.32}$$

At points of Σ_2 it is evident that $\alpha^{mm} = 0$ and, by virtue of the condition 2), $\alpha^{mj}_{y_k} = 0$ for $j = 1, \cdots, m$ and $k \neq m$, $\alpha^{mm}_{y_m} = 0$, and therefore $b = \beta^m - \alpha^{mj}_{x_j} = \beta^m < 0$. Suppose $u = 0$ on Σ_2. From (1.8.32) and the equation obtained from it by differentiating with respect to y_m, we may find all derivatives of u with respect to y_m on Σ_2 up to order μ inclusive. For w in a neighborhood of P_1 we take a polynomial in y_m of degree μ which on Σ_2 coincides with u, together with all its derivatives up to order μ inclusive. Clearly $L(w) - f = \Psi$, where $\Psi = O(y_m^\mu)$ and Ψ belongs to class $C_{(\mu)}$.

In order to construct w in the entire domain Ω_1, we consider a covering of the domain $\bar{\Omega}_1$ by domains ω_j ($j = 1, \cdots, N$) such that if ω_j contains points of Σ_2, then in the domain ω_j one may pass to local coordinates y and define a function w_j such as was constructed above in a neighborhood of P_1. In those regions ω_j not containing points of Σ_2 we set $w_j \equiv 0$. Suppose $\{\psi_j\}$ is a partition of unity corresponding to the covering of $\bar{\Omega}_1$ by the domains ω_j (see Chapter II, §1). Then the function $w = \Sigma_1^N \psi_j w_j$, defined in $\bar{\Omega}_1$, satisfies the required condition. In fact, in a neighborhood of Σ_2 we have $L(w) - f = O(y_m^\mu)$, since in the intersection of two domains ω_j, each containing points of Σ_2, the functions w_j differ only by a function which is equal to zero on Σ_2 together with its derivatives up to order μ. We may suppose that the ω_j contain points from only one of the sets Σ_2, $\Sigma_1 \cup \Sigma_0$, or Σ_3. We consider the function $\tilde{f} = f - L(\omega)$ in Ω_1. It is clear that $\tilde{f}$ vanishes on Σ_2 together with its derivatives up to order $\mu - 1$ inclusive, and belongs to class $C_{(\mu)}(\bar{\Omega}_1)$. In the domain Ω_1 we consider the equation

$$L_\varepsilon(u) = \varepsilon\Delta u + a\Delta u + L(u) = f_1, \tag{1.8.33}$$

where the function $a \in C_{(\mu)}(\bar{\Omega}_1)$, $a = 0$ in Ω and $a > 0$ in $\bar{\Omega}_1 \backslash \bar{\Omega}$, with $\epsilon = \text{const} > 0$, and the function $f_1 \equiv 0$ in those parts of $\Omega_1 \backslash \Omega$ whose boundary contains Σ_2, and $f_1 = \tilde{f}$ in the remaining parts of the domain Ω_1. Clearly $f_1 \in C_{(\mu)}(\Omega_1)$. We consider the solution $u_\epsilon(x)$ of the equation (1.8.33) in the domain

Ω_1 satisfying a condition of the form

$$u_\epsilon |_{S_1} = 0 \tag{1.8.34}$$

on the boundary S_1. Since the inequality

$$an_k n_k + a^{kj} n_k n_j > 0$$

is satisfied at all points of the boundary S_1, it follows that we may conclude on the basis of the assumptions of the theorem and of Lemmas 1.8.2 and 1.8.4 that the solution u_ϵ has derivatives in Ω_1 up to order μ inclusive which are bounded uniformly in ϵ. Consequently as $\epsilon \to 0$ the function $u_\epsilon(x) \to \widetilde{u}(x)$, the function $\widetilde{u}(x) \in C_{(\mu)}(\Omega_1)$, $\widetilde{u}(x)$ satisfies the equation $L(\widetilde{u}) = \widetilde{f}$ in Ω, and $\widetilde{u} = 0$ on Σ_3.

We shall show that $\widetilde{u} = 0$ on Σ_2. In the regions of $\Omega_1 \setminus \overline{\Omega}$ containing points of Σ_2 on their boundary, the function $\widetilde{u}(x)$ satisfies the equation

$$a\Delta\widetilde{u} + L(\widetilde{u}) = 0$$

and is moreover equal to zero on S_1. Since the points Σ_2 of the domain Ω are points of type Σ_1 for such domains $\Omega_1 \setminus \overline{\Omega}$, so by virtue of the uniqueness theorem 1.6.2, $\widetilde{u} \equiv 0$ in these domains, i.e. $\widetilde{u} = 0$ on Σ_2. Consequently $u = \widetilde{u} + w$ is a solution of the problem (1.1.4), (1.1.5) and belongs to the class $C_{(\mu)}(\overline{\Omega})$. The theorem is proved.

Condition 2) of Theorem 1.8.2 on the possibility of extending the coefficients of equation (1.1.4) into a neighborhood of Σ_2 with retention of smoothness and of the nonnegativity of the characteristic form $a^{kj}\xi_k\xi_j$ is essential, as is shown by the following example, taken from [64].

The function $u = y^r e^{\rho y}$, where r and ρ are positive constants, satisfies the equation

$$yu_{yy} + (1 - r)\, u_y - (1 + r + \rho y)\, \rho u = 0 \tag{1.8.35}$$

on the segment [0, 1] of the y axis. The coefficients of the equation are infinitely differentiable, the point $y = 0$ of the boundary of the domain $\Omega\{0 < y < 1\}$ belongs to Σ_2, and the point $y = 1$ belongs to Σ_3. It is clear that for sufficiently large ρ the coefficient of u in (1.8.35) is arbitrarily large in modulus and negative. However, if $r < \mu$ and r is not an integer, then u has no more than $[r]$ bounded derivatives, where $[r]$ is the integral part of r.

Equation (1.8.35) clearly may not be continued into a neighborhood of the point $y = 0$ with retention of smoothness and nonnegativity of the coefficient of the second derivative. Inequality (1.7.1) is not fulfilled for this equation.

The condition of Theorem 1.8.2 to the effect that no two of $\overline{\Sigma}_2$, $\overline{\Sigma}_3$ and $\Sigma_1 \cup \Sigma_0$ have common points is also essential, as the following example shows.

In the domain Ω of the (x_1, x_2) plane included between the circles $x_1^2 + x_2^2 = 1$ and $x_1^2 + x_2^2 = 4$ we consider the equation

$$\frac{\partial u}{\partial x_2} + cu = 0, \quad c = \text{const} < 0. \tag{1.8.36}$$

The points $(-2, 0)$, $(-1, 0)$, $(1, 0)$ and $(2,0)$ belong to Σ_0, the arc $x_1^2 + x_2^2 = 1$ for $x_2 > 0$ and the arc $x_1^2 + x_2^2 = 4$ for $x_2 < 0$ belong to Σ_1, and the remaining part of the boundary belongs to Σ_2. On Σ_2 we prescribe the following function:

$$u = 0 \text{ on } x_1^2 + x_2^2 = 1 \text{ for } x_2 < 0,$$
$$u = 1 \text{ on } x_1^2 + x_2^2 = 4 \text{ for } x_2 > 0.$$

The weak solution of this problem clearly has a discontinuity along the line $x_1 = 1$ for $x_2 < 0$ and along the line $x_1 \equiv -1$ for $x_2 < 0$.

We note that if the assumptions of Lemma 1.8.1 are fulfilled, if the boundary σ of the domain ω lies in Ω, $\sigma \cap E = \varnothing$, and also if the assumptions of Theorem 1.6.2 on uniqueness hold, then the weak solution of (1.1.4), (1.1.5) satisfies a Lipschitz condition in ω. In exactly the same way it follows from Lemma 1.8.2 that if $\sigma \cap E = \varnothing$ and the assumptions of Lemma 1.8.2 are fulfilled for a neighborhood of ω, then the weak solution $u(x)$ of (1.1.4), (1.1.5) belongs to class $C_{(\mu)}$ in ω.

The assumptions of Theorems 1.8.1 and 1.8.2 to the effect that $B_l < 0$ in a neighborhood of the set of points where $\det \| a^{kj} \| = 0$ for $l \leqslant \mu$ is always fulfilled if $c < -c_0 < 0$, where the constant c_0 is sufficiently large, depending on μ. The following example shows this assumption to be necessary. The function $u = r^{\alpha}$ $(\alpha > 0, \alpha - [\alpha] > 0, r^2 = \Sigma_{k=1}^m x_k^2)$ in the domain $\Omega = \{r < 1\}$ is a weak solution of (1.1.4), (1.1.5) for the equation

$$L(u) = r^2 \Delta u - [\alpha m + \alpha(\alpha - 2)] u = 0$$

with the condition $u = 1$ on Σ_3. This example shows that the equation $L(u) = 0$ with analytic coefficients and analytic function g, and arbitrarily large c_0, may have a weak solution with only a finite number of bounded derivatives, even under the condition that the equation $L(u) = 0$ is elliptic in a neighborhood of the boundary Σ.

We now examine the question: under which conditions on the equation $L(u) = f$ and on the boundary Σ will it be true that the boundary function g on a certain segment of the boundary does not influence the smoothness of the solution of (1.1.4), (1.1.5) at internal points?

For Lemma 1.8.2 it follows in particular that this is true when the equation $L(u) = f$ is elliptic in a neighborhood of the boundary, $c < -c_0 < 0$, and c_0 is sufficiently large at points where $\det \| a^{kj} \| = 0$.

LEMMA 1.8.5. *Suppose the function* $u_\epsilon(x)$ *satisfies the equation*

$$\epsilon \Delta u + L(u) = f,$$

in the domain ω, *which is the intersection of* Ω *with some neighborhood of the point* P *on the boundary, and that this equation in local coordinates* $y_1, \cdots, y_m$ *assumes the form*

$$L_\varepsilon(u) = L'_\varepsilon(u) + cu = f, \tag{1.8.37}$$

where

$$L'_\varepsilon(u) \equiv \varepsilon\mu^{kj} u_{y_k y_j} + \varepsilon\nu^k u_{y_k} + \alpha^{kj} u_{y_k y_j} + \beta^k u_{y_k},$$

$$\mu^{kj}\xi_k\xi_j \geqslant \mu_0 \sum_{k=1}^{m} \xi_k^2, \quad \mu_0 = \text{const} > 0, \quad \alpha^{kj}\xi_k\xi_j \geqslant 0.$$

(It is assumed that the boundary Σ *lies in the plane* $y_m = 0$, *and* $y_m > 0$ *for points in* Ω.) *Suppose* $u_\epsilon(x) \in C^{(\mu+2)}(\bar{\omega})$, $|u_\epsilon| \leqslant K_0$ *in* ω, *where* K_0 *is a constant, and that the coefficients of* (1.8.37) *and the function* f *belong to the class* $C_{(\mu)}(\omega)$, $\mu \geqslant 1$. *Assume that for certain positive constants* K_j *one of the following conditions holds*;

1) *The form* $\alpha^{kj}\xi_k\xi_j$ *satisfies* (1.7.1) *at points of* ω, *and in* ω *for some* $\kappa \geqslant 2$ *or* $\kappa = 0$ *and any* $\nu \leqslant \mu$, *either*

$$-y_m\beta^m + [(4m^\nu - 1)(\varkappa + 2\nu) + 2]\,\alpha^{mm} \leqslant K_1 y_m^2, \tag{1.8.38}$$

or

$$\alpha^{kj}\xi_k\xi_j \geqslant K_2\alpha^{mm} \sum_{j=1}^{m} \xi_j^2; \quad -y_m\beta^m \leqslant K_3(\alpha^{mm} + y_m^2); \tag{1.8.39}$$

2) *The form* $y_m\alpha^{kj}\xi_k\xi_j$ *satisfies* (1.7.1) *at points of* ω, *and everywhere in* ω *for some* $\kappa \geqslant 2$ *or* $\kappa = 0$ *and any* $\nu \leqslant \mu$, *either*

$$-y_m\beta^m + [(4m^\nu - 1)(\varkappa + 2\nu) + 2]\,\alpha^{mm} + 2(\varkappa + 2\nu)^{-1} M_1 y_m \leqslant K_4 y_m^2, \tag{1.8.40}$$

or

$$\alpha^{kj}\xi_k\xi_j \geqslant K_5\alpha^{mm} \sum_{j=1}^{m} \xi_j^2; \quad -y_m\beta^m + 2(\varkappa + 2\nu)^{-1} M_1 y_m \leqslant K_6\left(\alpha^{mm} + y_m^2\right). \tag{1.8.41}$$

Here the constant M_1 *is determined by the coefficients of* (1.1.4), $M_1 = M_0 + \gamma \max|\beta^m| \cdot m^\nu$, *where* M_0 *depends on* α^{kj} *and also on* β^k *for* $k \neq m$, *and* $\gamma = \text{const} > 1$.

Suppose $p^l \equiv \Sigma_{\bar{S}_l}(D_{\bar{S}_l}u_\epsilon)^2$, $\bar{S}_l = (s_1, \cdots, s_l)$, $s_j = 1, \cdots, m$, $D_{\bar{S}_l} = \partial/\partial x_{s_1} \cdots \partial/\partial x_{s_l}$, $V^\nu = \Sigma_{l=0}^{\nu} C_l y_m^{2l+\kappa} p^l$. *Then at points of* ω, *for* $\nu \leqslant \mu$ *and for certain positive constants* C_l *independent of* ϵ, *the functions* V^ν *satisfy the inequality*

$$\frac{1}{2} L'_\varepsilon (V^\nu) + (c + M_2) V^\nu + E_\nu V^{\nu-1} -$$

$$- a_0 \sum_{l=1}^{\nu} C_l \sum_{\bar{S}_l} y_m^{2l+\kappa} (\alpha^{kj} + \varepsilon\mu^{kj}) D_{\bar{S}_l} u_{y_k} D_{\bar{S}_l} u_{y_j} \geqslant - \widetilde{E}_\nu, \tag{1.8.42}$$

where a_0, E_ν *and* $\widetilde{E}_\nu$ *are positive constants independent of* ϵ, *and* M_2 *is determined by the coefficients* b^k, a^{kj}, *and their derivatives of first and second order, respectively.*

PROOF. Inequality (1.8.42) is proved by induction. First we obtain it for $\nu = 1$. Differentiating (1.8.37) with respect to y_s, multiplying it by u_{y_s} and summing with respect to s, we obtain for $p^1 = \Sigma_1^m u_{y_k}^2$ an equation of the form

$$\begin{aligned} &\frac{1}{2} L'_\varepsilon (p^1) + cp^1 + \left[- \alpha^{kj} u_{y_s y_k} u_{y_s y_j} + \alpha^{kj}_{y_s} u_{y_k y_j} u_{y_s}\right] \\ &+ \varepsilon\left[- \mu^{kj} u_{y_s y_k} u_{y_s y_j} + \mu^{kj}_{y_s} u_{y_k y_j} u_{y_s}\right] + \beta^k_{y_s} u_{y_k} u_{y_s} \\ &+ c_{y_s} u u_{y_s} + \varepsilon \nu^k_{y_s} u_{y_k} u_{y_s} = f_{y_s} u_{y_s}. \end{aligned} \tag{1.8.43}$$

Since the form $\mu^{kj}\xi_k\xi_j$ is positive definite, it follows that

$$\left|\mu^{kj}_{y_s} u_{y_k y_j} u_{y_s}\right| \leqslant \gamma_0 \mu^{kj} u_{y_s y_k} u_{y_s y_j} + R_1 p^1, \quad \gamma_0 = \text{const} < 1.$$

By R_j we denote constants not depending on ϵ. We estimate the terms of the form $\alpha^{kj}_{y_s} u_{y_k y_j} u_{y_s}$ in exactly the same way, if (1.7.1) holds for the form $\alpha^{kj}\xi_k\xi_j$ in ω. In the case when condition 2) is fulfilled, inequality (1.7.1) holds for the form $y_m \alpha^{kj}\xi_k\xi_j$. This implies the inequalities

$$y_m \left|\alpha^{kj}_{y_s} u_{y_k y_j}\right|^2 \leqslant M \alpha^{kj} u_{y_\rho y_k} u_{y_\rho y_j} \quad \text{for } s \neq m,$$

$$y_m^2 \left|\alpha^{kj}_{y_m} u_{y_k y_j}\right|^2 \leqslant R_2 y_m \alpha^{kj} u_{y_s y_k} u_{y_s y_j} + \gamma^2 \left(\alpha^{kj} u_{y_k y_j}\right)^2, \quad \gamma = \text{const} > 1.$$

Considering the latter inequality and using (1.8.37), we obtain

$$\begin{aligned} \left|y_m^2 \alpha^{kj}_{y_s} u_{y_k y_j} u_{y_s}\right| &\leqslant y_m \sum_{s=1}^{m} \left[R_3 \left(y_m \alpha^{kj} u_{y_\rho y_k} u_{y_\rho y_j}\right)^{1/2} \right. \\ &+ \left. \gamma \left|\alpha^{kj} u_{y_k y_j}\right|\right] \left|u_{y_s}\right| \leqslant \gamma_1 y_m^2 \alpha^{kj} u_{y_\rho y_k} u_{y_\rho y_j} + R_4 y_m p^1 \\ &+ \gamma_1 y_m^2 \varepsilon \left(\mu^{kj} u_{y_k y_j} + \nu^k u_{y_k}\right)^2 + R_5 \varepsilon p^1 + y_m R_6 p_1 + R_7, \end{aligned}$$

where $\gamma_1 = \text{const} < 1$, where R_4 depends only on α^{kj}, $R_6 \leq \gamma m \cdot \max|\beta^m| + R_8$, and R_8 depends on β^k for $k \neq m$.

Therefore by multiplying (1.8.43) by $y_m^{2+\kappa}$ and using the inequalities obtained above (for sufficiently small γ_0 and γ_1) to estimate the terms in square brackets, we obtain the following relation for $q^1 = y_m^{2+\kappa} p^1$.

$$\begin{aligned}
&\frac{1}{2} L'_\varepsilon (q^1) + (c + M_2) q^1 - a_1 y_m^{\varkappa+2} (\alpha^{kj} + \varepsilon\mu^{kj}) u_{y_s y_k} u_{y_s y_j} \\
&\quad - \frac{1}{2} (\varkappa + 2)(\beta^m + \varepsilon\nu^m) y_m^{\varkappa+1} p^1 - \frac{1}{2} (\alpha^{mm} + \varepsilon\mu^{mm})(\varkappa+2)(\varkappa+1) y_m^{\varkappa} p^1 \\
&\quad - (\varkappa + 2) y_m^{\varkappa+1} \alpha^{km} p^1_{y_k} - (\varkappa + 2) \varepsilon\mu^{km} y_m^{\varkappa+1} p^1_{y_k} \\
&\quad + y_m^{\varkappa+1} (R_4 + R_6) p^1 + R_5 \varepsilon y_m^{\varkappa} p^1 \geqslant - R_9,
\end{aligned} \tag{1.8.44}$$

where $R_4 + R_6 = 0$, if condition 1) of the lemma is satisfied; here the constant a_1 does not depend on ϵ and $0 < a_1 < 1$.

It is easy to see from (1.7.9) that

$$\begin{aligned}
y_m^{\varkappa+1} (\varkappa + 2) |\alpha^{km} p^1_{y_k}| &= 2 (\varkappa + 2) y_m^{\varkappa+1} |\alpha^{km} u_{y_s y_k} u_{y_s}| \\
&\leqslant \gamma_3 y_m^{\varkappa+2} \alpha^{kj} u_{y_s y_k} u_{y_s y_j} + \gamma_4 \alpha^{mm} p^1 y_m^{\varkappa},
\end{aligned} \tag{1.8.45}$$

where the positive constants $\gamma_3 < 1$, $\gamma_4 > 2(\kappa + 2)^2 \cdot m$.

In a similar way we obtain

$$(\varkappa + 2) y_m^{\varkappa+1} \varepsilon |\mu^{km} u_{y_s y_k} u_{y_s}| \leqslant \gamma_3 \varepsilon y_m^{\varkappa+2} \mu^{kj} u_{y_s y_k} u_{y_s y_j} + \varepsilon \gamma_4 \mu^{mm} y_m^{\varkappa} p^1. \tag{1.8.46}$$

Considering (1.8.45), (1.8.46) and choosing $\gamma_3 < a_1$, we obtain from (1.8.44) that

$$\begin{aligned}
&\frac{1}{2} L'_\varepsilon (q^1) + (c + M_2) q^1 - a_2 y_m^{\varkappa+2} (\varepsilon\mu^{kj} + \alpha^{kj}) u_{y_s y_k} u_{y_s y_j} \\
&\quad - y_m^{\varkappa+1} \left[\frac{1}{2} (\beta^m + \varepsilon\nu^m)(\varkappa + 2) - R_4 - R_6\right] p^1 \\
&\quad + \left[\left(\gamma_4 - \frac{1}{2} (\varkappa + 2)(\varkappa + 1)\right) (\alpha^{mm} + \varepsilon\mu^{mm}) + \varepsilon R_5\right] y_m^{\varkappa} p^1 \geqslant - R_{10},
\end{aligned} \tag{1.8.47}$$

where $a_2 = \text{const} > 0$. It is easy to see that for $q^0 = y_m^{\kappa} u_\epsilon^2$ with $\kappa \geqslant 2$ we have the relation

$$\begin{aligned}
&\frac{1}{2} L'_\varepsilon (q^0) - y_m^{\varkappa} (\alpha^{kj} + \varepsilon\mu^{kj}) u_{y_k} u_{y_j} + (c + M_2) q^0 \\
&\quad - 2\varkappa y_m^{\varkappa-1} (\alpha^{km} + \varepsilon\mu^{km}) u u_{y_k} \geqslant - R_{11}.
\end{aligned}$$

Since according to (1.7.9) for $\kappa \geqslant 2$ we have

$$|2 y_m^{\varkappa-1} (\alpha^{km} + \varepsilon\mu^{km}) u u_{y_k}| \leqslant \gamma_5 y_m^{\varkappa} (\alpha^{mm} + \varepsilon\mu^{mm}) (\alpha^{kj} + \varepsilon\mu^{kj}) u_{y_k} u_{y_j} + R_{12},$$

it follows by choosing the constant γ_5 sufficiently small that

$$\frac{1}{2} L_\varepsilon'(q^0) - a_3 y_m^\varkappa (\alpha^{kj} + \varepsilon\mu^{kj}) u_{y_k} u_{y_j} + (c + M_2) q^0 \geqslant - R_{13}. \tag{1.8.48}$$

From (1.8.47) and (1.8.48) we deduce an inequality for V^1 of the form

$$\begin{aligned} &\frac{1}{2} L_\varepsilon'(V^1) + (c + M_2) V^1 - a_4\{C_0 y_m^\varkappa(\alpha^{kj} + \varepsilon\mu^{kj}) u_{y_k} u_{y_j} \\ &\quad + y_m^{\varkappa+2}(\varepsilon\mu^{kj} + \alpha^{kj}) u_{y_s y_k} u_{y_s y_j}\} \\ &+ \varepsilon y_m^\varkappa\Big[R_5 + \Big(\gamma_4 - \frac{1}{2}(\varkappa + 2)(\varkappa + 1)\Big)\mu^{mm}\Big] p^1 + y_m^\varkappa\Big[-\frac{1}{2}(\varkappa + 2)(\beta^m + \varepsilon\nu^m) y_m \\ &\quad + \Big(\gamma_4 - \frac{1}{2}(\varkappa + 2)(\varkappa + 1)\Big)\alpha^{mm} + (R_4 + R_6) y_m\Big] p^1 \geqslant -R_{14}. \end{aligned} \tag{1.8.49}$$

Considering conditions (1.8.38), (1.8.39) and conditions (1.8.40), (1.8.41), and choosing the constant C_0 sufficiently large, we obtain from (1.8.49) the relation

$$\begin{aligned} &\frac{1}{2} L_\varepsilon'(V^1) + (c + M_2) V^1 - a_5 [C_0 y_m^\varkappa (\alpha^{kj} \\ &+ \varepsilon\mu^{kj}) u_{y_k} u_{y_j} + y_m^{\varkappa+2}(\varepsilon\mu^{kj} + \alpha^{kj}) u_{y_s y_k} u_{y_s y_j}] \geqslant - R_{15}. \end{aligned} \tag{1.8.50}$$

Furthermore assuming that (1.8.42) is valid for some $\nu < \mu$ $(\nu \geqslant 1)$ we deduce, exactly as we derived (1.8.50), that (1.8.42) is also true for $\nu + 1$. The lemma is thereby proved.

REMARK 2. Suppose the following assumption is put in place of conditions 1) and 2) of Lemma 1.8.5:

3) Inequality (1.7.1) is valid for the form $\alpha^{kj}\xi_k\xi_j$ or for the form $y_m\alpha^{kj}\xi_k\xi_j$, and in ω,

$$\alpha^{kj}\xi_k\xi_j \geqslant K_1 \alpha^{mm} \sum_{j=1}^{m} \xi_j^2 \text{ and } \alpha^{mm} \geqslant K_0 y_m^s, \tag{1.8.51}$$

where s is a nonnegative integer. Then an inequality of the form (1.8.42) is valid for the functions

$$V^\nu = \sum_{l=0}^{\nu} C_l y_m^{l(s+1)} p^l.$$

The proof of this assertion is analogous to the proof of Lemma 1.8.5.

THEOREM 1.8.3. *Suppose the boundary* Σ *of the domain* Ω *consists of two closed nonintersecting sets* Σ' *and* $\Sigma \backslash \Sigma'$ *(either of them may be empty). Assume that for some sufficiently small* δ *the assumptions of Theorem* 1.8.2 *are fulfilled in regard to the boundary* $\Sigma\backslash\Sigma'$, *the coefficients of the equation* (1.1.4), *and the functions* f *and* g *in the domain* $\Omega\backslash G_\delta'$ *(where* G_δ' *is the* δ*-neighborhood of the set* Σ'*); also assume* *that* G'_δ *may be covered by a finite number of domains* $\{\Omega'_j\}$ *such that the assumptions of Lemma* 1.8.5 *regarding the coefficients* α^{kj} *and* β^k *are fulfilled for each of the domains* $\omega_j = \Omega'_j \cap \Omega$. *Assume, finally, that the function* g *is bounded on* Σ' *and that* $c_1 < -c_0 < 0$ *with a constant* c_0 *sufficiently large, depending on* α^{ki} *and* β^k. *Then the generalized solution of the problem* (1.1.4), (1.1.5) *belongs to class* $C_{(\mu)}(\Omega \backslash G'_\delta)$ *for any* $\delta > 0$, *and the estimate*

$$\sum_{l=0}^{\mu} \rho^{2l+\varkappa} \sum_{\bar{S}_l} (D_{\bar{S}_l} u)^2 < C_1 \tag{1.8.52}$$

is fulfilled, where $C_1 = \text{const}$, ρ *is the distance from the point* x *to the boundary* Σ', *and* $\kappa \geqslant 2$ *or* $\kappa = 0$.

PROOF. Suppose the domain Ω_0 is such that $\Omega_0 \supset \Omega, \Omega_0 \supset (\Sigma \backslash \Sigma') \cap (\Sigma_2 \cup \Sigma_1 \cup \Sigma_0)$, and the boundary S_0 of the domain Ω_0 contains $\Sigma_3 \cup \Sigma'$ and $\Omega_0 \in A^{(\mu+2)}$. We consider the solution $u_\epsilon(x)$ of the equation

$$L_\varepsilon(u) = \varepsilon\Delta u + a(x)\Delta u + L(u) = f_1 \text{ in } \Omega_0 \tag{1.8.53}$$

with the condition

$$u = g_n \quad \text{on } S_0, \tag{1.8.54}$$

where $g_n = 0$ on $S_0 \backslash \Sigma'$, g_n is infinitely differentiable on Σ', $g_n \to g$ in $L_2(\Sigma')$ as $n \to \infty$, g_n are bounded uniformly in n; $a(x) \in C_{(\mu)}(\Omega_0)$, $a = 0$ in Ω, and $a > 0$ in $\overline{\Omega}_0 \backslash \overline{\Omega}$. The function f_1 is constructed from f exactly as in the proof of Theorem 1.8.2. We may suppose that $f_1 = f$ outside a neighborhood of $(\Sigma \backslash \Sigma') \cap \Sigma_2$.

Let the domains Ω'_j $(j = 1, \cdots, N)$ and Ω''_j $(j = N+1, \cdots, N_0)$ cover $\overline{\Omega}_0$ and the functions $\{\psi_j\}$, $j = 1, \cdots, N_0$ form a partition of unity corresponding to this covering of Ω_0. We may suppose that the domains Ω''_j do not contain points of Σ'. In each of the domains $\omega_j = \Omega'_j \cap \Omega$ we consider functions V_j^ν, $\nu \leqslant \mu$, constructed in Lemma 1.8.5, and in the domains $\Omega''_j \cap \Omega_0$ we set $V_j^\nu = \Sigma_{l=0}^{\nu} C_l p^l$, where $p^l \equiv \Sigma_{\bar{S}_l} (D_{\bar{S}_l} u_\epsilon)^2$ and C_l are positive constants to be chosen below.

We consider the function

$$W^\nu = \sum_{j=0}^{N_0} V_j^\nu \psi_j, \quad \nu \leqslant \mu,$$

defined in Ω_0. In the same manner as Lemma 1.8.2 was proved, we obtain that in the domain Ω''_j the functions V_j^ν satisfy the relation

$$\frac{1}{2} L'_\varepsilon(V_j^\nu) + \frac{1}{2} a(x)\Delta V_j^\nu + (c + M_2) V_j^\nu + E_\nu V_j^{\nu-1} - a_0 \sum_{l=0}^{\nu} C_l \sum_{\bar{S}_l} (a^{ks} + \varepsilon\delta^{ks}) D_{\bar{S}_l} u_{x_k} D_{\bar{S}_l} u_{x_s} \geqslant -\tilde{E}_\nu, \tag{1.8.55}$$

where $M_2, E_\nu, a_0, \tilde{E}_\nu, C_l$ are certain positive constants independent of ϵ, and $\delta^{ks} = 0$ for $k \neq s$, $\delta^{ks} = 1$ for $k = s$.

Multiplying relation (1.8.42) by ψ_j $(j = 1, \cdots, N)$ and summing over j from 1 to N, and then multiplying (1.8.55) by ψ_j and summing over j from $N + 1$ to N_0, we obtain

$$\sum_{j=1}^{N_0} \frac{1}{2} [\psi_j L'_\varepsilon(V_j^\nu) + \psi_j a(x)\Delta V_j^\nu] + (c + M_2) W^\nu + E_\nu W^{\nu-1} - \sum_{j=1}^{N_0} \tilde{a}\psi_j \Phi_j^\nu \geqslant -R_1, \tag{1.8.56}$$

where for $j = 1, \cdots, N$

$$\Phi_j^\nu = \sum_{l=0}^{\nu} y_m^{2l+\varkappa} C_l \sum_{\bar{S}_l} (a^{ks} + \varepsilon\mu^{ks}) D_{\bar{S}_l} u_{y_k} D_{\bar{S}_l} u_{y_s} \text{ in } \Omega'_j \cap \Omega_0$$

and for $j = N + 1, \cdots, N_0$

$$\Phi_j^\nu = \sum_{l=0}^{\nu} C_l \sum_{\bar{S}_l} (a^{ks} + \varepsilon\delta^{ks}) D_{\bar{S}_l} u_{x_k} D_{\bar{S}_l} u_{x_s} \text{ in } \Omega''_j \cap \Omega_0.$$

By R_s we denote constants independent of ϵ. It is easy to see that

$$\sum_{j=1}^{N_0} \psi_j L'_\varepsilon(V_j^\nu) = L'_\varepsilon(W^\nu) - \sum_{j=1}^{N_0} [L'_\varepsilon(\psi_j) V_j^\nu] - 2\sum_{j=1}^{N} (a^{ks} + \varepsilon\mu^{ks}) V_{jy_k}^\nu \psi_{jy_s} - 2\sum_{j=N+1}^{N_0} (a^{ks} + \varepsilon\delta^{ks}) V_{jx_k}^\nu \psi_{jx_s}. \tag{1.8.57}$$

We estimate the last three terms of (1.8.57). It is clear that

$$|L'_\varepsilon(\psi_j) V_j^\nu| \leqslant R_2 W^\nu.$$

Since for $k \neq m$ and $j = 1, \cdots, N$

$$V_{jy_k}^\nu = \sum_{l=0}^{\nu} C_l y_m^{2l+\varkappa} p_{y_k}^l = \sum_{l=0}^{\nu} 2C_l y_m^{2l+\varkappa} D_{\bar{S}_l} u D_{\bar{S}_l} u_{y_k},$$

$$V_{jy_m}^\nu = \sum_{l=0}^{\nu} \left[2C_l y_m^{2l+\varkappa} D_{\bar{S}_l} u D_{\bar{S}_l} u_{y_m} + C_l (2l + \varkappa) y_m^{2l+\varkappa-1} p^l\right],$$

the following inequality holds for $1 \leqslant j \leqslant N$:

$$2\left|(\alpha^{ks}+\varepsilon\mu^{ks})V^{\nu}_{jy_k}\psi_{jy_s}\right| \leqslant \left|2\sum_{l=0}^{\nu} C_l(\alpha^{mk}+\varepsilon\mu^{mk})(2l+\varkappa)y_m^{2l+\varkappa-1}p^l\psi_{jy_k}\right|$$
$$+\sum_{l=0}^{\nu}\gamma y_m^{2l+\varkappa}C_l(\alpha^{ks}+\varepsilon\mu^{ks})D_{\overline{S}_l}u_{y_k}D_{\overline{S}_l}u_{y_s}\psi_j \tag{1.8.58}$$
$$+\sum_{l=0}^{\nu}C_l\frac{1}{\gamma}(\alpha^{ks}+\varepsilon\mu^{ks})\psi_{jy_k}\psi_{jy_s}\psi_j^{-1}y_m^{2l+\varkappa}p^l.$$

The functions ψ_j are nonnegative, and therefore according to (1.7.2) we have $|\psi_{jy_k}|^2 \leqslant R_3\psi_j$, $k=1,\cdots,m$. From this it follows that the last term in (1.8.58) does not surpass R_4W^{ν}.

For sufficiently small γ the second term on the right of (1.8.58) does not surpass $\tilde{a}\,\psi_j\Phi_j^{\nu}$. If the inequality $\alpha^{kj}\xi_k\xi_j \geqslant K_1\alpha^{mm}\sum_{k=1}^{m}\xi_k^2$ is fulfilled in the domain $\Omega_j'\cap\Omega_0$, then we use (1.7.10) to estimate the first term on the right side of (1.8.58). We thus obtain

$$\sum_{l=0}^{\nu}C_l\left|\alpha^{mk}(2l+\varkappa)\,y_m^{2l+\varkappa-1}p^l\psi_{jy_k}\right|$$
$$\leqslant\sum_{l=0}^{\nu}C_l\sqrt{2\alpha^{mm}}\sqrt{\alpha^{ks}\psi_{jy_k}\psi_{jy_s}}\,(2l+\varkappa)\,y_m^{2l+\varkappa-1}p^l$$
$$\leqslant\sum_{l=0}^{\nu}R_5C_l\sqrt{2\alpha^{mm}}\sqrt{\psi_j}\,(2l+\varkappa)\,y_m^{2l+\varkappa-1}p^l \tag{1.8.59}$$
$$\leqslant\sum_{l=0}^{\nu}\{C_l\gamma\alpha^{mm}y_m^{2(l-1)+\varkappa}p^l\psi_j+R_6C_ly_m^{2l+\varkappa}\varphi_jp^l\}$$
$$\leqslant\sum_{l=0}^{\nu}C_l\gamma\alpha^{mm}y_m^{2(l-1)+\varkappa}\psi_jp^l+R_7W^{\nu},$$

where ϕ_j are functions with support in Ω_j', $0\leqslant\phi_j\leqslant 1$, and $\phi_j\equiv 1$ at points where $\psi_j\neq 0$.

In the same way we obtain

$$\left|C_l\varepsilon\mu^{mk}(2l+\varkappa)\,y_m^{2l+\varkappa-1}p^l\psi_{jy_k}\right|\leqslant C_l\gamma\varepsilon\mu^{mm}y_m^{2(l-1)+\varkappa}\psi_jp^l+\varepsilon R_8W^{\nu}. \tag{1.8.60}$$

If (1.8.38) or (1.8.40) holds in $\Omega_j'\cap\Omega_0$ and, consequently, $\alpha^{mm}=0$ and $\alpha^{mk}=0$ on Σ', then to estimate the first term on the right side of (1.8.58) we use the relation

$$C_l\left|\alpha^{mk}(2l+\varkappa)\,y_m^{2l+\varkappa-1}p^l\psi_{jy_k}\right|\leqslant R_9C_ly_m^{2l+\varkappa}\varphi_jp^l\leqslant R_{10}W^{\nu}.$$

For the last term in (1.8.57) we have the estimate

$$|(a^{ks}+\varepsilon\delta^{ks}) V^{\nu}_{jx_k}\psi_{jx_s}| \leqslant R_{11} \sum_{l=0}^{\nu} C_l (a^{ks}+\varepsilon\delta^{ks}) \psi_{jx_k}\psi_{jx_s} p^l \psi_j^{-1}$$

$$+\gamma \sum_{l=0}^{\nu} C_l \sum_{\overline{S}_l} (a^{ks}+\varepsilon\delta^{ks}) D_{\overline{S}_l} u_{x_k} D_{\overline{S}_l} u_{x_s} \psi_j \leqslant R_{12} W^{\nu} + \gamma\psi_j \Phi^{\nu}_j.$$

Substituting into (1.8.56) the expression for $\Sigma_{j=1}^{N_0} \psi_j L'_\epsilon(V^\nu_j)$ given by the right side of (1.8.57) and bearing in mind the estimates derived above, we obtain

$$\begin{aligned}
&\frac{1}{2}\left[L'_\varepsilon(W^{\nu})+a\Delta W^{\nu}\right]+(c+M_2)W^{\nu}+R_{13}W^{\nu}+E_{\nu}W^{\nu-1}\\
&\quad+(\gamma-\widetilde{a})\sum_{j=1}^{N_0}\psi_j\Phi^{\nu}_j+\gamma\varepsilon\sum_{j=1}^{N}\sum_{l=1}^{\nu}C_l y_m^{2(l-1)+\varkappa}\mu^{mm}\psi_j p^l\\
&\quad+\gamma\sum_{j=1}^{N}\sum_{l=1}^{\nu}C_l y_m^{2(l-1)+\varkappa}a^{mm}\psi_j p^l \geqslant -R_1.
\end{aligned} \tag{1.8.61}$$

The last summation on the left of (1.8.61) is carried through only those values of j for which (1.8.39) or (1.8.41) is fulfilled in Ω'_j. Therefore, choosing γ sufficiently small, we obtain that the sum of the three last terms on the left of (1.8.61) is nonpositive. Thus in Ω_0 we obtain

$$\frac{1}{2}\left[L'_\varepsilon(W^{\nu})+a\Delta W^{\nu}\right]+(c+M_2+R_{14})W^{\nu}+E_{\nu}W^{\nu-1} \geqslant -R_1, \tag{1.8.62}$$

for W^ν when $\nu \leqslant \mu$; here R_{14} depends on ν and on the functions a^{kj}.

From the maximum principle for (1.8.53) it follows that W^0 is bounded uniformly in ϵ. Furthermore, from (1.8.62) it follows that W^ν is bounded in Ω_0 uniformly in ϵ and n, if this is valid for $W^{\nu-1}$ and $c + M_2 + R_{14} < 0$ in Ω_0. In fact, if W^ν takes on its greatest value inside Ω_0, then from (1.8.62) it follows that

$$W^{\nu} \leqslant -(c+M_2+R_{14})^{-1}(R_1+E_{\nu}W^{\nu-1}).$$

It is clear that $W^\nu = 0$ on Σ'. On the remaining part of the boundary S_0 the condition

$$a^{kj}n_k n_j + a n_k n_k > 0$$

holds and hence (1.8.23) is valid. As a consequence, if W^ν assumes its greatest value on $S_0 \setminus \Sigma'$, then from (1.8.23) it follows that $W^\nu \leqslant R_{15}$. Thus by induction we obtain the boundedness of W^ν for $\nu \leqslant \mu$ uniformly in ϵ and n. From this we obtain the assertion of Theorem 1.8.3 as in the proof of Theorem 1.8.2.

It is easy to see that the hypotheses of Theorem 1.8.3 will be fulfilled, in particular, if $\Sigma' \subset \Sigma_1$ and if the coefficients a^{kj} or Φa^{kj} ($\Phi = 0$ on Σ_1, $\Phi > 0$ in Ω, grad $\Phi \neq 0$) may be extended to a neighborhood of Σ_1 in such a manner that

$a^{kj} \in C_{(2)}, a^{kj}\xi_k\xi_j \geqslant 0$, or $\Phi a^{kj} \in C_{(2)}$ and $\Phi a^{kj}\xi_k\xi_j \geqslant 0$.

REMARK 3. Theorem 1.8.3 is also true in the case when in place of conditions 1) and 2) of Lemma 1.8.5, condition (1.8.51) is fulfilled for some of the domains Ω_j', as long as we also assume that if $s > 1$, then such domains Ω_j' cover a connected component of Σ'. We denote it by $\widetilde{\Sigma}'$. In this case in place of (1.8.52) we have

$$\sum_{l=1}^{\mu} \rho^{2l+\varkappa} \cdot \rho_1^{l(s+1)} \sum_{\bar{s}_l} (D_{\bar{s}_l} u)^2 < C_1,$$

where ρ_1 is the distance from the point x to $\widetilde{\Sigma}'$, and ρ is the distance from the point x to $\Sigma' \setminus \widetilde{\Sigma}'$.

REMARK 4. It is easy to see that Theorem 1.8.3 is not true if in place of condition (1.8.39) or (1.8.41) we merely assume that the boundary of Σ' belongs to Σ_3. In fact, suppose that Ω contains the square $\{0 < x_1 < 1, 0 < x_2 < 1\}$, in which $L(u) \equiv \partial^2 u/\partial x_2^2 + cu = 0$, $c = \text{const} < 0$. The segment $0 \leqslant x_1 \leqslant 1$ of the line $x_2 = 0$, and also the segment $0 \leqslant x_1 \leqslant 1$ of the line $x_2 = 1$, belong to Σ_3. By virtue of the uniqueness theorem, the generalized solution $u(x)$ in the square is a linear combination of the functions g given on the sides, so that the smoothness of the solution u inside Ω depends on the smoothness of the boundary functions g. The same assertion is also true if $L(u) \equiv \partial u/\partial x_2 + cu$, $c = \text{const} < 0$, in the same square, which means that the segment $0 \leqslant x_1 \leqslant 1$ of the line $x_2 = 0$ belongs to Σ_2.

REMARK 5. In Theorem 1.8.3, suppose the boundary Σ' includes an $(m-1)$-dimensional closed manifold $\widetilde{\Sigma}'$ such that $\widetilde{\Sigma}' \subset \Sigma_0 \cup \Sigma_1$ and there exists a sequence of smooth surfaces $\widetilde{\Sigma}^n$ such that $\widetilde{\Sigma}^n \subset \Omega$ and $\widetilde{\Sigma}^n \to \widetilde{\Sigma}'$ as $n \to \infty$ in the sense that the measure of the set bounded by $\widetilde{\Sigma}^n \cup \widetilde{\Sigma}'$ tends to zero as $n \to \infty$. We assume that for the domain Ω^n whose boundary consists of $(\Sigma \setminus \widetilde{\Sigma}') \cup \widetilde{\Sigma}^n$ the portion $\widetilde{\Sigma}^n$ is of type $\Sigma_0 \cup \Sigma_1$. Exactly as was shown in Theorem 1.1.2 (the maximum principle), it may be shown that if p^l takes on its greatest value on $\widetilde{\Sigma}^n$, then p^l is uniformly bounded with respect to n, and consequently, under the assumptions of Theorem 1.8.3, the weak solution of (1.1.4), (1.1.5) has bounded derivatives of order $\mu - 3$ in a neighborhood of $\widetilde{\Sigma}'$ as well.

We note that condition 2) of Lemma 1.8.5 does not assume the possibility of extending the coefficients a^{kj} to a neighborhood of Σ' with retention of the conditions $a^{kj}\xi_k\xi_j \geqslant 0$ and $a^{kj} \in C_{(2)}$, and, in particular, it includes the case when, on Σ', $\alpha^{mm} \equiv a^{kj}\Phi_{x_k}\Phi_{x_j}$ is of order Φ, where the equation $\Phi = 0$ defines the boundary Σ' and $\operatorname{grad} \Phi \neq 0$ on Σ'. The conditions also admit the possibility that $(\Sigma_1 \cup \Sigma_0) \cap \overline{\Sigma}_3$ is nonempty.

We now establish theorems on the smoothness of weak solutions of (1.1.4), (1.1.5) in the closed region $\Omega \cup \Sigma$. We shall show that in many cases the conditions of Theorem 1.8.2 may be weakened.

LEMMA 1.8.6. *In the domain* ω, *the intersection of a neighborhood of some point* P_1 *on the boundary* Σ *with* Ω, *suppose the function* $u_\epsilon(x)$ *satisfies*

$$L_\epsilon(u) = f \text{ and } u_\epsilon = 0 \text{ on } \Sigma.$$

Using local coordinates $y_1, \cdots, y_m$ *in which the boundary* Σ *lies in the plane* $y_m = 0$ *and* $y_m > 0$ *in* ω, *this equation has the form*

$$L_\varepsilon(u) \equiv L'_\varepsilon(u) + cu = f, \tag{1.8.63}$$

where

$$L'_\varepsilon(u) \equiv \varepsilon\mu^{kj}u_{y_k y_j} + \alpha^{kj}u_{y_k y_j} + \beta^k u_{y_k},$$

$$\mu^{mj} = 0 \text{ for } j \neq m,\ \mu^{mm} = 1,\ \alpha^{mj} = 0 \text{ for } j \neq m,$$

$$\mu^{kj}\xi_k\xi_j \geqslant \mu_0 \sum_{k=1}^{m} \xi_k^2,\quad \mu_0 = \text{const} > 0,\quad \alpha^{kj}\xi_k\xi_j \geqslant 0.$$

Suppose $u_\epsilon \subset C^{(\mu+2)}(\bar{\omega})$, $|u_\epsilon| \leqslant K_0$ *in* ω, $K_0 = \text{const} > 0$, *and the coefficients of the equation and the function* f *belong to the class* $C_{(\mu)}(\omega)$. *Assume that for some constants* $K_j > 0$ *the following conditions hold.*

1) $\alpha^{mm} \leqslant K_1 y_m$; $\beta^m = \text{const} > 0$,

$$\begin{aligned} &\left| D'_{\bar{S}_l}\alpha^{mm}_{y_m} \right| \leqslant K_3 y_m,\quad 1 \leqslant l \leqslant \mu - 1, \\ &\left| D'_{\bar{S}_l}\alpha^{mm} \right| \leqslant K_2 y_m^2,\quad 2 \leqslant l \leqslant \mu, \end{aligned} \tag{1.8.64}$$

where $D'_{\bar{S}_l}v$ *denotes the derivative*

$$\frac{\partial}{\partial y_{s_1}} \cdots \frac{\partial}{\partial y_{s_l}} v$$

of order l, *in which* $\bar{S}_l = (s_1, \cdots, s_l)$ *and* $s_j \neq m$ *for* $j = 1, \cdots, l$.

2) *The form* $\Sigma_{k,j=1}^{m-1} \alpha^{kj}\xi_k\xi_j$ *satisfies* (1.7.1) *in* ω, *i. e.*

$$\left| \sum_{k,j=1}^{m-1} \alpha^{kj}_{y_s} v_{y_k y_j} \right|^2 \leqslant M \sum_{k,j=1}^{m-1} \alpha^{kj} v_{y_\rho y_k} v_{y_\rho y_j},$$
$$s = 1, \ldots, m;\ M = \text{const} > 0. \tag{1.8.65}$$

3) *Either the function* α^{mm}, *extended to be zero outside* ω, *has bounded derivatives of second order in a neighborhood of* Σ, *or the function* $y_m\alpha^{mm}$, *extended to be zero outside* ω, *belongs to class* $C_{(2)}$ *in a neighborhood of* Σ *and*

$\beta^m = \beta_0$, where β_0 is a sufficiently large number depending on α^{mm}. (In the first case one may assume in place of the condition $\beta^m = \text{const} > 0$ that $\beta^m = 0$ on Σ.)

Set

$$p^{\varkappa,l} = \sum_{\overline{S}_l} y_m^{2\varkappa} \left(\frac{\partial^\varkappa}{\partial y_m^\varkappa} D'_{\overline{S}_l} u_\varepsilon \right)^2, \quad \varkappa \geqslant 0;$$
$$Z^\nu = \sum_{l+\varkappa \leqslant \nu} C_{\varkappa,l} p^{\varkappa,l}; \quad Q_\varepsilon(v, w) = (\varepsilon\mu^{kj} + \alpha^{kj}) v_{y_k} w_{y_j}, \tag{1.8.66}$$

where $C_{\kappa,l}$ are some constants.

Then the following inequality is valid in ω for the functions Z^ν with $\nu \leqslant \mu$ and for certain constants $C_{\kappa,l}$ independent of ϵ:

$$\frac{1}{2} L'_\varepsilon (Z^\nu) + (c + M_1) Z^\nu + E_\nu Z^{\nu-1}$$
$$- a_0 \sum_{l+\varkappa \leqslant \nu} C_{\varkappa,l} y_m^{2\varkappa} Q_\varepsilon \left(\frac{\partial^\varkappa}{\partial y_m^\varkappa} D'_{\overline{S}_l} u, \frac{\partial^\varkappa}{\partial y_m^\varkappa} D'_{\overline{S}_l} u \right) \geqslant - \widetilde{E}_\nu, \tag{1.8.67}$$

where $a_0, M_1, E_\nu, \widetilde{E}_\nu$ are positive constants independent of ϵ, and M_1 depends on β^k and α^{kj} and their derivatives of first and second order respectively.

PROOF. We prove (1.8.67) by induction in a manner analogous to the proof of Lemma 1.8.5. For this purpose we first obtain (1.8.67) for $\nu = 1$. It is easy to see that for

$$h_1 = y_m^2 u_{y_m}^2 + \sum_{\rho=1}^{m-1} u_{y_\rho}^2 = \sum_{\varkappa+l=1} p^{\varkappa,l}$$

we have the relation

$$\frac{1}{2} L'_\varepsilon (h_1) - \sum_{\rho=1}^{m-1} Q_\varepsilon (u_{y_\rho}, u_{y_\rho}) - y_m^2 Q_\varepsilon (u_{y_m}, u_{y_m})$$
$$+ \sum_{\rho, k, j=1}^{m-1} (\varepsilon\mu_{y_\rho}^{kj} + \alpha_{y_\rho}^{kj}) u_{y_k y_j} u_{y_\rho}$$
$$+ \sum_{\rho=1}^{m-1} \alpha_{y_\rho}^{mm} u_{y_m y_m} u_{y_\rho} + y_m^2 \alpha_{y_m}^{mm} u_{y_m y_m} u_{y_m} \tag{1.8.68}$$
$$+ \sum_{k,j=1}^{m-1} y_m^2 (\varepsilon\mu_{y_m}^{kj} + \alpha_{y_m}^{kj}) u_{y_k y_j} u_{y_m} + c h_1 + M_1 h_1 - \beta^m y_m u_{y_m}^2$$
$$- (\varepsilon + \alpha^{mm}) u_{y_m}^2 - 4 (\varepsilon + \alpha^{mm}) y_m u_{y_m y_m} u_{y_m} \geqslant - R_1.$$

By R_j we denote constants independent of ϵ. We estimate the individual terms of (1.8.68). Clearly

$$|2(\varepsilon + a^{mm}) y_m u_{y_m y_m} \cdot u_{y_m}| \leqslant \gamma(\varepsilon + a^{mm})\, y_m^2 (u_{y_m y_m})^2$$
$$+ \frac{1}{\gamma}(\varepsilon + a^{mm})\, u_{y_m}^2 \leqslant \gamma y_m^2 Q_\varepsilon(u_{y_m}, u_{y_m}) + \frac{1}{\gamma}(\varepsilon + a^{mm})\, u_{y_m}^2.$$

Since the form $\mu^{kj}\xi_k\xi_j$ is positive definite and satisfies condition 2), it follows that

$$\left|\sum_{k,j,\rho=1}^{m-1} (\varepsilon\mu_{y_\rho}^{kj} + a_{y_\rho}^{kj})\, u_{y_k y_j} \cdot u_{y_\rho}\right| \leqslant \gamma \sum_{\rho=1}^{m-1}\left[\sum_{k,j=1}^{m-1} (\varepsilon\mu_{y_\rho}^{kj} + a_{y_\rho}^{kj})\, u_{y_k y_j}\right]^2 + \frac{1}{\gamma}\sum_{\rho=1}^{m-1} u_{y_\rho}^2$$
$$\leqslant \gamma R_0 \sum_{\rho=1}^{m-1} Q_\varepsilon(u_{y_\rho}, u_{y_\rho}) + R_2 h_1.$$

Furthermore, according to (1.7.2) and the condition $|a_{y_\rho y_\rho}^{mm}| \leqslant K_3 y_m^2$ we have

$$\left|\sum_{\rho=1}^{m-1} a_{y_\rho}^{mm} u_{y_m y_m} u_{y_\rho}\right| \leqslant \gamma \sum_{\rho=1}^{m-1} \max |a_{y_\rho y_\rho}^{mm}| \cdot a^{mm} u_{y_m y_m}^2 + R_3 h_1$$
$$\leqslant \gamma_1 y_m^2 Q_\varepsilon(u_{y_m}, u_{y_m}) + R_3 h_1.$$

Using condition 2), we obtain the estimate

$$y_m^2 \left|\sum_{k,j=1}^{m-1} (\varepsilon\mu_{y_m}^{kj} + a_{y_m}^{kj})\, u_{y_k y_j} u_{y_m}\right| \leqslant \gamma y_m^2 \sum_{\rho=1}^{m} Q_\varepsilon(u_{y_\rho}, u_{y_\rho}) + R_4 h_1.$$

If the first assumption in condition 3) is fulfilled, then, applying (1.7.2) to the function a^{mm} extended to be zero in a neighborhood of Σ for $y_m < 0$, we obtain

$$y_m^2 |a_{y_m}^{mm} u_{y_m y_m} \cdot u_{y_m}| \leqslant \gamma y_m^2 a^{mm} (u_{y_m y_m})^2 + R_5 h_1$$
$$\leqslant \gamma y_m^2 Q_\varepsilon(u_{y_m}, u_{y_m}) + R_5 h_1.$$

If the second assumption in condition 3) is fulfilled, then (1.7.2) is applicable to the function $y_m a^{mm}$ in ω. We have

$$|y_m|^3 \left|a_{y_m}^{mm}\right|^2 \leqslant M y_m^2 a^{mm} + M^0 y_m |a^{mm}|^2, \quad M, M^0 = \text{const} > 0.$$

Therefore, considering that $a^{mm} \leqslant K_1 y_m$, we obtain

$$|y_m^2 a_{y_m}^{mm} u_{y_m y_m} \cdot u_{y_m}| \leqslant \gamma y_m^3 (a_{y_m}^{mm} u_{y_m y_m})^2 + \frac{1}{2\gamma} y_m u_{y_m}^2$$
$$\leqslant \gamma M y_m^2 a^{mm} (u_{y_m y_m})^2 + \gamma M^0 y_m (a^{mm})^2 (u_{y_m y_m})^2 + \frac{1}{2\gamma} y_m (u_{y_m})^2$$
$$\leqslant \gamma_1 y_m^2 Q_\varepsilon(u_{y_m}, u_{y_m}) + R_6 y_m u_{y_m}^2,$$

where R_6 is determined by the function a^{mm}.

Using the estimates obtained above, choosing the constants γ and γ_1 sufficiently small, and assuming that $R_6 < \beta_0$, we deduce from (1.8.68) that

$$\frac{1}{2} L'_\varepsilon(h_1) - a_0\left[\sum_{\rho=1}^{m-1} Q_\varepsilon(u_{y_\rho}, u_{y_\rho}) + y_m^2 Q_\varepsilon(u_{y_m}, u_{y_m})\right] + (c + M_2) h_1 + R_7(\varepsilon + \alpha^{mm}) u_{y_m}^2 \geqslant - R_8. \tag{1.8.69}$$

Since $h_0 = u_\epsilon^2$ clearly satisfies the inequality

$$\frac{1}{2} L'_\varepsilon(h_0) + ch_0 - Q_\varepsilon(u, u) \geqslant - R_9, \tag{1.8.70}$$

on choosing the constant $C_{0,0}$ sufficiently large we obtain from (1.8.69) and (1.8.70) a relation for $Z^1 = h_1 + C_{0,0}h_0$ of the form

$$\frac{1}{2} L'_\varepsilon(Z^1) - a_1\left[C_{0,0}Q_\varepsilon(u, u) + y_m^2 Q_\varepsilon(u_{y_m}, u_{y_m}) + \sum_{\rho=1}^{m-1} Q_\varepsilon(u_{y_\rho}, u_{y_\rho})\right] + (c + M_2) Z^1 \geqslant - R_{10}. \tag{1.8.71}$$

The proof of (1.8.67) now proceeds by induction. We calculate the expression $\frac{1}{2}L'_\epsilon(p^{\kappa,l})$ for $\kappa + l = \nu$, carry through estimates analogous to those obtained for $\nu = 1$, and hence, assuming that (1.8.67) holds for $Z^{\nu-1}$ and choosing $C_{\kappa,l}$ for $\kappa + l \leqslant \nu - 1$ sufficiently large, derive an inequality for the Z^ν. In this way the assertion of Lemma 1.8.6 is obtained.

THEOREM 1.8.4. *Suppose the boundary* Σ *of* Ω *consists of two closed non-intersecting sets* Σ' *and* $\Sigma \setminus \Sigma'$ *(either of them may be empty). Let the assumptions of Theorem* 1.8.2 *be valid for the boundary* $\Sigma \setminus \Sigma'$ *and for the coefficients of* (1.1.4) *and the functions* f *and* g *in the domain* $\Omega \setminus G'_\delta$. *Here* G'_δ *is the* δ*-neighborhood of the set* Σ', *and we assume that* G'_δ *may be covered by a finite number of domains* $\{\Omega'_j\}$, $j = 1, \cdots, N$, *such that for each of the domains* $\omega_j = \Omega \cap \Omega'_j$ *the assumptions of Lemma* 1.8.6 *are satisfied relative to the coefficients* α^{kj}, β^k, c, *and the function* f, *but with the condition* $\beta^m > \text{const} > 0$ *in* ω_j *in place of the condition* $\beta^m = \text{const} > 0$. *Suppose that* $g = 0$ *on* Σ' *and that* $c < -c_0 < 0$, *where constant* c_0 *is sufficiently large, depending on* α^{kj} *and* β^k. *Then the weak solution* $u(x)$ *of* (1.1.4), (1.1.5) *belongs to the class* $C_{(\mu)}(\Omega \setminus G'_\delta)$ *for any* $\delta > 0$. *Also, the following estimate holds for* $u(x)$ *in each of the domains* ω_j, *where the local coordinates* $y_1, \cdots, y_m$ *are such that* Σ' *lies in the plane* $y_m = 0$:

$$\sum_{k+l\leqslant\mu} y_m^{2k}\left(\frac{\partial^k}{\partial y_m^k} D'_{\bar{s}_l} u\right)^2 \leqslant C_1, \quad C_1 = \text{const}. \tag{1.8.72}$$

PROOF. Theorem 1.8.4 may be proved in exactly the same way as Theorem

1.8.3. The only difference is in the construction of the operator (1.8.53) in the neighborhood of Σ'. In the domain Ω_0 constructed in the proof of Theorem 1.8.3 we consider the equation

$$L_\varepsilon(u) = \varepsilon P(u) + a(x)\Delta u + \frac{1}{\tilde{\beta}^m} L(u) = \frac{f_1}{\tilde{\beta}^m} \tag{1.8.73}$$

with the condition $u = 0$ on S_0, where $P(u)$ is an elliptic operator in $\bar{\Omega}_0$, $P(u) = \Delta u$ in $\Omega_0 \backslash G'_\delta$ for some $\delta > 0$, and in some neighborhood of Σ' the operator $P(u) = u_{y_m y_m} + P_1(u)$, where $P_1(u)$ is an elliptic operator of second order given on Σ' and extended inside Ω_0 with coefficients not depending on y_m. Also the functions $\tilde{\beta}^m = \beta^m$ in the δ-neighborhood of Σ', $\tilde{\beta}^m = 1$ outside the 2δ-neighborhood of Σ' and $\tilde{\beta}^m > \text{const} > 0$ if $\beta^m > 0$ on Σ'. However, if $\beta^m = 0$ on Σ', then $\tilde{\beta}^m \equiv 1$.

We note that the condition of Lemma 1.8.5 that $\alpha^{mj} = 0$ for $j \neq m$ for the operator L' in a neighborhood of Σ' may always be fulfilled with an appropriate choice of local coordinates, if, for some s, the function

$$\frac{a^{kl}F_{x_l}}{a^{sl}F_{x_l}}, \quad k = 1, \ldots, m,$$

belongs to the class $C_{(\mu)}(\omega_j)$. Here a^{kl} are the coefficients of the operator $L(u)$, $F = 0$ is the equation of the boundary Σ', and $\operatorname{grad} F \neq 0$.

THEOREM 1.8.5. *Suppose all assumptions of Theorem 1.8.4 are fulfilled. Furthermore suppose that within each of the domains ω_j, one of the following conditions is fulfilled in a neighborhood of Σ': either*

$$\begin{gathered} \left|D'_{\bar{s}_l}\alpha^{mm}\right| \leqslant K_3 y_m^2, \quad 0 \leqslant l \leqslant \mu, \\ \left|D'_{\bar{s}_l}\alpha^{mm}_{y_m}\right| \leqslant K_4 y_m, \quad 0 \leqslant l \leqslant \mu - 1; \quad \beta^m > \text{const} > 0, \end{gathered} \tag{1.8.74}$$

or

$$\alpha^{mm}_{y_m} \neq 0, \quad \frac{\beta^m}{\max\left|\alpha^{mm}_{y_m}\right|} > 1, \tag{1.8.75}$$

where K_3 and K_4 are certain positive constants. Then the weak solution of (1.1.4), (1.1.5) *belongs to class $C_{(\mu-1)}(\Omega)$ and to class $C_{(\mu)}(\Omega \backslash G'_\delta)$ for every $\delta > 0$ if conditions* (1.8.75) *are satisfied, and to class $C_{(k)}(\Omega)$ and to class $C_{(\mu)}(\Omega \backslash G'_\delta)$ for every $\delta > 0$, $k = [\mu/2]$, if conditions* (1.8.74) *are satisfied.*

PROOF. We first assume that in ω_j conditions (1.8.74) are satisfied. Then, using an estimate of the form (1.8.72) with constant C_1 independent of ϵ for the solution u_ϵ of (1.8.73) which is equal to zero on the boundary S_0, and then passing to the limit as $\epsilon \to 0$ in (1.8.73), we obtain

$$\alpha^{mm} u_{y_m y_m} + \beta^m u_{y_m} = F_1 \tag{1.8.76}$$

at interior points of ω_j, where F_1 is a bounded function in ω_j, in accordance with (1.8.72). From (1.8.72) it also follows that $\alpha^{mm} u_{y_m y_m}$ is bounded in ω_j, since $\alpha^{mm} \leqslant K_3 y_m^2$ in ω_j. From (1.8.76) it follows that u_{y_m} is bounded in ω_j.

Furthermore, applying the operator $D'_{\overline{S}_l}$ with $l \leqslant \mu - 2$ to the right and left sides of (1.8.76), we obtain the equation

$$\alpha^{mm} D'_{\overline{S}_l} u_{y_m y_m} + \beta^m D'_{\overline{S}_l} u_{y_m} = F_2,$$

where F_2 contains derivatives of u with respect to $y_1, \cdots, y_{m-1}$, and terms of the form $D'_{\overline{S}_{l-\gamma}} \alpha^{mm} D'_{\overline{S}_\gamma} u_{y_m y_m}$, $l - \gamma > 0$, $l \leqslant \mu - 2$. Since by assumption $|D'_{\overline{S}_l} \alpha^{mm}| \leqslant K_3 y_m^2$, by virtue of (1.8.72) the function F_2 is bounded in ω_j, and consequently $D'_{\overline{S}_l} u_{y_m}$ is bounded in ω_j for $l \leqslant \mu - 2$.

Applying sequentially the operators $D'_{\overline{S}_l} \partial^\nu / \partial y_m^\nu$, $\nu = 1, \cdots, \mu - 2, l + \nu \leqslant \mu - 2$, to (1.8.76) and multiplying the equations so obtained by y_m^ν, we obtain equations which, together with (1.8.74) and (1.8.72), imply that the quantities

$$y_m^\nu D'_{\overline{S}_l} \frac{\partial^{\nu+1} u}{\partial y_m^{\nu+1}} \quad \text{for } l + \nu \leqslant \mu - 2$$

are bounded in ω_j.

The proof of the boundedness in ω_j of all derivatives up to order $[\mu/2]$ is carried out by induction on κ. We assume that for some $\kappa \geqslant 1$ the estimate

$$\left| y_m^\nu D'_{\overline{S}_l} \frac{\partial^{\nu+\kappa} u}{\partial y_m^{\nu+\kappa}} \right| < C_2 = \text{const}, \qquad l + \nu \leqslant \mu - 2\kappa, \tag{1.8.77}$$

holds in ω_j, and then show that the same estimate is also valid for $\kappa + 1$. For this purpose we apply the operator $D'_{\overline{S}_l} \partial^{\nu+\kappa} / \partial y_m^{\nu+\kappa}$, $l + \nu \leqslant \mu - 2(\kappa + 1)$, to the equation (1.8.76) and multiply it by y_m^ν.

The equations so obtained, together with the condition (1.8.74) and the induction assumption (1.8.77), yield that an estimate of the form (1.8.77) holds also for $\kappa + 1$.

From (1.8.72) and the estimate (1.8.77) proved for $\nu = 0$, $\kappa \leqslant [\mu/2]$, it follows that $u(x) \in C_{(\mu-1)}(\omega_j)$. Thus the theorem is proved in the case of condition (1.8.74).

If $\alpha^{mm}_{y_m} \neq 0$ on Σ', then (1.8.76) in ω_j may be written in the form

$$y_m u_{y_m y_m} + q(y_1, \ldots, y_m) u_{y_m} = \Psi, \tag{1.8.78}$$

where the function Ψ is bounded in ω_j and $q > q_0 = \text{const} > 0$. In ω_j we fix the point $P^0 = (y_1^0, \cdots, y_m^0)$, $y_m^0 > 0$ and consider the segment $\{0 < y_m \leq y_m^0, y_1 = y_1^0, \cdots, y_{m-1} = y_{m-1}^0\}$. Multiplying (1.8.78) by

$$y_m^{-1} \exp\left[\int_{y_m^0}^{y_m} q(y_1^0, \ldots, y_{m-1}^0, s)\, s^{-1} ds\right]$$

and integrating it with respect to y_m from $y_m = \rho$ to $y_m = y_m^0$, $0 < \rho < y_m^0$, we obtain

$$|u_{y_m}(P^0)| \leq \exp\left[\int_{y_m^0}^{\rho} q(y_1^0, \ldots, y_{m-1}^0, s)\, s^{-1} ds\right] \cdot |u_{y_m}(y_1^0, \ldots, y_{m-1}^0 \rho)| \tag{1.8.79}$$
$$+ \max_{\omega_j} |\Psi| \int_{\rho}^{y_m^0} z^{-1} \exp\left[\int_{y_m^0}^{z} q(y_1^0, \ldots, y_{m-1}^0, s)\, s^{-1} ds\right] dz.$$

It is easy to see that

$$\exp\left[\int_{y_m^0}^{\rho} q(y_1^0, \ldots, y_{m-1}^0, s)\, s^{-1} ds\right] \leq \exp\left[\int_{y_m^0}^{\rho} q_0 s^{-1} ds\right] = \rho^{q_0}(y_m^0)^{-q_0},$$

$$\int_{\rho}^{y_m^0} z^{-1} \exp\left[\int_{y_m^0}^{z} q(y_1^0, \ldots, y_{m-1}^0, s)\, s^{-1} ds\right] dz \leq \int_{\rho}^{y_m^0} z^{q_0 - 1} (y_m^0)^{-q_0} dz$$
$$= q_0^{-1}(1 - \rho^{q_0}(y_m^0)^{-q_0}).$$

From (1.8.79) it follows that

$$|u_{y_m}(P^0)| \leq (y_m^0)^{-q_0} \rho^{q_0} |u_{y_m}(y_1^0, \ldots, y_{m-1}^0, \rho)| \tag{1.8.80}$$
$$+ \max |\Psi|\, q_0^{-1}(1 - (y_m^0)^{-q_0} \rho^{q_0}).$$

Since, according to condition (1.8.75), $q_0 > 1$, when we let ρ tend to zero in (1.8.80) and consider (1.8.72) we obtain

$$|u_{y_m}(P^0)| \leq \max |\Psi| q_0^{-1}. \tag{1.8.81}$$

Furthermore, applying the operator $D'_{\bar{S}_l}$ to (1.8.78) successively for $l = 1, \cdots, \mu - 2$, we obtain equations for the $D_{\bar{S}_l} u_{y_m}$, from which we obtain (successively for $l = 1, \cdots, \mu - 2$) the boundedness of $|D'_{\bar{S}_l} u_{y_m}|$ in ω_j exactly as the estimate (1.8.81) was obtained.

We assume that we have proved, for some $\kappa > 1$, the boundedness in ω_j of the derivatives of the form $D'_{\bar{s}_l}\partial^\kappa u/\partial y_m^\kappa$ for $l+\kappa \leqslant \mu-1$. We show that this implies the boundedness of the derivatives $D'_{\bar{s}_l}\partial^\nu u/\partial y_m^\nu$ for $\nu=\kappa+1, l+\kappa+1 \leqslant \mu-1$. For this we first apply the operator $\partial^\kappa/\partial y_m^\kappa$ to the equation (1.8.78) and obtain for $v=\partial^{\kappa+1}u/\partial y_m^{\kappa+1}$ an equation of the form

$$y_m(v)_{y_m} + (q+\varkappa)v = \Psi_\varkappa, \tag{1.8.82}$$

where Ψ_κ is a bounded function. From this equation we obtain the boundedness of $\partial^{\kappa+1}u/\partial y_m^{\kappa+1}$ in ω_j exactly as the estimate (1.8.81) was obtained. Afterwards, applying to (1.8.82) successively the operators $D'_{\bar{s}_l}$ for $l=1, \cdots, \mu-\kappa-2$, we obtain the boundedness of derivatives of the form $D'_{\bar{s}_l}\partial^{\kappa+1}u/\partial y_m^{\kappa+1}$, $l+\kappa \leqslant \mu-2$. With this the theorem is proved.

REMARK 6. Theorem 1.8.5 yields conditions under which there exists a weak solution of the problem (1.1.4), (1.1.5) smooth in a neighborhood of Σ_1. In this process it is not assumed that (1.1.4) may be extended to a neighborhood of Σ_1 with retention of smoothness of its coefficients and of nonnegativity of the form $a^{kj}\xi_k\xi_j$, as was done in Theorem 1.8.2. An analogous theorem was obtained in the paper of Kohn and Nirenberg [64]. However, in that paper, in distinction to the considerations given above, the process of obtaining a smooth solution of (1.1.4), (1.1.5) as the limit as $\epsilon \to 0$ of solutions of elliptic equations was carried through by adjoining to (1.1.4), in a neighborhood of Σ_1, an elliptic operator of higher order with small parameter ϵ, the order of the operator depending on the smoothness of the data of the problem (the number μ). And outside the neighborhood of Σ_1 they adjoin an elliptic operator of second order, also with parameter ϵ. The results of the paper [64] will be examined in the next section.

LEMMA 1.8.7. *Suppose the assumptions of Lemma* 1.8.6 *are satisfied, except that in place of condition* 1) *we have in* ω_j *for some constants* K_j *and* $s \geqslant 1$ *the relations*

$$\alpha^{mm} \geqslant K_0 y_m^s, \tag{1.8.83}$$

$$\alpha^{mm} \leqslant K_1 y_m, \quad |D'_{\bar{s}_l}\alpha^{mm}| \leqslant K_2 y_m^{s+1} \quad \text{for } 2 \leqslant l \leqslant \mu,$$

$$|D'_{\bar{s}_l}\alpha^{mm}_{y_m}| \leqslant K_3 y_m^{\frac{s+1}{2}} \quad \text{for } 1 \leqslant l \leqslant \mu-1, \quad \beta^m = \text{const}.$$

Let

$$p^{\varkappa,l} = y_m^{\varkappa(s+1)} \sum_{\bar{s}_l} \left(\frac{\partial^\varkappa}{\partial y_m^\varkappa} D'_{\bar{s}_l} u_\varepsilon\right)^2$$

and, as before, $Z^\nu = \Sigma_{l+\kappa\le\nu} C_{\kappa,l} p^{\kappa,l}$. *Then at points of* ω_j *we have the following inequality for* Z^ν *with* $\nu \le \mu$:

$$\frac{1}{2} L'_\varepsilon(Z^\nu) + (c + M_1) Z^\nu + E_\nu Z^{\nu-1} - a_0 \sum_{l+\varkappa\le\nu} C_{\varkappa l} y_m^{\varkappa(s+1)} Q_\varepsilon \left(\frac{\partial^\varkappa}{\partial y_m^\varkappa} D'_{\overline{s}_l} u, \frac{\partial^\varkappa}{\partial y_m^\varkappa} D'_{\overline{s}_l} u \right) \ge -\widetilde{E}_\nu, \tag{1.8.84}$$

where $C_{\kappa,l}, a_0, M_1, E_\nu, \widetilde{E}_\nu$ *are constants independent of* ϵ, *with* M_1 *depending on* β^k, α^{kj} *and their derivatives of first and second orders respectively.*

PROOF. Inequality (1.8.84) is proved exactly as the analogous inequality (1.8.67). We first obtain it for $\nu = 1$. Suppose $h_1 = y_m^{s+1} u_{y_m}^2 + \Sigma_{\rho=1}^{m-1} u_{y_\rho}^2$. We have

$$\frac{1}{2} L'_\varepsilon(h_1) + ch_1 + M_1 h_1 - \sum_{\rho=1}^{m-1} Q_\varepsilon(u_{y_\rho}, u_{y_\rho}) - y_m^{s+1} Q_\varepsilon(u_{y_m}, u_{y_m})$$
$$+ \left[\sum_{k,j,\rho=1}^{m-1} (\varepsilon\mu_{y_\rho}^{kj} + \alpha_{y_\rho}^{kj}) u_{y_k y_j} u_{y_\rho} + \sum_{\rho=1}^{m-1} \alpha_{y_\rho}^{mm} u_{y_m y_m} u_{y_\rho} \right]$$
$$+ y_m^{s+1} \left[\alpha_{y_m}^{mm} u_{y_m y_m} u_{y_m} + \sum_{k,j=1}^{m-1} (\varepsilon\mu_{y_m}^{kj} + \alpha_{y_m}^{kj}) u_{y_k y_j} u_{y_m} \right]$$
$$- \frac{(s+1)}{2} \beta^m y_m^s u_{y_m}^2 - (\varepsilon + \alpha^{mm}) y_m^{s-1} \frac{s(s+1)}{2} u_{y_m}^2$$
$$- 2(\varepsilon + \alpha^{mm})(s+1) y_m^s u_{y_m y_m} u_{y_m} \ge -R_1.$$

The estimate for the last term, and also for the terms in square brackets, is carried out exactly as in the proof of Lemma 1.8.6. We estimate the terms $-((s+1)/2)\beta^m y_m^s u_{y_m}^2 + R_2 y_m^s u_{y_m}^2$ by bearing in mind that in (1.8.70) for $h^0 = u_\epsilon^2$ we have a term of the form $-Q_\epsilon(u, u)$. Since $\alpha^{mm} \ge K_0 y_m^s$, it follows that in the equation for $h_1 + C_{0,0} h^0$ for sufficiently large $C_{0,0}$ the sum

$$-\frac{(s+1)}{2} \beta^m y_m^s u_{y_m}^2 + R_1 y_m^s u_{y_m}^2 - C_{0,0} \alpha^{mm} u_{y_m}^2$$

is nonpositive. The remaining portions of the proof of Lemma 1.8.7 repeat the proof of Lemma 1.8.6.

THEOREM 1.8.6. *Suppose the boundary* Σ *consists of two closed nonintersecting sets* Σ' *and* $\Sigma \backslash \Sigma'$ *(either of them may be empty). Assume that the hypotheses of Theorem* 1.8.2 *are fulfilled with regard to the boundary* $\Sigma\backslash\Sigma'$ *and to the coefficients of* (1.1.4) *and the functions* f *and* g *in the domain* $\Omega \backslash G'_\delta$, *where* G'_δ *is the* δ*-neighborhood of the set* Σ'. *Assume also that* G'_δ *may be covered by a finite number of domains* Ω'_j, $j = 1, \cdots, N$, *such that in each of the domains* $\omega_j = \Omega \cap \Omega'_j$ *either the assumptions of Lemma* 1.8.6 *or the assumptions of Lemma* 1.8.7 *are satisfied* *with regard to the coefficients* α^{kj}, β^k, c *and the function* f. *Suppose* $g = 0$ *on* Σ', *and* $c < -c_0 < 0$ *in* G'_δ, *with the constant* c_0 *sufficiently large depending on* α^{kj}, β^k, *and their derivatives of first and second orders. Then the weak solution* $u(x)$ *of* (1.1.4), (1.1.5) *belongs to class* $C_{(\mu)}(\Omega \backslash G'_\delta)$ *for any* $\delta > 0$. *Furthermore, in each of the domains* ω_j *the following estimate holds for* $u(x)$, *where* $y_1, \cdots, y_m$ *are local coordinates such that* Σ' *lies in the plane* $y_m = 0$ *with* $y_m > 0$ *in* Ω:

$$\sum_{k+l\leqslant\mu} y_m^\nu \left(\frac{\partial^k}{\partial y_m^k} D'_{\overline{S}_l} u\right)^2 \leqslant C_1 = \text{const}, \tag{1.8.85}$$

where $\nu = 2k$ *if the point* $(y_1, \cdots, y_m)$ *belongs to a* ω_j *for which the assumptions of Lemma* 1.8.6 *are satisfied, and* $\nu = k(s+1)$ *if the assumptions of Lemma* 1.8.7 *are fulfilled in* ω_j.

The proof of this theorem is carried out exactly as the proof of Theorem 1.8.4.

THEOREM 1.8.7. *Suppose the assumptions of Theorem* 1.8.6 *are fulfilled, with* $\Sigma' = \Sigma_1^1 \cup \Sigma_2^1$, *where those domains* ω_j *whose boundaries contain points of* Σ_1^1 *satisfy the conditions of Theorem* 1.8.5 *with* (1.8.75), *and those* ω_j *whose boundaries contain points of* Σ_2^1 *satisfy the assumptions of Lemma* 1.8.7, *as well as the condition*

$$\beta^m + (\nu - 1)\,\alpha^{mm}_{y_m} < 0 \quad \textit{for} \quad \nu \leqslant \mu. \tag{1.8.86}$$

Then the weak solution $u(x)$ *of* (1.1.4), (1.1.5) *belongs to the class* $C_{(\mu-1)}(\Omega)$ *and to the class* $C_{(\mu)}(\Omega \backslash G'_\delta)$ *for any* $\delta > 0$.

PROOF. In view of Theorem 1.8.5, it is sufficient to prove that the solution $u(x)$ in a neighborhood of Σ_2^1 belongs in the class $C_{(\mu-1)}$. Clearly $\Sigma_2^1 \subset \Sigma_2$. We first show that in ω_j the functions $u_{\epsilon y_m}$ are uniformly bounded, where u_ϵ is the solution of an equation analogous to (1.8.73) equal to zero on the boundary S_0, with $\widetilde{\beta}^m = |\beta^m|$ in the δ-neighborhood of Σ_2^1. For the proof of Theorem 1.8.6 we obtained the estimate

$$\sum_{k+l\leqslant\mu} y_m^{k(s+1)} \left(\frac{\partial^k}{\partial y_m^k} D'_{\overline{S}_l} u_\epsilon\right)^2 \leqslant C_1 \tag{1.8.87}$$

for u_ϵ in a neighborhood of Σ_2^1, where the constant C_1 does not depend on ϵ. From equation (1.8.73) with consideration taken of (1.8.87), we obtain

$$(\alpha^{mm} + \epsilon|\beta^m|)\,u_{\epsilon y_m y_m} + \beta^m u_{\epsilon y_m} = \Psi_\epsilon, \tag{1.8.88}$$

where the function Ψ_ϵ is bounded uniformly in ϵ. According to estimate (1.8.87), the $u_{\epsilon y_m}$ are bounded uniformly in ϵ for $y_m \geqslant \delta_1 > 0$.

We consider the $u_{\epsilon y_m}$ for $0 \leqslant y_m \leqslant \delta_1$. If $u_{\epsilon y_m}$ takes its greatest positive or least negative value for $0 \leqslant y < \delta_1$, then at this point $u_{\epsilon y_m} u_{\epsilon y_m y_m} \leqslant 0$ and, since $\beta^m < 0$, it follows from (1.8.88) that

$$|u_{\epsilon y_m}| \leqslant \max |\Psi_\epsilon| \cdot |\beta^m|^{-1},$$

i.e. the function $p^1 = \Sigma_{j=1}^m u_{\epsilon x_j}^2$ is bounded uniformly in ϵ.

Furthermore, considering that p^1 is bounded in a neighbrohood of Σ_2^1 uniformly with respect to ϵ, and for $y_m \geqslant \delta$ all derivatives up to order μ inclusive are bounded uniformly in ϵ, we may show exactly as in Lemma 1.8.7 that the sum

$$\sum_{1 \leqslant l+\nu+1 \leqslant \mu} C_{\nu+1,l}\, y_m^{\nu(s+1)} \left(\frac{\partial^{\nu+1}}{\partial y_m^{\nu+1}} D'_{\overline{S}_l} u_\epsilon\right)^2$$

is bounded in a neighborhood of Σ_2^1 uniformly in ϵ.

The proof of the boundedness of all derivatives up to order $\mu - 1$ inclusive in a neighborhood of Σ_2^1 proceeds by induction on κ. We assume that for some $\kappa \geqslant 1$ the estimate

$$\sum_{1 \leqslant l+\nu+\kappa \leqslant \mu} C_{\nu+\kappa,l}\, y_m^{\nu(s+1)} \left(\frac{\partial^{\nu+\kappa}}{\partial y_m^{\nu+\kappa}} D'_{\overline{S}_l} u_\epsilon\right)^2 < C_2 \tag{1.8.89}$$

holds, where the constant C_2 does not depend on ϵ. We show that all derivatives of u_ϵ of order $\kappa + 1$ are bounded in ω_j uniformly in ϵ. From this we obtain, exactly as we proved Lemma 1.8.7, that an estimate of the form (1.8.89) is valid for $\kappa + 1$ as well. For this we apply the operator $\partial^\kappa/\partial y_m^\kappa$ to the equation (1.8.88). In ω_j we have

$$(\alpha^{mm} + \epsilon|\beta^m|)\left(\frac{\partial^{\kappa+1} u_\epsilon}{\partial y_m^{\kappa+1}}\right)_{y_m} + (\beta^m + \kappa\alpha^{mm}_{y_m} + \kappa\epsilon|\beta^m|_{y_m})\frac{\partial^{\kappa+1} u_\epsilon}{\partial y_m^{\kappa+1}} = \Psi_\epsilon^\kappa, \tag{1.8.90}$$

where the functions Ψ_ϵ^κ according to (1.8.89) are bounded uniformly in ϵ. Since by assumption of Theorem 1.8.7 we have $\beta^m + \kappa\alpha^{mm}_{y_m} < 0$, it follows from (1.8.90), in exactly the same way that we proved the boundedness of $u_{\epsilon y_m}$, that the functions $(\partial^{\kappa+1}/\partial y_m^{\kappa+1}) u_\epsilon$ are bounded in a neighborhood of Σ_2^1 uniformly in ϵ.

From this and from (1.8.89) it follows that all derivatives of u_ϵ of order $\kappa + 1$ are uniformly bounded, which implies that an estimate of the form (1.8.89) is valid also for $\kappa + 1$.

Thus we obtain the uniform (in ϵ) boundedness in a neighborhood of Σ_2^1 of all derivatives of u_ϵ to order $\mu - 1$ inclusive. The theorem is thereby proved.

Condition (1.8.83) of Theorem 1.8.7 may be omitted. Namely, the following result holds:

THEOREM 1.8.8. *Suppose all the assumptions of Theorem* 1.8.7 *are satisfied with the exception of* (1.8.83), *which in Theorem* 1.8.7 *was assumed fulfilled for points lying in a neighborhood of* Σ_2^1. *Then the generalized solution of* (1.1.4), (1.1.5) *belongs to class* $C_{(\mu)}$ *in* $\Omega\setminus G'_\delta$, *to class* $C_{(\mu-1)}$ *in a neighborhood of* Σ_1^1, *and to class* $C_{(\mu-3)}$ *in a neighborhood of* Σ_2^1.

PROOF. In the domain Ω_0 defined in the proof of Theorems 1.8.3 and 1.8.4 we consider the equation

$$\varepsilon\widetilde{L}(u) + a(x)\Delta u + L(u) = f_1, \tag{1.8.91}$$

where $\widetilde{L}$ is elliptic everywhere in $\overline{\Omega}_0\setminus\Sigma_2^1$, $\widetilde{L}(u)\equiv P(u)$ outside the δ-neighborhood of Σ_2^1 (the operator P is defined in the proof of Theorem 1.8.4), and in the $\delta/2$-neighborhood of Σ_2^1

$$\widetilde{L}(u) \equiv \widetilde{a}(y_m)\, u_{y_m y_m} + P_1(u),$$

where $P_1(u)$ is an elliptic operator given on Σ_2^1 and extended inside Ω with coefficients independent of y_m, $\widetilde{a}(y_m) = y_m^2$ for $y_m \leqslant \delta/4$ and $\widetilde{a}(y_m) = 1$ for $\delta/3 \leqslant y_m \leqslant \delta/2$, $0 < \widetilde{a} \leqslant 1$ for $0 < y_m \leqslant \delta/2$. In the domain Ω_0 the equation (1.8.91) has a solution, according to Theorem 1.8.7, equal to zero on the boundary S_0. This solution belongs to class $C_{(\mu-1)}(\Omega_0)$. On the boundary Σ_2^1 we may express all derivatives of u_ϵ up to order $\mu - 1$ by use of equation (1.8.91) and the equations obtained from it by differentiating with respect to $y_1, \cdots, y_m$. It is easy to see that these derivatives are bounded uniformly in ϵ. Beyond this, the proof of Theorem 1.8.8 proceeds according to the same scheme as in Theorem 1.8.4, with this difference, that in those regions ω_j whose boundaries contain points of Σ_2^1 we set $Z^\nu = \Sigma_{l\leqslant\nu} C_l \Sigma_{\overline{S}_l}(D_{\overline{S}_l}u_\epsilon)^2$.

§9. On conditions for the existence of a solution of the first boundary value problem in the spaces of S. L. Sobolev

In this section we shall give conditions under which a solution of the problem (1.1.4), (1.1.5) with square integrable weak derivatives exists. If, under these conditions, the hypotheses of the theorem yielding uniqueness of weak solutions of this problem are fulfilled, then the conditions guarantee a certain amount of smoothness of the weak solution of (1.1.4), (1.1.5).

The exposition of this section basically follows the papers of Kohn and Nirenberg [64, 147].

For simplicity we shall assume that the coefficients of the equation

$$L(u) = a^{kj}u_{x_k x_j} + b^k u_{x_k} + cu = f \tag{1.9.1}$$

are infinitely differentiable (or sufficiently smooth) functions in $\Omega \cup \Sigma$, that the boundary function g in (1.1.5) is equal to zero, and that the domain $\Omega \in A^{(n)}$ for sufficiently large n.

As in §8, we denote

$$D_{\overline{S}_l} u = \frac{\partial}{\partial x_{s_1}} \cdots \frac{\partial}{\partial x_{s_l}} u, \quad \overline{S}_l = (s_1, \ldots, s_l), \quad s_j = 1, \ldots, m,$$

$$D'_{\overline{S}_l} u = \frac{\partial}{\partial y_{s_1}} \cdots \frac{\partial}{\partial y_{s_l}} u,$$

with the condition that $s_j \neq m$. By $W_2^k(\Omega)$ we denote the space of functions for which there exist weak derivatives in the domain Ω up to order k inclusive, and

$$\| u \|_{k;\Omega}^2 = \int_\Omega \sum_{l=0}^{k} \sum_{\overline{S}_l} (D_{\overline{S}_l} u)^2 \, dx < \infty.$$

In the case $k = 0$ we shall also use the notation $\|u\|_{0;\Omega} = \|u\|_0$. In a neighborhood ω of a point P_1 on the boundary Σ we introduce local coordinates $y_1, \cdots, y_m$ so that the boundary Σ lies in the plane $y_m = 0$ and $y_m > 0$ in Ω. Let G_δ^0 be the δ-neighborhood of the set Σ_0, G_δ^2 the δ-neighborhood of the set Σ_2, and let δ be so small that G_δ^0 and G_δ^2 may be covered by domains ω in which there exist local coordinates $y_1, \cdots, y_m$ with the properties indicated above. We define

$$\begin{aligned} ||| u |||_N^2 = &\int_{G_\delta^2} \Bigg[\sum_{\overline{S}_l; l \leqslant N} \left| D'_{\overline{S}_l} u \right|^2 + \sum_{\substack{\overline{S}_l \\ l+k \leqslant N,\ k \geqslant 1}} | y_m |^{k-1} \left(\frac{\partial^k}{\partial y_m^k} D'_{\overline{S}_l} u \right)^2 \Bigg] dy \\ &+ \int_{G_\delta^0} \sum_{\overline{S}_l,\ l+k \leqslant N} | y_m |^k \left| \frac{\partial^k}{\partial y_m^k} D'_{\overline{S}_l} u \right|^2 dy + \int_{\Omega \setminus (G_\delta^0 \cup G_\delta^2)} \sum_{\overline{S}_l, l \leqslant N} (D_{\overline{S}_l} u)^2 \, dx. \end{aligned} \tag{1.9.2}$$

THEOREM 1.9.1 (SEE [64]). *Assume the following conditions on the boundary* Σ *and on the coefficients in* (1.9.1):

1) $$(\Sigma_1 \cup \Sigma_0) \cap \overline{\Sigma_2 \cup \Sigma_3} = \varnothing. \tag{1.9.3}$$

2) *The coefficient* $c < -c_0 < 0$ *in* Ω *and the constant* c_0 *is sufficiently large, depending on* a^{kj}, b^k, *and their derivatives of order* 3 *and less.*

3) *At points of* Σ_2,

$$\gamma = 1 + \frac{1}{2} \mu \frac{\alpha}{\tilde{b}^2} > 0, \tag{1.9.4}$$

where $\tilde{b} = (b^k - a^{kj}_{x_j}) F_{x_k}$, $\alpha = (b^k - a^{kj}_{x_j})(a^{rs} F_{x_r} F_{x_s})_{x_k}$, μ *is a positive integer, the equation* $F = 0$ *defines the boundary* Σ, *and* grad F *has the direction of the interior normal to* Σ *(clearly,* $\tilde{b} < 0$ *and* $\alpha \leqslant 0$ *on* Σ_2, *and the functions* $\tilde{b}$ *and* α *are invariant relative to nondegenerate changes of independent variables, as is seen from Lemma* 1.1.1).

4) *At points of* Σ_2

$$\frac{\sqrt{L_0(a)}}{|\tilde{b}|} \leqslant K_1 \frac{\gamma}{\mu}, \tag{1.9.5}$$

where K_1 *is a constant depending only on the dimension of the space* R^m, *and* $L_0 \equiv a^{kj}\partial^2/\partial x_k \partial x_j$, $a(x) = a^{kj}F_{x_k}F_{x_j}$. (*If the point* x_0 *is an interior point of* Σ_2 *or the limit of interior points of* Σ_2, *it is easy to show that at* x_0 *the left side of* (1.9.5) *vanishes, so that condition* (1.9.5) *is satisfied.*)

Then for any function $f \in W_2^\mu(\Omega)$ *there exists a solution* $u(x)$ *of the problem* (1.1.4), (1.1.5) *with* $g = 0$, *and the estimate*

$$|||u|||_\mu \leqslant C_1 \|f\|_{\mu:\Omega} \tag{1.9.6}$$

is satisfied, where the constant C_1 *does not depend on* u, *and* μ *is a given positive integer. Moreover,* $u \in W_2^\mu$ *in* $\Omega \backslash (G_\delta^0 \cup G_\delta^2)$, $u \in W_2^{\mu/2}$ *for* μ *even and* $u \in W_2^{(\mu+1)/2}$ *for* μ *odd in a neighborhood of* Σ_2; *finally,* $u \in W_2^{\mu/2}$ *for* μ *even and* $u \in W_2^{(\mu-1)/2}$ *for* μ *odd in a neighborhood of* Σ_0.

Below we shall give a proof of Theorem 1.9.1 in the simplest case, when $\Sigma = \Sigma_3$. The solution of (1.1.4), (1.1.5) in this case, as in Theorem 1.8.2, will be obtained as the limit as $\epsilon \to 0$ of solutions of the elliptic equation

$$\epsilon \Delta u + L(u) = f \text{ in } \Omega \tag{1.9.7}$$

with boundary condition

$$u = 0 \text{ on } \Sigma. \tag{1.9.8}$$

In the case when Σ contains a set Σ_1, the equation analogous to (1.9.7) in the proof of Theorem 1.9.1 in [64] involves the adjoining of an elliptic operator of order 2μ with small parameter ϵ in a neighborhood of Σ_1. We shall not give here the complete proof of the assertions of Theorem 1.9.1 because of its complexity. Certain results regarding the smoothness of the solution of (1.1.4), (1.1.5), near to those of Theorem 1.9.1, were proved by other methods in §8. These other methods used a regularization of the type (1.9.7), (1.9.8) throughout.

THEOREM 1.9.2. *Suppose* $\Sigma = \Sigma_3$, $f \in W_2^\mu(\Omega)$, *that the coefficient* $c < -c_0 < 0$ *in* Ω *and the constant* c_0 *is sufficiently large depending on* a^{kj}, b^k *and their derivatives of order* 2 *and less, and that the coefficients of* (1.9.1) *and the boundary of* Ω *are infinitely differentiable* (*or sufficiently smooth*). *Then there exists a solution of* (1.1.4), (1.1.5) *with* $g = 0$ *in the class* $W_2^\mu(\Omega)$, *and*

$$\|u\|_{\mu;\Omega} \leqslant C_2 \|f\|_{\mu;\Omega}, \tag{1.9.9}$$

where C_2 *does not depend on* u, *and* μ *is a positive integer.*

First we shall prove certain auxiliary results.

We introduce some notation. For any functions u and v in $C_{(1)}(\Omega)$ we set

$$-Q(u, v) \equiv \int_\Omega \Big[-a^{kj} v_{x_k} u_{x_j} - \varepsilon v_{x_k} u_{x_k} + \frac{1}{2}(b^k - a^{kj}_{x_j})(u_{x_k} v - v_{x_k} u) + \frac{1}{2}(2c - b^k_{x_k} + a^{kj}_{x_k x_j}) uv \Big] dx. \tag{1.9.10}$$

In the following, the symbols C_j will be used to denote constants depending only on a^{ks}, b^k and their derivatives up to order two inclusive, and K_j will denote any constants independent of the parameter ϵ. Their numeration is retained only within the confines of the proof of a single lemma or a single theorem.

LEMMA 1.9.1. *Let* ϕ *be an infinitely differentiable function with support contained in a small neighborhood* U *of a point* $x_0 \in \Sigma$ *(so that in* U *we may introduce local coordinates* $y_1, \cdots, y_m$ *with* Σ *lying in the plane* $y_m = 0$, *and* $y_m > 0$ *in* Ω*). Then for* $l \geqslant 1$,

$$\begin{aligned} Q(\varphi D'_{\bar{S}_l} u, \varphi D'_{\bar{S}_l} u) - (-1)^l\, Q(u, D'_{\bar{S}_l} \varphi^2 D'_{\bar{S}_l} u) \\ \leqslant C_1 \Big\{ \sum_{\bar{S}_{l-1}} \|\psi D'_{\bar{S}_{l-1}} u_{y_m}\|_0^2 + \sum_{\bar{S}_l} \|\psi D'_{\bar{S}_l} u\|_0^2 \Big\} \\ + K_1 \Big\{ \sum_{\bar{S}_\rho, \rho \leqslant l-1} \|\psi D'_{\bar{S}_\rho} u\|_0^2 + \sum_{\bar{S}_\rho, \rho \leqslant l-2} \|\psi D'_{\bar{S}_\rho} u_{y_m}\|_0^2 \Big\}, \end{aligned} \tag{1.9.11}$$

where $u \in C^{(l+1)}(\bar{\Omega})$, $\psi \in C^\infty(\bar{\Omega})$, $\psi \geqslant 1$ *on* supp ϕ, *and* $\psi \equiv 0$ *outside* U.

PROOF. We denote the expression on the right of (1.9.11) with constants C_s and K_ρ in place of C_1 and K_1 by the symbol $R(C_s, K_\rho)$. Let $(v, w) = \int_\Omega vw\, dx$. It may easily be seen that

$$Q(u, v) = \sum_j (M_j u, L_j v),$$

where M_j and L_j are first order differential operators which in local coordinates $y_1, \cdots, y_m$ have the form

$$M_j \equiv a_j \frac{\partial}{\partial y_m} + \sum_{k=1}^{m-1} b_j^k \frac{\partial}{\partial y_k} + c_j,$$

$$L_j \equiv \alpha_j \frac{\partial}{\partial y_m} + \sum_{k=1}^{m-1} \gamma_j^k \frac{\partial}{\partial y_k} + r_j, \tag{1.9.12}$$

where a_j, α_j, b_j^k and γ_j^k do not depend on c.

We consider the operators $A = \phi D^l_{\bar{S}_l}$ and $A^* = (-1)^l D^l_{\bar{S}_l}\phi$. The following identity holds:

$$\begin{aligned} 2[Q(Au, Au) - Q(u, A^*Au)] &= [Q(Au, Au) - Q(u, A^*Au)] \\ &+ [Q(A^*Au, u) - Q(u, A^*Au)] + [Q(Au, Au) - Q(A^*Au, u)] \\ &= -\{(M_jAu, [A, L_j]u) + (M_ju, [L_j, A^*]Au)\} \\ &- \{(L_jAu, [A, M_j]u) + ([M_j, A^*]Au, L_ju)\} \\ &+ [Q(A^*Au, u) - Q(u, A^*Au)], \end{aligned} \tag{1.9.13}$$

where by definition $[A, B] \equiv AB - BA$ for any two operators A and B. Here the summation of the repeated index j is assumed. We estimate the terms in the last part of (1.9.13).

Obvious transformations lead to the identity

$$\begin{aligned} &(M_jAu, [A, L_j]u) + (M_ju, [L_j, A^*]Au) \\ &\equiv ([A, M_j]u, [L_j, A]u) - ((A - A^*)M_ju, [L_j, A]u) \\ &+ (M_ju, [L_j, A^* - A]Au) + (M_ju, [[L_j, A], A]u). \end{aligned} \tag{1.9.14}$$

Further using the representation (1.9.12) for the operators M_j and L_j, we obtain

$$\begin{aligned} ([A, M_j]u, [L_j, A]u) = &-([A, a_j]u_{y_m} - a_jA_{(y_m)}u \\ &+ \sum_{k=1}^{m-1}\left[A, b_j^k \frac{\partial}{\partial y_k}\right]u + [A, c_j]u, [A, \alpha_j]u_{y_m} \\ &- \alpha_jA_{(y_m)}u + \sum_{k=1}^{m-1}\left[A, \gamma_j^k \frac{\partial}{\partial y_k}\right]u + [A, r_j]u), \end{aligned}$$

where the operator $A_{(y_m)}$ is defined by the equation $A_{(y_m)} \equiv \phi_{y_m} D^l_{\bar{S}_l}$. The following estimates are evident:

$$\|[A, a_j]u_{y_m}\|_0^2 + \|[A, \alpha_j]u_{y_m}\|_0^2 \leqslant C_2 \sum_{\bar{S}_{l-1}} \|\psi D'_{\bar{S}_{l-1}} u_{y_m}\|_0^2 + K_2 \sum_{\bar{S}_p,\ p \leqslant l-2} \|\psi D'_{\bar{S}_p} u_{y_m}\|_0^2;$$

$$\|a_jA_{(y_m)}u\|_0^2 + \sum_{k=1}^{m-1}\left\{\left\|\left[A, b_j^k \frac{\partial}{\partial y_k}\right]u\right\|_0^2 + \left\|\left[A, \gamma_j^k \frac{\partial}{\partial y_k}\right]u\right\|_0^2\right\} + \|\alpha_jA_{(y_m)}u\|_0^2 \leqslant$$

$$\leqslant C_3 \sum_{\bar{S}_l} \|\psi D'_{\bar{S}_l} u\|_0^2 + K_3 \sum_{\bar{S}_\rho,\ \rho \leqslant l-1} \|\psi D'_{\bar{S}_\rho} u\|_0^2;$$

$$\|[A, c_j]\, u\|_0^2 + \|[A, r_j]\, u\|_0^2 \leqslant K_4 \sum_{\bar{S}_\rho,\ \rho \leqslant l-1} \left\|\psi D'_{\bar{S}_\rho} u\right\|_0^2.$$

From the estimates obtained above it follows that $|([A, M_j]\, u,\ [L_j, A]\, u)| \leqslant R(C_4, K_5)$. In an analogous manner we estimate the integral $((A - A^*)M_j u,\ [L_j, A]\, u)$. We further examine the integrals $(M_j u,\ [[L_j, A], A]\, u)$ and $(M_j u,\ [L_j, A^* - A]\, Au)$, on the right of (1.9.14). Using representation (1.9.12), we obtain

$$\left(M_j u,\ [[L_j, A], A]\, u\right) = \left(M_j u,\ [[\alpha_j, A], A]\, u_{y_m} + [\alpha_j A_{(y_m)}, A]\, u \right. \tag{1.9.15}$$
$$\left. + [[r_j, A], A]\, u + [\alpha_j, A]\, A_{(y_m)} u + \sum_{k=1}^{m-1} \left[\left[\gamma_j^k \frac{\partial}{\partial y_k}, A\right], A\right] u\right).$$

We transform the integrals on the right of (1.9.15) by integration by parts and recall that $[[\alpha_j, A], A]$ is a differential operator of order $2l - 2$ containing differentiations only with respect to the variables $y_1, \cdots, y_{m-1}$. Also recalling that $[\alpha_j A_{(y_m)}, A]$, $[[r_j, A], A]$, $[\alpha_j, A] A_{(y_m)}$ and $\Sigma_{k=1}^{m-1} [[\gamma_j^k \partial/\partial y_k, A], A]$ are differential operators of order no greater than $2l - 1$, also containing differentiations only with respect to the variables $y_1, \cdots, y_{m-1}$, we obtain the estimate

$$|(M_j u,\ [[L_j, A], A]\, u)| \leqslant R(C_5, K_6)$$

from (1.9.15). We estimate the integral $(M_j u,\ [L_j, A^* - A] Au)$ in exactly the same way. As a consequence,

$$|(M_j Au,\ [A, L_j]\, u) + (M_j u,\ [L_j, A^*]\, Au)| \leqslant R(C_6, K_7).$$

The two remaining terms in the last part of (1.9.13) may be estimated in an analogous manner; and from these results the assertion of the lemma follows.

LEMMA 1.9.2. *Suppose condition* 2) *of Theorem* 1.9.1 *is satisfied, and let* $u_\varepsilon(x)$ *be a solution of the equation*

$$L_\varepsilon(u) \equiv \varepsilon \Delta u + L(u) = f \text{ in } \Omega, \quad \varepsilon = \text{const} > 0, \tag{1.9.16}$$

with condition

$$u = 0 \quad on \quad \Sigma, \tag{1.9.17}$$

and suppose $u_\varepsilon \in C^{(\mu+2)}(\bar{\Omega})$; $f \in W_2^\mu(\Omega)$. *Then for any* $l \leqslant \mu$ *the estimate*

$$c_0 \left\|\varphi D'_{\bar{S}_l} u_\varepsilon\right\|_0^2 \leqslant K_8 \sum_{\bar{S}_\rho,\ \rho \leqslant l} \left\|\psi D'_{\bar{S}_\rho} f\right\|_0^2 + C_7 \left\{\sum_{\bar{S}_l} \left\|\psi D'_{\bar{S}_l} u_\varepsilon\right\|_0^2 + \sum_{\bar{S}_{l-1}} \left\|\psi D'_{\bar{S}_{l-1}} u_{\varepsilon y_m}\right\|_0^2\right\} +$$

$$+ K_9 \left\{ \sum_{\bar{S}_\rho,\ \rho \leqslant l-1} \| \psi D'_{\bar{S}_\rho} u_\varepsilon \|_0^2 + \sum_{\bar{S}_\rho,\ \rho \leqslant l-2} \| \psi D'_{\bar{S}_\rho} u_{\varepsilon y_m} \|_0^2 \right\} \tag{1.9.18}$$

holds, where the functions ϕ *and* ψ *are defined as in Lemma* 1.9.1. *If* $\phi \in C_0^\infty(\Omega)$, *then for* $l \leqslant \mu$

$$\begin{aligned} c_0 \| \varphi D_{\bar{S}_l} u_\varepsilon \|_0^2 &\leqslant K_{10} \sum_{\bar{S}_\rho,\ \rho \leqslant l} \| \psi D_{\bar{S}_\rho} f \|_0^2 \\ &+ C_8 \sum_{\bar{S}_l} \| \psi D_{\bar{S}_l} u_\varepsilon \|_0^2 + K_{11} \sum_{\bar{S}_\rho,\ \rho \leqslant l-1} \| \psi D_{\bar{S}_\rho} u_\varepsilon \|_0^2, \end{aligned} \tag{1.9.19}$$

where $\psi \in C_0^\infty(\Omega)$ *and* $\psi \geqslant 1$ *on the support of* ϕ.

PROOF. For any smooth functions u and v equal to zero on Σ we have the identity

$$-(L_\varepsilon(u), v) = Q(u, v). \tag{1.9.20}$$

By virtue of assumption 2) of Theorem 1.9.1, we may suppose that in Ω

$$c_0 + b^k_{x_k} - a^{kj}_{x_k x_j} \geqslant 0.$$

Therefore from (1.9.10) it follows that

$$Q(v, v) \geqslant \frac{c_0}{2} \|v\|_0^2 \tag{1.9.21}$$

for any smooth function v equal to zero on Σ. In (1.9.21) we substitute $v = \phi D'_{\bar{S}_l} u_\varepsilon$, thus obtaining

$$\frac{c_0}{2} \| \varphi D'_{\bar{S}_l} u_\varepsilon \|_0^2 \leqslant Q\left(\varphi D'_{\bar{S}_l} u_\varepsilon, \varphi D'_{\bar{S}_l} u_\varepsilon \right). \tag{1.9.22}$$

To estimate the right side of (1.9.22) we use the inequality (1.9.11) of Lemma 1.9.1. We have

$$Q\left(\varphi D'_{\bar{S}_l} u_\varepsilon, \varphi D'_{\bar{S}_l} u_\varepsilon \right) \leqslant \left| Q\left(u_\varepsilon, D'_{\bar{S}_l} \varphi^2 D'_{\bar{S}_l} u_\varepsilon \right) \right| + R(C_1, K_1). \tag{1.9.23}$$

From (1.9.20) it follows that

$$\left| Q\left(u_\varepsilon, D'_{\bar{S}_l} \varphi^2 D'_{\bar{S}_l} u_\varepsilon \right) \right| = \left| \left(f, D'_{\bar{S}_l} \varphi^2 D'_{\bar{S}_l} u_\varepsilon \right) \right| \leqslant \left\{ \| \varphi D'_{\bar{S}_l} u_\varepsilon \|_0^2 + \| \varphi D'_{\bar{S}_l} f \|_0^2 \right\}. \tag{1.9.24}$$

Estimates (1.9.22)–(1.9.24) imply (1.9.18).

Inequality (1.9.19) is proved in a similar way. In this case it is necessary to reason that if $\phi \in C_0^\infty(\Omega)$, then one may obtain an estimate of the form

$$Q\left(\varphi D_{\bar{S}_l} u, \varphi D_{\bar{S}_l} u\right) \leqslant (-1)^l Q\left(u, D_{\bar{S}_l} \varphi^2 D_{\bar{S}_l} u\right) + C_9 \sum_{\bar{S}_l} \left\|\psi D_{\bar{S}_l} u\right\|_0^2 + K_{12} \sum_{\bar{S}_\rho, \rho \leqslant l-1} \left\|\psi D_{\bar{S}_\rho} u\right\|_0^2 \tag{1.9.25}$$

just as Lemma 1.9.1 was obtained, except that certain simplifications may be introduced. The lemma is thereby proved.

PROOF OF THEOREM 1.9.2. The solution of the problem (1.1.4), (1.1.5) with $g = 0$ will be obtained as the limit as $\epsilon \to 0$ of solutions of the elliptic equation (1.9.16) with condition (1.9.17). We first assume that $f \in C^\infty(\Omega \cup \Sigma)$. Let $\{\Omega_j\}$, $j = 1, \cdots, N$, be a finite covering of $\Omega \cup \Sigma$ such that Ω_1 contains no points of the boundary, and Ω_j for $j > 1$ contains points of Σ and is so small that one may introduce local coordinates $y_1, \cdots, y_m$ in Ω_j such that the boundary Σ lies in the plane $y_m = 0$ and $y_m > 0$ for points of Ω. Let $\{\phi_j\}$, $\{\chi_j\}$ and $\{\psi_j\}$ be partitions of unity (see Chapter II, §1) corresponding to the covering $\{\Omega_j\}$ with $\chi_j \geqslant \gamma =$ const > 0 on the support of ϕ_j, and $\psi_j \geqslant \gamma$ on the support of χ_j, $j = 1, \cdots, N$. We shall prove that

$$\|u_\varepsilon\|_{\mu;\Omega}^2 \leqslant C_1 \|f\|_{\mu;\Omega}^2, \tag{1.9.26}$$

where the constant C_1 does not depend on ϵ. From the estimate (1.9.26) it follows that one may find a sequence $\epsilon \to 0$, such that $u_\epsilon(x) \to u(x)$ weakly in the space $W_2^\mu(\Omega)$ and such that (1.9.9) is valid for the limit function $u(x)$, i.e. for $\mu \geqslant 2$, $u(x)$ is a solution of (1.1.4), (1.1.5) with the required properties. From the Sobolev imbedding theorem (see [125, 84, 139] and also Chapter II, §1) it follows that $u(x) \in C^{(k)}(\Omega)$ if $2(\mu - k) > m$. We note that if f belongs to $W_2^\mu(\Omega)$ then we may approximate it in the norm of $W_2^\mu(\Omega)$ by functions f_n in the class $C^\infty(\Omega \cup \Sigma)$, and the solution of (1.1.4), (1.1.5) is obtained as the limit of u_ϵ as $1/n$ and ϵ tend to zero through some sequence.

Thus for the proof of Theorem 1.9.2 it is sufficient to obtain the estimate (1.9.26). Clearly the inequality

$$\begin{aligned} \|u_\varepsilon\|_{\mu;\Omega}^2 &\leqslant N \sum_j^N \|\varphi_j u_\varepsilon\|_{\mu;\Omega}^2 \\ &\leqslant 2N \sum_{\bar{S}_\mu} \left\|\varphi_1 D_{\bar{S}_\mu} u_\varepsilon\right\|_0^2 + K_1 \|u_\varepsilon\|_{\mu-1;\Omega}^2 + 2N \sum_{j>1} \sum_{l+r=\mu} \sum_{\bar{S}_l} \left\|\varphi_j \frac{\partial^r}{\partial y_m^r} D'_{\bar{S}_l} u_\varepsilon\right\|_0^2 \end{aligned} \tag{1.9.27}$$

is valid. According to Lemma 1.9.2, estimate (1.9.19) holds for the integral $\|\phi_1 D_{\bar{S}_\mu} u_\epsilon\|_0^2$ with the function $\psi = \psi_1/\gamma$. We now estimate the last summation in the relation (1.9.27). We show that for any $\delta_1 > 0$ and $l \leqslant \mu$,

$$\sum_{\bar{S}_\rho, \rho+r=l} \left\| \varphi_j \frac{\partial^r}{\partial y_m^r} D'_{\bar{S}_\rho} u_\varepsilon \right\|_0^2 \leqslant \left(\delta_1 + \frac{C_1}{c_0 \delta_1}\right) \sum_{\bar{S}_\rho, r+\rho=l} \left\| \psi_j \frac{\partial^r}{\partial y_m^r} D'_{S_\rho} u_\varepsilon \right\|_0^2 + K_2 \left[\|f\|_{l;\Omega}^2 + \|u_\varepsilon\|_{l-1;\Omega}^2 \right]. \tag{1.9.28}$$

For this purpose we write (1.9.16) in $\omega_j = \Omega_j \cap \Omega$ for $j > 1$ using local coordinates $y_1, \cdots, y_m$. We have

$$L_\varepsilon(u) \equiv \varepsilon \mu^{kj} u_{y_k y_j} + \varepsilon \nu^k u_{y_k} + \alpha^{kj} u_{y_k y_j} + \beta^k u_{y_k} + cu = f.$$

According to the definition of the form $Q(u, v)$, for any smooth function u with support in ω_j, $j > 1$, we have

$$((\alpha^{mm} + \varepsilon\mu^{mm}) u_{y_m}, u_{y_m}) \leqslant C_2 \left\{ Q(u, u) + 2 \left| \sum_{k=1}^{m-1} ((\alpha^{mk} + \varepsilon\mu^{mk}) u_{y_k}, u_{y_m}) \right| + \sum_{k, \rho=1}^{m-1} ((\alpha^{k\rho} + \varepsilon\mu^{k\rho}) u_{y_k}, u_{y_\rho}) \right\} + K_3 \|u\|_0^2. \tag{1.9.29}$$

Since $\Sigma = \Sigma_3$, in ω_j we have

$$\alpha^{mm} > \frac{1}{2} \max_{\bar{\omega}_j} \alpha^{mm} > 0, \tag{1.9.30}$$

if the domain Ω_j is sufficiently small. Therefore, bearing in mind inequality (1.9.29) for $u = \phi_j u_\epsilon$, we obtain

$$\begin{aligned} \|\varphi_j u_{\varepsilon y_m}\|_0^2 &\leqslant C_3 \left[((\alpha^{mm} + \varepsilon\mu^{mm}) (\varphi_j u_\varepsilon)_{y_m}, (\varphi_j u_\varepsilon)_{y_m}) + \|\psi_j u_\varepsilon\|_0^2 \right] \\ &\leqslant C_4 \left[Q(\varphi_j u_\varepsilon, \varphi_j u_\varepsilon) + 2 \left| \sum_{k=1}^{m-1} ((\alpha^{mk} + \varepsilon\mu^{mk}) (\varphi_j u_\varepsilon)_{y_k}, (\varphi_j u_\varepsilon)_{y_m}) \right| \right. \\ &\quad \left. + \sum_{k, \rho=1}^{m-1} ((\alpha^{k\rho} + \varepsilon\mu^{k\rho}) (\varphi_j u_\varepsilon)_{y_k}, (\varphi_j u_\varepsilon)_{y_\rho}) \right] + K_4 \|\psi_j u_\varepsilon\|_0^2. \end{aligned}$$

From this it follows that

$$\|\varphi_j u_{\varepsilon y_m}\|_0^2 \leqslant \delta_1 \|\psi_j u_{\varepsilon y_m}\|_0^2 + \frac{C_5}{\delta_1} \sum_{\bar{S}_1} \|\chi_j D'_{\bar{S}_1} u_\varepsilon\|_0^2 + C_4 Q(\varphi_j u_\varepsilon, \varphi_j u_\varepsilon) + K_5 \|\psi_j u_\varepsilon\|_0^2. \tag{1.9.31}$$

Using (1.9.20), it is not difficult to show that

$$Q(\varphi_j u_\varepsilon, \varphi_j u_\varepsilon) \leqslant K_6 \{\|\psi_j f\|_0^2 + \|\psi_j u_\varepsilon\|_0^2\} + \delta_1 \sum_{\bar{S}_1} \|\psi_j D'_{\bar{S}_1} u_\varepsilon\|_0^2 + \delta_1 \|\psi_j u_{\varepsilon y_m}\|_0^2, \tag{1.9.32}$$

where the constant K_6 depends on δ_1.

From inequality (1.9.18) taken for $l = 1$ we deduce that

$$\|\chi_j D'_{\bar{S}_1} u_\varepsilon\|_0^2 + \|\varphi_j D'_{\bar{S}_1} u_\varepsilon\|_0^2 \leqslant K_7 \Big\{\sum_{\bar{S}_1} \|\psi_j D'_{\bar{S}_1} f\|_0^2 + \|\psi_j f\|_0^2 + \|\psi_j u_\varepsilon\|_0^2\Big\}$$
$$+ \frac{C_6}{c_0}\Big\{\sum_{\bar{S}_1} \|\psi_j D'_{\bar{S}_1} u_\varepsilon\|_0^2 + \|\psi_j u_{\varepsilon y_m}\|_0^2\Big\}. \qquad (1.9.33)$$

From estimates (1.9.31), (1.9.32) and (1.9.33) it follows that

$$\sum_{\bar{S}_1} \|\varphi_j D'_{\bar{S}_1} u_\varepsilon\|_0^2 + \|\varphi_j u_{\varepsilon y_m}\|_0^2 \leqslant \Big(2\delta_1 + \frac{C_7}{c_0\delta_1}\Big)\Big\{\|\psi_j u_{\varepsilon y_m}\|_0^2 + \sum_{\bar{S}_1} \|\psi_j D'_{\bar{S}_1} u_\varepsilon\|_0^2\Big\}$$
$$+ K_8\{\|f\|_{1;\,\Omega}^2 + \|u_\varepsilon\|_0^2\}. \qquad (1.9.34)$$

This means that for $l = 1$ inequality (1.9.28) is satisfied.

We show by induction that (1.9.28) is valid for any $l \leqslant \mu$. Let it be true for $l = l_0$. We show that this relation holds also for $l = l_0 + 1$. From (1.9.18) it follows that

$$\|\chi_j D'_{\bar{S}_{l_0+1}} u_\varepsilon\|_0^2 + \|\varphi_j D'_{\bar{S}_{l_0+1}} u_\varepsilon\|_0^2 \leqslant \frac{C_8}{c_0}\Big[\sum_{\bar{S}_{l_0+1}} \|\psi_j D'_{\bar{S}_{l_0+1}} u_\varepsilon\|_0^2 + \sum_{\bar{S}_{l_0}} \|\psi_j D'_{\bar{S}_{l_0}} u_{\varepsilon y_m}\|_0^2\Big]$$
$$+ K_9(\|f\|_{l_0+1;\,\Omega}^2 + \|u_\varepsilon\|_{l_0;\,\Omega}^2).$$
$$(1.9.35)$$

We consider the inequality (1.9.29) for $u = \phi_j D'_{\bar{S}_{l_0}} u_\varepsilon$. Bearing in mind condition (1.9.30), we obtain the following, exactly as inequality (1.9.31) was obtained:

$$\sum_{\bar{S}_{l_0}} \|\varphi_j D'_{\bar{S}_{l_0}} u_{\varepsilon y_m}\|_0^2 \leqslant C_9 \sum_{\bar{S}_{l_0}} Q\big(\varphi_j D'_{\bar{S}_{l_0}} u_\varepsilon,\ \varphi_j D'_{\bar{S}_{l_0}} u_\varepsilon\big)$$
$$+ \delta_1 \sum_{\bar{S}_{l_0}} \|\psi_j D'_{\bar{S}_{l_0}} u_{\varepsilon y_m}\|_0^2 + \frac{C_{10}}{\delta_1} \sum_{\bar{S}_{l_0+1}} \|\chi_j D'_{\bar{S}_{l_0+1}} u_\varepsilon\|_0^2 \qquad (1.9.36)$$
$$+ K_{10}\Big\{\sum_{\bar{S}_p,\ p\leqslant l_0} \|\psi_j D'_{\bar{S}_p} u_\varepsilon\|_0^2 + \sum_{\bar{S}_p,\ p\leqslant l_0-1} \|\psi_j D'_{\bar{S}_p} u_{\varepsilon y_m}\|_0^2\Big\}.$$

From Lemma 1.9.1 follows the inequality

$$\sum_{\bar{S}_{l_0}} Q\big(\varphi_j D'_{\bar{S}_{l_0}} u_\varepsilon,\ \varphi_j D'_{\bar{S}_{l_0}} u_\varepsilon\big) \leqslant \big|Q\big(u_\varepsilon,\ D'_{\bar{S}_{l_0}} \varphi_j^2 D'_{\bar{S}_{l_0}} u_\varepsilon\big)\big| + K_{11}\|\psi_j u_\varepsilon\|_{l_0;\,\Omega}^2. \qquad (1.9.37)$$

Since $D'_{\bar{S}_{l_0}} \phi_j^2 D'_{\bar{S}_{l_0}} u_\varepsilon = 0$ on Σ, we may deduce from (1.9.20), (1.9.37) that

$$\sum_{\bar{S}_{l_0}} Q\left(\varphi_j D'_{\bar{S}_{l_0}} u_\varepsilon, \varphi_j D'_{\bar{S}_{l_0}} u_\varepsilon\right) \leqslant K_{12}\{\|f\|^2_{l_0;\,\Omega} + \|\psi_j u_\varepsilon\|^2_{l_0;\,\Omega}\}.$$

From (1.9.36) and the last inequality it follows that

$$\sum_{\bar{S}_{l_0}} \|\varphi_j D'_{\bar{S}_{l_0}} u_{\varepsilon y_m}\|_0^2 \leqslant \left(\delta_1 + \frac{C_{11}}{\delta_1 c_0}\right)\left\{\sum_{\bar{S}_{l_0}} \|\psi_j D'_{\bar{S}_{l_0}} u_{\varepsilon y_m}\|_0^2 + \|\psi_j D'_{\bar{S}_{l_0+1}} u_\varepsilon\|_0^2\right\}$$
$$+ K_{13}\{\|f\|^2_{l_0+1;\,\Omega} + \|\psi_j u_\varepsilon\|^2_{l_0;\,\Omega}\}. \tag{1.9.38}$$

We now apply the operator $\phi_j D'_{\bar{S}_\rho} \partial^{r-2}/\partial y_m^{r-2}$ for $r \geqslant 2$, $\rho + r = l_0 + 1$, to the equation $L_\epsilon(u) = f$. We obtain

$$(a^{mm} + \varepsilon\mu^{mm})\varphi_j D'_{\bar{S}_\rho}\frac{\partial^r u_\varepsilon}{\partial y_m^r} = -2\sum_{k=1}^{m-1}(a^{km} + \varepsilon\mu^{km})\varphi_j D'_{\bar{S}_\rho}\frac{\partial^r u_\varepsilon}{\partial y_k \partial y_m^{r-1}}$$
$$-\sum_{s,k=1}^{m-1}(a^{ks} + \varepsilon\mu^{ks})\varphi_j D'_{\bar{S}_\rho}\frac{\partial^r u_\varepsilon}{\partial y_m^{r-2}\partial y_k \partial y_s} + \varphi_j D'_{\bar{S}_\rho}\frac{\partial^{r-2} f}{\partial y_m^{r-2}} + A_{l_0} u_\varepsilon, \tag{1.9.39}$$

where A_{l_0} is a linear differential operator of order l_0 whose coefficients are equal to zero outside supp ϕ_j. Bearing condition (1.9.30) in mind we deduce from (1.9.39) that for $\rho + r = l_0 + 1$,

$$\left\|\varphi_j D'_{\bar{S}_\rho}\frac{\partial^r u_\varepsilon}{\partial y_m^r}\right\|_0^2 \leqslant C_{12}\left[\sum_{\bar{S}_\gamma;\ \gamma+r-1=l_0+1}\left\|\varphi_j D'_{\bar{S}_\gamma}\frac{\partial^{r-1} u_\varepsilon}{\partial y_m^{r-1}}\right\|_0^2\right.$$
$$\left.+ \sum_{\bar{S}_\gamma;\ \gamma+r-2=l_0+1}\left\|\varphi_j D'_{\bar{S}_\gamma}\frac{\partial^{r-2} u_\varepsilon}{\partial y_m^{r-2}}\right\|_0^2\right]$$
$$+ K_{14}(\|f\|^2_{l_0+1;\,\Omega} + \|u_\varepsilon\|^2_{l_0;\,\Omega}). \tag{1.9.40}$$

Therefore if it is assumed that an estimate of the form

$$\sum_{\bar{S}_\rho;\ \rho+r=l_0+1}\left\|\varphi_j D'_{\bar{S}_\rho}\frac{\partial^r u_\varepsilon}{\partial y_m^r}\right\|_0^2 \leqslant \left(\delta_1 + \frac{C_{14}}{\delta_1 c_0}\right)\sum_{\bar{S}_\rho;\ \rho+r=l_0+1}\left\|\psi_j D'_{\bar{S}_\rho}\frac{\partial^r u_\varepsilon}{\partial y_m^r}\right\|_0^2$$
$$+ K_{15}(\|f\|^2_{l_0+1;\,\Omega} + \|u_\varepsilon\|^2_{l_0;\,\Omega}) \tag{1.9.41}$$

is valid for $r \leqslant r_0$, then from (1.9.40) it follows that an estimate of the same form (1.9.41) is also valid for $r = r_0 + 1$. Since (1.9.41) is satisfied for $r = 1$ because of (1.9.38) and (1.9.35), it follows that (1.9.41) holds for all ρ and r such that $\rho + r \leqslant l_0 + 1$, i.e. (1.9.28) holds for $l = l_0 + 1$.

We use inequality (1.9.28) to estimate the terms in the last summation in (1.9.27). Assuming then that the constant δ_1 is chosen sufficiently small and the constant c_0, according to condition 2) of Theorem 1.9.1, is sufficiently large, we use (1.9.27) to obtain the estimate

$$\|u_\varepsilon\|^2_{\mu;\,\Omega} \leqslant C_{15}\{\|f\|^2_{\mu;\,\Omega} + \|u_\varepsilon\|^2_{\mu-1;\,\Omega}\}. \tag{1.9.42}$$

In exactly the same way we may obtain the estimate

$$\|u_\varepsilon\|^2_{l;\,\Omega} \leqslant C_{16}\{\|f\|^2_{l;\,\Omega} + \|u_\varepsilon\|^2_{l-1;\,\Omega}\} \tag{1.9.43}$$

for $l \leqslant \mu$. Since

$$\|u_\varepsilon\|^2_0 \leqslant C_{17}\,\|f\|^2_0 \tag{1.9.44}$$

is clearly true, estimate (1.9.26) follows from (1.9.42)–(1.9.44). This proves the theorem.

If $\mu = 1$, then Theorem 1.9.2 yields conditions for the existence of a weak solution of the problem (1.1.4), (1.1.5) in the sense of integral identity (1.5.3).

CHAPTER II

ON THE LOCAL SMOOTHNESS OF WEAK SOLUTIONS AND HYPOELLIPTICITY OF SECOND ORDER DIFFERENTIAL EQUATIONS

Chapter II is basically concerned with the study of local smoothness of weak solutions of second order equations with nonnegative characteristic form and, in particular, conditions for hypoellipticity.

As we already said in the Introduction, conditions for hypoellipticity for the class of second order equations of the form (9) were first given in the paper by Hörmander [55], and for general second order equations in [113] and in Chapter II, §§5, 6. Hörmander's proof uses certain results from the theory of Lie algebras, and special function spaces. Another proof of Hörmander's theorem on hypoellipticity, based on the theory of pseudodifferential operators, is given in §5 of this chapter.

In §6 examples are given of hypoelliptic second order equations with nonnegative characteristic form not representable in the form (9); in this section results on hypoellipticity for general second order equations are given. In §5 it is also shown that the conditions of Hörmander's theorem may be weakened; analogous theorems are proved in §6 for general second order equations. In §§5 and 6 the hypoellipticity of equations is proved as a consequence of a priori estimates of Schauder type. These estimates are used in §7 to construct a solution of the first boundary value problem in nonsmooth domains for hypoelliptic equations by the method of M. V. Keldyš [63]. In §§1 and 2 basic results on the spaces $\mathcal{H}_s$ and on pseudodifferential operators are developed, and a detailed bibliography of papers dealing with these questions is given. In §8, necessary and sufficient conditions for the hypoellipticity of second order equations with analytical coefficients are given.

§1. The spaces $\mathcal{H}_s$

This section is of an auxiliary nature. We discuss here those properties of the spaces $\mathcal{H}_s$ which will be necessary for our treatment.

We first give some notation, definitions, and theorems from the theory of distributions which will be needed later. Let Ω be a domain in m-dimensional space R^m and $u(x)$ a continuous complex valued function in Ω. The support of a function $u(x)$ in Ω is the closure of the set $\{x; x \in \Omega, u(x) \neq 0\}$. The support of the function u is denoted by supp u. By $C_0^\infty(\Omega)$ we denote the set of all infinitely differentiable functions in Ω which have compact support in Ω. For any set A in R^m, we understand by $C_0^\infty(A)$ the set of all real functions ϕ in $C_0^\infty(R^m)$ such that $\operatorname{supp} \phi \subset A$. The following theorem on partitions of unity holds.

THEOREM 2.1.1. *Let $\Omega_1, \cdots, \Omega_k$ be open sets and let K be a compact set such that $\bigcup_1^k \Omega_j \supset K$. There exist functions $\phi_j \in C_0^\infty(\Omega_j)$ such that $\phi_j \geqslant 0$ and $\Sigma_1^k \phi_j \leqslant 1$, with $\Sigma_1^k \phi_j \equiv 1$ on K.*

By α we shall denote a multi-index $\alpha_1, \cdots, \alpha_m$, where the α_j are nonnegative integers; their sum $\Sigma_1^m \alpha_j$ we denote by $|\alpha|$, and the product $\alpha_1! \cdots \alpha_m!$ by $\alpha!$. We also introduce the notation

$$D_j = -i\frac{\partial}{\partial x_j}, \quad D^\alpha = D_1^{\alpha_1} \ldots D_m^{\alpha_m}.$$

Here $i = \sqrt{-1}$. We set $x^\alpha = x_1^{\alpha 1} \cdots x_m^{\alpha m}$. A functional $u(\phi)$, defined on the functions ϕ in $C_0^\infty(\Omega)$, is called a distribution in Ω if $u(\phi)$ satisfies the conditions: 1) $u(c_1\phi_1 + c_2\phi_2) = c_1 u(\phi_1) + c_2 u(\phi_2)$, where c_1 and c_2 are complex numbers; $\phi_1, \phi_2 \in C_0^\infty(\Omega)$; 2) $u(\phi_n) \to 0$ as $n \to \infty$, if the sequence ϕ_n converges to zero in the following sense: for any multi-index α, $\sup_\Omega |D^\alpha \phi_n| \to 0$ as $n \to \infty$ and there exists a compact set K in Ω which contains the supports of all functions ϕ_n.

The linear space of functions C_0^∞ in which the convergence of sequences to zero is defined in this way is denoted by $D(\Omega)$. This definition of distribution in Ω is equivalent to the following: a distribution in Ω is a functional $u(\phi)$ defined on functions ϕ in $C_0^\infty(\Omega)$ satisfying the condition 1) and such that for any compact set $K \subset \Omega$ there exist constants C and k such that

$$|u(\varphi)| \leqslant C \sum_{|\alpha| \leqslant k} \sup |D^\alpha \varphi|, \quad \varphi \in C_0^\infty(K).$$

The set of all distributions in Ω will be denoted by $D'(\Omega)$. Clearly the distributions in $D'(\Omega)$ form a vector space with the natural definitions of the operations of addition and multiplication by complex numbers:

$$(a_1 u_1 + a_2 u_2)\varphi = a_1 u_1(\varphi) + a_2 u_2(\varphi); \ u_1, u_2 \in D'(\Omega),$$
$$\varphi \in C_0^\infty(\Omega); \ a_1, a_2 = \text{const}.$$

We shall say that a distribution $u(\phi)$ is the limit in $D'(\Omega)$ of distributions $u_j(\phi)$

as $j \to \infty$ if for any function $\phi \in C_0^\infty(\Omega)$

$$\lim_{j\to\infty} u_j(\varphi) = u(\varphi).$$

It can be proved that if the sequence of distributions $u_j(\phi)$ is such that for any function $\phi \in C_0^\infty(\Omega)$ the limit

$$\lim_{j\to\infty} u_j(\varphi) = u(\varphi)$$

exists, then $u(\phi)$ is a distribution in $D'(\Omega)$. We shall say that two distributions u_1 and u_2 in $D'(\Omega)$ are equal in a neighborhood of the point $x_0 \in \Omega$ if $u_1(\phi) = u_2(\phi)$ for all functions $\phi \in C_0^\infty(\Omega_1)$, where Ω_1 is some neighborhood of the point x_0.

Suppose, for some point x, that there does not exist a neighborhood where the distribution u is equal to zero. The set of all such points x is called the support of the distribution $u(\phi)$.

From this definition it follows that $u = 0$ in the complement of $\operatorname{supp} u$ in Ω, i.e. $u(\phi) = 0$ if $\phi \in C_0^\infty(\Omega)$ and $\operatorname{supp} u \cap \operatorname{supp} \phi = \varnothing$. Here and everywhere the symbol $\varnothing$ denotes the empty set.

The derivative $D_k u$ of the distribution $u \in D'(\Omega)$ is defined by the equation

$$D_k u(\varphi) = -u(D_k\varphi) \quad \text{for} \quad \varphi \in C_0^\infty(\Omega).$$

It is clear that $D^\alpha u(\phi) = (-1)^{|\alpha|} u(D^\alpha \phi)$, $\phi \in C_0^\infty(\Omega)$. If $u \in D'(\Omega)$ and $a \in C^\infty(\Omega)$, then the product of $u(\phi)$ by a is defined by the formula

$$(au)(\varphi) = u(a\varphi), \quad \varphi \in C_0^\infty(\Omega).$$

Let $p(\xi)$ be a polynomial in m variables $\xi_1, \cdots, \xi_m$ with complex coefficients. We denote by $P(D)$ the differential operator obtained from $p(\xi)$ by replacing ξ_j by D_j. It is clear that

$$P(D) e^{i\langle x,\xi\rangle} = p(\xi) e^{i\langle x,\xi\rangle},$$

where $\langle x, \xi\rangle = x_1\xi_1 + \cdots + x_m\xi_m$. We denote

$$p^{(\alpha)}(\eta) = \frac{\partial^{|\alpha|} p(\eta)}{\partial\eta_1^{\alpha_1} \cdots \partial\eta_m^{\alpha_m}} = i^{|\alpha|} D_\eta^\alpha p(\eta).$$

The generalized Leibniz formula

$$P(D)(au) = \sum_\alpha \frac{1}{\alpha!} (D^\alpha a)(P^{(\alpha)}(D) u) \tag{2.1.1}$$

holds.

The convolution $u * \phi$ of two continuous functions u and ϕ, one of which has compact support in R^m, is defined by the formula

$$(u*\varphi)(x) = \int u(x-y)\varphi(y)\,dy = \int u(y)\varphi(x-y)\,dy = (\varphi*u)(x).$$

(Here and in the following, if no integration domain is explicitly indicated, it is understood to be the entire space R^m.)

If $u \in D'(R^m)$ and $\phi \in C_0^\infty(R^m)$, then $u * \phi$ denotes the function defined by the equation

$$(u * \phi)(x) = u_y(\phi(x-y)), \tag{2.1.2}$$

where u_y means that u acts on $\phi(x-y)$ as a function of y with x fixed.

By the vector sum of two sets A and B in R^m we understand the set

$$A + B = \{x + y;\ x \in A;\ y \in B\}.$$

THEOREM 2.1.2. *If $u \in D'(R^m)$ and $\phi \in C_0^\infty(R^m)$, then $u * \phi \in C_0^\infty(R^m)$ and* $\operatorname{supp}(u * \phi) \subset \operatorname{supp} u + \operatorname{supp} \phi$. *Moreover,*

$$D^\alpha(u*\varphi) = (D^\alpha u)*\varphi = u*(D^\alpha\varphi).$$

THEOREM 2.1.3. *Let $\phi \in C_0^\infty(R^m)$, $\int \phi\,dx = 1$, $\phi \geqslant 0$,* $\operatorname{supp}\phi = \{x, |x| \leqslant 1\}$, *and $\phi_\epsilon(x) = \epsilon^{-m}\phi(x/\epsilon)$, where $\epsilon = \text{const} > 0$. If $u \in D'(R^m)$, then $u * \phi_\epsilon \in C^\infty(R^m)$ and*

$$\operatorname{supp}(u*\varphi_\varepsilon) \subset \operatorname{supp} u + \{x;\ |x| \leqslant \varepsilon\}.$$

As $\epsilon \to 0$ the functions $u * \phi_\epsilon \to u$ in $D'(R^m)$.

The Fourier transform $\hat{f}$ of a function $f \in \mathfrak{L}_1(R^m)$ is defined by the formula

$$\hat{f}(\xi) = \int e^{-i\langle x,\xi\rangle} f(x)\,dx. \tag{2.1.3}$$

If $\hat{f}(\xi)$ also belongs to $\mathfrak{L}_1(R^m)$, then the inversion formula

$$f(x) = (2\pi)^{-m}\int e^{i\langle x,\xi\rangle}\hat{f}(\xi)\,d\xi \tag{2.1.4}$$

holds. By S we denote the set of all functions $\phi \in C^\infty(R^m)$ such that

$$\sup_x |x^\beta D^\alpha\varphi(x)| < \infty$$

for any multi-indices α and β. Functions in the class S form a locally compact topological space, if one introduces the following system of seminorms on S (see [144]):

$$p_{\alpha,\beta}(\varphi) = \sup_{x\in R^m} |x^\beta D^\alpha\varphi(x)|. \tag{2.1.5}$$

It is clear that $C_0^\infty(R^m) \subset S$.

THEOREM 2.1.4. *The Fourier transform maps* S *continuously into* S. *The formula* (2.1.4) *for the inverse Fourier transform is valid for functions in* S. *The Fourier transform of* $D_j\phi$ *is* $\xi_j\hat{\phi}(\xi)$ *and the Fourier transform of* $x_j\phi$ *is* $-D_j\hat{\phi}(\xi)$.

THEOREM 2.1.5. *If* ϕ *and* ψ *are functions in* S, *then*

$$\int \hat{\varphi}\psi\, dx = \int \varphi\hat{\psi} dx, \tag{2.1.6}$$

$$\int \hat{\varphi}\bar{\hat{\psi}} d\xi = (2\pi)^m \int \varphi\bar{\psi} dx, \tag{2.1.7}$$

$$\widehat{(\varphi * \psi)} = \hat{\varphi}\cdot\hat{\psi}, \tag{2.1.8}$$

$$\widehat{(\varphi \cdot \psi)} = (2\pi)^{-m}\hat{\varphi}*\hat{\psi}. \tag{2.1.9}$$

A continuous linear functional on S is called a distribution in the space S'. Since $C_0^\infty \subset S$ and since the convergence of a sequence in $D(R^m)$ implies its convergence in S, each distribution in S' yields a continuous linear functional on $D(R^m)$, i.e. is a distribution in $D'(R^m)$.

It is easy to show that the set C_0^∞ is dense in S and that S' may be continuously imbedded in $D'(R^m)$.

If $u \in S'$, then the Fourier transform $\hat{u}$ of the distribution $u(\phi)$ is defined by the equation

$$\hat{u}(\varphi) = u(\hat{\varphi}), \quad \varphi \in S. \tag{2.1.10}$$

For each distribution in S' the inversion formula for the Fourier transform is valid in the form

$$\hat{\hat{u}} = (2\pi)^m \check{u},$$

where $\check{\phi} = \phi(-x)$ for $\phi \in S$ and $\check{u}(\phi) = u(\check{\phi})$ for distributions in S'.

THEOREM 2.1.6. *If the distribution* $u(\phi)$ *in* $D'(R^m)$ *has compact support, then* $u \in S'$ *and the Fourier transform of* u *is defined for all complex values of* ξ *by the formula*

$$\hat{u}(\xi) = u_x(e^{-i\langle x,\xi\rangle}),$$

and $\hat{u}(\xi)$ *is an entire analytic function of* ξ.

The proofs of Theorems 2.1.1–2.1.6 may be found in the books [40], [56], [120] or [126].

In the function space S we introduce the scalar product

$$(u,v)_s = (2\pi)^{-m}\int (1 + |\xi|^2)^s \hat{u}(\xi)\overline{\hat{v}(\xi)}\, d\xi, \tag{2.1.11}$$

where s is any real number. This scalar product generates a norm

$$\|u\|_s^2 = (u, u)_s = (2\pi)^{-m} \int (1 + |\xi|^2)^s |\hat{u}(\xi)|^2 d\xi. \tag{2.1.12}$$

The closure of the space S with respect to this norm is called the space $\mathcal{H}_s$. Clearly $\mathcal{H}_s$ is a separable Hilbert space.

THEOREM 2.1.7. *The space* $\mathcal{H}_s$ *is isomorphic to the subspace of distributions* u *in* S', *of which the Fourier transform* $\hat{u}(\xi)$ *is an ordinary function and*

$$\|u\|_s^2 = (2\pi)^{-m} \int (1 + |\xi|^2)^s |\hat{u}(\xi)|^2 d\xi < \infty. \tag{2.1.13}$$

PROOF. To each element u in $\mathcal{H}_s$ belonging to S we associate a distribution in S' by

$$u(\varphi) = \int u\varphi dx; \quad u, \varphi \in S.$$

If the sequence u_n is fundamental in $\mathcal{H}_s$, then to it there corresponds a distribution $u(\phi)$ equal to $\lim_{n\to\infty} u_n$ in S'. This limit exists, since $\lim_{n\to\infty} \int u_n \phi \, dx$ exists as $n \to \infty$ for each $\phi \in S$. In fact, using Parseval's equation and the Schwarz inequality, we obtain

$$\left| \int (u_n - u_{n'}) \varphi dx \right| = |(u_n - u_{n'}, \bar{\varphi})_0|$$
$$\leqslant \| u_n - u_{n'} \|_s \| \bar{\varphi} \|_{-s}. \tag{2.1.14}$$

The right side of (2.1.14) tends to zero as $n, n' \to \infty$, since the sequence u_n is fundamental. We shall show that the distribution equal to $\lim_{n\to\infty} u_n$ has a Fourier transform $\hat{u}(\xi)$ such that condition (2.1.13) is fulfilled.

Since

$$\| u_n - u_{n'} \|_s^2 = (2\pi)^{-m} \int (1 + |\xi|^2)^s |\hat{u}_{n'} - \hat{u}_n|^2 d\xi,$$

it follows that the functions $\hat{u}_n$ form a fundamental sequence in the space of functions with norm given by the right side of (2.1.12). The completeness of this space follows from the completeness of $\mathfrak{L}_2(R^m)$. Thus there exists a function $\hat{u}(\xi)$ such that condition (2.1.13) is fulfilled, and

$$\int (1 + |\xi|^2)^s |\hat{u}(\xi) - \hat{u}_n(\xi)|^2 d\xi \to 0 \quad \text{as} \quad n \to \infty. \tag{2.1.15}$$

We shall show that the function $\hat{u}(\xi)$ is the Fourier transform of the distribution $\lim_{n\to\infty} u_n$.

According to (2.1.6) and the definition of the Fourier transform for distributions,

$$\int \hat{u}_n \varphi dx = \int u_n \hat{\varphi} dx = u_n(\hat{\varphi}) = \hat{u}_n(\varphi).$$

Passing to the limit as $n \to \infty$ in these equations and recalling (2.1.15), we obtain

$$\int \hat{u} \varphi dx = u(\hat{\varphi}) = \hat{u}(\varphi).$$

This means that $\hat{u}(\xi)$ is the Fourier transform of the distribution $u(\phi) = \lim_{n\to\infty} u_n(\phi)$. Now let $u(\phi)$ be a distribution in S' such that its Fourier transform is a function $\hat{u}(\xi)$ satisfying the condition (2.1.13). We construct a sequence of functions $\hat{u}_n(\xi)$ such that $\hat{u}_n(\xi) \in C_0^\infty(R^m)$ and

$$\int (1 + |\xi|^2)^s |\hat{u}(\xi) - \hat{u}_n(\xi)|^2 d\xi \to 0 \quad \text{as} \quad n \to \infty.$$

Then the functions u_n, defined by the formulas

$$u_n(x) = (2\pi)^{-m} \int e^{i\langle x,\xi\rangle} \hat{u}_n(\xi)\, d\xi,$$

belong to the space S and form a fundamental sequence in the space $\mathcal{H}_s$. From the equation

$$\int \hat{u}_n \varphi dx = \int u_n \hat{\varphi} dx = u_n(\hat{\varphi})$$

it follows that for any function $\phi \in S$

$$\lim_{n\to\infty} u_n(\hat{\varphi}) = \int \hat{u} \varphi dx = \hat{u}(\varphi) = u(\hat{\varphi}).$$

This means that the sequence u_n converges in S' to the distribution $u(\phi)$ considered by us, and is fundamental in $\mathcal{H}_s$.

Thus to each element u in $\mathcal{H}_s$ there corresponds, in a one-to-one manner, a distribution $u(\phi)$ in S' such that

$$\| u \|_s^2 = (2\pi)^{-m} \int (1 + |\xi|^2)^s |\hat{u}(\xi)|^2 d\xi,$$

where $\hat{u}(\xi)$ is the Fourier transform of the distribution $u(\phi)$, and $\|u\|_s$ is the norm of the element u in $\mathcal{H}_s$. The theorem is proved.

The following properties of the spaces $\mathcal{H}_s$ may easily be proved.

1) $\mathcal{H}_t \subset \mathcal{H}_s$ if $s < t$, and

$$\| u \|_s \leqslant \| u \|_t. \tag{2.1.16}$$

2)
$$|(u, v)_s| \leqslant \| u \|_s \| v \|_s. \tag{2.1.17}$$

This inequality is derived by applying the Schwarz inequality to the integral on the right side of (2.1.11).

3) If $u \in \mathcal{H}_{t+|p|}$, then

$$\| D^p u \|_t \leqslant \| u \|_{t+|p|}. \tag{2.1.18}$$

4) If $u \in \mathcal{H}_{s+t}$ and $v \in \mathcal{H}_{s-t}$, then (2.1.11) defines a scalar product $(u, v)_s$ for them with the following inequalities valid:

$$|(u, v)_s| \leqslant \|u\|_{s+t} \|v\|_{s-t}, \tag{2.1.19}$$

$$|(u, v)_s| \leqslant \frac{1}{2}\left(\mu \|u\|_{s+t}^2 + \frac{1}{\mu}\|v\|_{s-t}^2\right), \tag{2.1.20}$$

where μ is an arbitrary number. Inequality (2.1.19) is sometimes called the generalized Schwarz inequality.

It is easy to see that if $s = 0$, then the space $\mathcal{H}_0$ coincides with the space $\mathcal{L}_2(R^m)$. If s is a positive integer, then $\mathcal{H}_s$ coincides with the Sobolev space W_2^s consisting of functions whose generalized derivatives up to order s inclusive belong to $\mathcal{L}_2(R^m)$ (see [125]).

THEOREM 2.1.8. (IMBEDDING THEOREM OF S. L. SOBOLEV.) *If $u \in \mathcal{H}_{l+k}(R^m)$, where $2l > m$ and k is a nonnegative integer, then $u \in C^{(k)}(R^m)$.*

PROOF. Functions u in S satisfy the relation

$$|D^\alpha u(x)| = (2\pi)^{-m}\left|\int \xi^\alpha \hat{u}(\xi) e^{i\langle x,\xi\rangle} d\xi\right|$$

$$\leqslant (2\pi)^{-m}\int (1 + |\xi|^2)^{|\alpha|/2} |\hat{u}(\xi)| d\xi.$$

Hence

$$\|u(x)\|_{C^{(k)}} \equiv \sum_{|\alpha| \leqslant k} \max_{x \in R^m} |D^\alpha u(x)| \leqslant C_1 \int (1 + |\xi|^2)^{\frac{k}{2}} |\hat{u}(\xi)| d\xi,$$

where the constant C_1 does not depend on u. From this estimate and the Schwarz inequality it follows that

$$\|u(x)\|_{C^{(k)}} \leqslant C_1 \left\{\int (1 + |\xi|^2)^{-l} d\xi\right\}^{1/2}$$

$$\times \left\{\int (1 + |\xi|^2)^{k+l} |\hat{u}(\xi)|^2 d\xi\right\}^{1/2} \leqslant C_2 \|u\|_{l+k}, \tag{2.1.21}$$

since $2l > m$.

If $u \in \mathcal{H}_{l+k}(R^m)$ there exists a sequence of functions $u_n \in S$ such that $\|u - u_n\|_{l+k} \to 0$ as $n \to \infty$ and $\|u_n - u_{n'}\|_{l+k} \to 0$ as $n, n' \to \infty$. From (2.1.21) it follows that $\|u_n - u_{n'}\|_{C(k)} \to 0$ as $n, n' \to \infty$, and hence there exists a function u' such that $\|u_n - u'\|_{C(k)} \to 0$ as $n \to \infty$. Evidently $u \equiv u'$ in R^m.

THEOREM 2.1.9. *The space $\mathcal{H}_s$ is dual to the space $\mathcal{H}_{-s}$ with respect to the scalar product of the space $\mathcal{H}_0$. This means that if $v \in \mathcal{H}_{-s}$, then*

$$l(u) = (u, v)_0 = (2\pi)^{-m} \int \hat{u}(\xi) \overline{\hat{v}(\xi)} d\xi$$

is a bounded linear functional in $\mathcal{H}_s$ *and the norm of the functional* $l(u)$ *is equal to* $\|v\|_{-s}$, *i. e. the relation*

$$\|v\|_{-s} = \sup_u \frac{(u, v)_0}{\|u\|_s} \tag{2.1.22}$$

holds.

Conversely, each bounded linear functional on $\mathcal{H}_s$ *may be represented in the form* $(u, v)_0$, *where the function* $v \in \mathcal{H}_{-s}$ *is uniquely determined, and the norm of the functional is equal to* $\|v\|_{-s}$.

PROOF. We consider $l(u) = (u, v)_0$ for $v \in \mathcal{H}_{-s}$, $u \in \mathcal{H}_s$. According to (2.1.19) $|l(u)| \leqslant \|u\|_s \|v\|_{-s}$ and therefore $l(u)$ is a bounded linear functional on $\mathcal{H}_s$.

By definition of the norm $l(u)$,

$$\|l\| = \sup_u \frac{(u, v)_0}{\|u\|_s} \tag{2.1.23}$$

which implies

$$\|l\| = \sup_u \frac{(2\pi)^{-\frac{m}{2}} \int [(1+|\xi|^2)^{s/2} \hat{u}(\xi)]\, \overline{[(1+|\xi|^2)^{-\frac{s}{2}} \hat{v}(\xi)]}\, d\xi}{\left[\int (1+|\xi|^2)^s |\hat{u}(\xi)|^2\, d\xi\right]^{1/2}}. \tag{2.1.24}$$

Applying the Schwarz inequality, we obtain

$$\|l\| \leqslant \|v\|_{-s}. \tag{2.1.25}$$

Substituting in the right side of (2.1.23) the distribution u whose Fourier transform is equal to $\hat{v}(\xi)(1 + |\xi|^2)^{-s}$, we obtain

$$\|l\| \geqslant \|v\|_{-s}. \tag{2.1.26}$$

From (2.1.25) and (2.1.26) we obtain (2.1.22).

Suppose now that $l(u)$ is a bounded linear functional in $\mathcal{H}_s$. By the Riesz theorem, $l(u) = (u, v_1)_s$ and $\|l\| = \|v_1\|_s$, where $u, v_1 \in \mathcal{H}_s$. According to the definition (2.1.11) of scalar product in $\mathcal{H}_s$

$$l(u) = (2\pi)^{-m} \int (1 + |\xi|^2)^s \hat{u}(\xi) \overline{\hat{v}_1(\xi)}\, d\xi$$

$$= (2\pi)^{-m} \int \hat{u}(\xi) \overline{[(1 + |\xi|^2)^s \hat{v}_1(\xi)]}\, d\xi = (u, v)_0,$$

where v is the distribution whose Fourier transform is equal to $(1 + |\xi|^2)^s \hat{v}_1(\xi)$. Clearly $\|l\| = \|v\|_{-s}$, since $v \in \mathcal{H}_{-s}$ and $\|v\|_{-s} = \|v_1\|_s$. Consequently $\|l\| = \|v\|_{-s}$. The theorem is thereby proved.

THEOREM 2.1.10. *Each distribution* u *in* $D'(R^m)$ *which has compact support belongs to* $\mathcal{H}_{-s}$ *for some* $s \in R^1$.

PROOF. Let $\psi \in C_0^\infty(R^m)$ and $\psi \equiv 1$ on the support of u. Also let ϕ be any function in $C_0^\infty(R^m)$. Then $u(\phi) = u(\psi\phi + (1 - \psi)\phi) = u(\psi\phi)$. Therefore we need consider u only applied to functions ϕ in $C_0^\infty(R^m)$ such that $\operatorname{supp}\phi \subset K$, where K is some compact set in R^m. We assume that u does not belong to $\mathcal{H}_{-s}$ for any $s \in R^1$, i.e. u is not representable in the form $u(\phi) = (u, \phi)_0$ for $u \in \mathcal{H}_{-s}$ for any s. This means that there exists a sequence of functions $\phi_n \in C_0^\infty(K)$ such that

$$|u(\varphi_n)| > n\|\varphi_n\|_n, \quad n = 1, 2, \ldots .$$

The functions $\psi_n = (n\|\phi_n\|)^{-1}\phi_n$ have support contained in K, and $\|\psi_n\|_s \to 0$ as $n \to \infty$ for every $s = 1, 2, \cdots$, since $\|\psi_n\|_s \leqslant 1/n$ if $n > s$. From Theorem 2.1.8 and estimate (2.1.21) it follows that the ψ_n converge to zero uniformly, together with all their derivatives. However $u(\psi_n) = u(\phi_n/n\|\phi_n\|_n) \geqslant 1$, so that $u(\psi_n)$ do not converge to zero as $n \to \infty$. This contradiction shows that $|u(\phi)| \leqslant C\|\phi\|_s$ for some s, i.e. $u \in \mathcal{H}_{-s}$.

THEOREM 2.1.11. *A set of elements* $\{u_n\}$ *of the space* $\mathcal{H}_t$ *which have support in some compact set* K *and which are bounded uniformly in the norm of the space* $\mathcal{H}_t$ *is compact in* $\mathcal{H}_s$ *if* $t > s$.

PROOF. It is sufficient to prove Theorem 2.1.11 for the case when all the $u_n \in C_0^\infty(K)$. In fact, for each u_n in $\mathcal{H}_t$ there is a function w_n in S such that

$$\|u_n - w_n\|_t \leqslant 1/n.$$

Let $\psi \in C_0^\infty(R^m)$ and $\psi \equiv 1$ on K. Then $\psi u_n = u_n$ and

$$\|\psi u_n - \psi w_n\|_t \leqslant C\,\|u_n - w_n\|_t \leqslant \frac{C}{n}, \tag{2.1.27}$$

where C does not depend on n. Estimate (2.1.27) follows from Theorem 2.2.1, which we shall prove in the next section. From (2.1.27) it follows that the sequence ψw_n is fundamental in $\mathcal{H}_s$ if the sequence u_n has this same property, and conversely. For the set of elements $\{\psi w_n\}$ the assumptions of Theorem 2.1.11 are fulfilled:

$$\|\psi w_n\|_t \leqslant C_1 \text{ and } \operatorname{supp}\psi w_n \subset K_1,$$

where K_1 is a bounded set in R^m. Moreover, $\psi w_n \in C_0^\infty(R^m)$.

Thus in proving the theorem we may assume that $u_n \in C_0^\infty(R^m)$, $\|u_n\|_t \leqslant C_1$ and $\operatorname{supp} u_n \subset K$. We show that a subsequence, fundamental in $\mathcal{H}_s$, may be extracted from the sequence $\{u_n\}$. Suppose u_n is a sequence converging weakly in $\mathcal{H}_t$, i.e. $(u_n, v)_t \to (u, v)_t$ for any v in $\mathcal{H}_t$. We estimate $\|u_n - u_{n'}\|_s$. Let $\epsilon > 0$ be an arbitrary number. Then for all n and n' we have

$$(2\pi)^{-m} \int\limits_{|\xi|>N} |\hat{u}_n - \hat{u}_{n'}|^2 (1+|\xi|^2)^s \, d\xi$$
$$\leqslant (2\pi)^{-m} \int (1+N^2)^{s-t} |\hat{u}_n - \hat{u}_{n'}|^2 (1+|\xi|^2)^t \, d\xi$$
$$\leqslant (2\pi)^{-m} 4C_1^2 (1+N^2)^{s-t} < \varepsilon, \tag{2.1.28}$$

if $N \geqslant N_1(\epsilon)$.

Let $\{\xi^k\}$ be a finite set of points of the space $R^m(\xi_1, \cdots, \xi_m)$ such that $|\xi^k| \leqslant N$. By virtue of the weak convergence of u_n in $\mathcal{H}_t$, for each point ξ^k the sequence

$$\hat{u}_n(\xi^k) = \int u_n(x) e^{-i\langle x, \xi^k \rangle} \psi(x) \, dx$$

converges as $n \to \infty$, where $\psi \in C_0^\infty(R^m)$ and $u_n \psi \equiv u_n$ for any n, since clearly $\psi(x) e^{-i\langle x, \xi^k \rangle} \in \mathcal{H}_{-t}$. We choose the points ξ^k so that for any ξ ($|\xi| \leqslant N$) there is a ξ^k such that

$$|\hat{u}_n(\xi) - \hat{u}_n(\xi^k)| = \left| \int u_n \psi (e^{-i\langle x, \xi \rangle} - e^{-i\langle x, \xi^k \rangle}) \, dx \right|$$
$$\leqslant \|u_n\|_t \, \|\psi (e^{-i\langle x, \xi \rangle} - e^{-i\langle x, \xi^k \rangle})\|_{-t} \leqslant \delta,$$

where $\delta > 0$ is a constant to be chosen later. From this it follows that $|\hat{u}_n(\xi) - \hat{u}_{n'}(\xi)| \leqslant 3\delta$ for $|\xi| \leqslant N$, if n and n' are sufficiently large. Therefore

$$(2\pi)^{-m} \int\limits_{|\xi| \leqslant N} |\hat{u}_n - \hat{u}_{n'}|^2 (1+|\xi|^2)^s \, d\xi \leqslant \varepsilon, \tag{2.1.29}$$

if n and n' are sufficiently large and the constant $\delta = \delta(\epsilon)$ is chosen to be sufficiently small.

From (2.1.28) and (2.1.29) it follows that $\|u_n - u_{n'}\|_s^2 \leqslant 2\epsilon$, if n and n' are sufficiently large, i.e. the sequence u_n is fundamental in $\mathcal{H}_s$.

THEOREM 2.1.12. *Let* $s_1 < s_2 < s_3$. *Then for any* $\epsilon > 0$ *and any* $u \in \mathcal{H}_{s_3}$ *the inequality*

$$\|u\|_{s_2}^2 \leqslant \varepsilon \|u\|_{s_3}^2 + C_\varepsilon \|u\|_{s_1}^2 \tag{2.1.30}$$

holds, where

$$C_\varepsilon = C(s_1, s_2, s_3) \, \varepsilon^{-\frac{s_3 - s_1}{s_3 - s_2}}$$

and $C(s_1, s_2, s_3) = \text{const} > 0$, *depending on* s_1, s_2 *and* s_3.

PROOF. For any $M > 0$ we have

$$(1+|\xi|^2)^{s_2} = (1+|\xi|^2)^{s_1} \left[M (1+|\xi|^2)^{s_2 - s_1} \cdot \frac{1}{M} \right]. \tag{2.1.31}$$

If we use the elementary inequality (see, for example, [125])

$$ab \leqslant \frac{1}{p} a^p + \frac{1}{q} b^q; \ \frac{1}{p} + \frac{1}{q} = 1, \tag{2.1.32}$$

with the values

$$a = M(1 + |\xi|^2)^{s_2 - s_1}; \ b = \frac{1}{M}; \ p = \frac{s_3 - s_1}{s_2 - s_1},$$

we obtain

$$M(1 + |\xi|^2)^{s_2 - s_1} \cdot \frac{1}{M} \leqslant \frac{1}{p} M^{\frac{s_3 - s_1}{s_2 - s_1}} (1 + |\xi|^2)^{s_3 - s_1} + \frac{1}{q} M^{-\frac{s_3 - s_1}{s_3 - s_2}}. \tag{2.1.33}$$

From (2.1.31) and estimate (2.1.33) it follows that

$$(1 + |\xi|^2)^{s_2} \leqslant \varepsilon (1 + |\xi|^2)^{s_3} + \frac{1}{q} (p\varepsilon)^{-\frac{s_2 - s_1}{s_3 - s_2}} (1 + |\xi|^2)^{s_1}, \ \text{where} \quad \varepsilon = p^{-1} M^{\frac{s_3 - s_1}{s_2 - s_1}}. \tag{2.1.34}$$

From (2.1.34) and the definition of the norm in $\mathcal{H}_s$ it follows that

$$\|u\|^2_{s_2} \leqslant \varepsilon \|u\|^2_{s_3} + \frac{1}{q} (p\varepsilon)^{-\frac{s_2 - s_1}{s_3 - s_2}} \|u\|^2_{s_1}.$$

Thus Theorem 2.1.12 is proved.

The space $\mathcal{H}_s$ is widely used in the study of problems in partial differential equations. An extensive literature is devoted to the study of these spaces (see, for example, [56, 125, 139]).

By $\mathcal{H}_s^{\mathrm{loc}}(\Omega)$ we denote the set of distributions in $D'(\Omega)$ such that $\psi u \in \mathcal{H}_s$ for any ψ in $C_0^\infty(\Omega)$.

§2. Some properties of pseudodifferential operators

In the following we shall use pseudodifferential operators with symbols belonging to a certain special class. At the present time there is a fairly wide literature devoted to the theory of pseudodifferential operators (see [22, 23, 53, 54, 65, 137] and others). However, for the convenience of the reader we shall present here not only new results, but also those known parts of the theory of pseudodifferential operators which we shall need.

A pseudodifferential operator is an operator P defined on functions u in the class S, which has the form

$$Pu(x) = (2\pi)^{-m} \int p(x, \xi) \hat{u}(\xi) e^{i\langle x, \xi\rangle} d\xi, \tag{2.2.1}$$

where $\hat{u}(\xi)$ is the Fourier transform of the function $u(x)$, and $p(x, \xi)$ is a function called the symbol of the operator P.

We assume that the symbol of the operators we consider satisfies the following conditions:

a) The function $p(x, \xi)$ may be represented in the form

$$p(x, \xi) = p_1(\xi) + p_2(x, \xi), \tag{2.2.2}$$

where $p_1(\xi)$ is an infinitely differentiable function of ξ defined for all ξ and $p_2(x, \xi)$ is an infinitely differentiable function of x and ξ defined for all $x \in R^m$ and $\xi \in R^m$ such that for each ξ the function $p_2(x, \xi)$, considered as a function of x, has support belonging to some compact set $K \subset R^m$, i.e $p_2(x, \xi) \equiv 0$ for $x \in R^m \setminus K$ and $\xi \in R^m$.

b) There exists a $\sigma \in R^1$ such that for any multi-indices α and β and any $x \in R^m$ and $\xi \in R^m$ the estimates

$$\left| \frac{\partial^{|\alpha|}}{\partial \xi^\alpha} p_1(\xi) \right| \leqslant C_\alpha (1 + |\xi|^2)^{\frac{1}{2}(\sigma - |\alpha|)}, \tag{2.2.3}$$

$$\left| \frac{\partial^{|\alpha|}}{\partial \xi^\alpha} D^\beta p_2(x, \xi) \right| \leqslant C_{\alpha,\beta} (1 + |\xi|^2)^{\frac{1}{2}(\sigma - |\alpha|)} \tag{2.2.4}$$

hold for constants C_α and $C_{\alpha\beta}$ depending only on α and β. It is easy to see that the operator P transforms a function u in S into a function Pu also in S.

Theorems 2.2.1–2.2.5, 2.2.7 and 2.2.8 proved below are also valid for a wider class of symbols. In particular, the proof of these theorems proceeds without change also in the case when $p_2(x, \xi)$ is an infinitely differentiable function of the variables x and ξ in the class S with respect to the variable x such that

$$(1 + |x|^2)^{k/2} \left| \frac{\partial^{|\alpha|}}{\partial \xi^\alpha} D^\beta p_2(x, \xi) \right| \leqslant C_{k,\alpha,\beta} (1 + |\xi|^2)^{\frac{1}{2}(\sigma - |\alpha|)} \tag{2.2.5}$$

for any k and any multi-indices α and β, $x \in R^m$ and $\xi \in R^m$, where the constants $C_{k,\alpha,\beta}$ depend only on k, α and β.

First we prove the following auxiliary propositions.

LEMMA 2.2.1. *For any* $s \in R^1$ *and any* ξ *and* η *in* R^m,

$$(1 + |\xi|^2)^s (1 + |\eta|^2)^{-s} \leqslant 2^{|s|} (1 + |\xi - \eta|^2)^{|s|}. \tag{2.2.6}$$

PROOF. From the inequality

$$|\xi|^2 = |\eta + (\xi - \eta)|^2 \leqslant 2|\xi - \eta|^2 + 2|\eta|^2$$

it follows that

$$1 + |\xi|^2 \leqslant 2(1 + |\xi - \eta|^2)(1 + |\eta|^2).$$

In the same way we obtain

$$1 + |\eta|^2 \leqslant 2(1 + |\xi - \eta|^2)(1 + |\xi|^2).$$

Therefore

$$(1+|\xi|^2)^{|s|}(1+|\eta|^2)^{-|s|} \leqslant 2^{|s|}(1+|\xi-\eta|^2)^{|s|},$$
$$(1+|\eta|^2)^{|s|}(1+|\xi|^2)^{-|s|} \leqslant 2^{|s|}(1+|\xi-\eta|^2)^{|s|}.$$

The desired inequality follows from this.

LEMMA 2.2.2. *For any* θ *such that* $0 \leqslant \theta \leqslant 1$ *and any* $s \in R^1$ *the inequalities*

$$(1+|\eta+\theta(\xi-\eta)|^2)^{-|s|} \leqslant 4^{|s|}(1+|\xi-\eta|^2)^{|s|}(1+|\eta|^2)^{-|s|} \tag{2.2.7}$$

and

$$(1+|\eta+\theta(\xi-\eta)|^2)^{|s|} \leqslant [(1+|\eta|^2)^{|s|}+(1+|\xi|^2)^{|s|}] \tag{2.2.8}$$

hold for any ξ *and* η *in* R^m.

PROOF. If $|\eta-\xi| \leqslant \frac{1}{2}|\eta|$, it is clear that

$$(1+|\eta+\theta(\xi-\eta)|^2) \geqslant 1+4^{-1}|\eta|^2,$$
$$(1+|\eta+\theta(\xi-\eta)|^2)^{-|s|} \leqslant 4^{|s|}(1+|\eta|^2)^{-|s|}.$$

However, if $|\eta-\xi| > \frac{1}{2}|\eta|$, then

$$(1+|\eta+\theta(\xi-\eta)|^2)^{-|s|} \leqslant \frac{(1+4^{-1}|\eta|^2)^{|s|}}{(1+4^{-1}|\eta|^2)^{|s|}}$$
$$\leqslant 4^{|s|}(1+|\eta-\xi|^2)^{|s|}(1+|\eta|^2)^{-|s|}.$$

From these inequalities, (2.2.7) easily follows. Inequality (2.2.8) follows from the fact that either $|\eta+\theta(\xi-\eta)| \leqslant |\xi|$ or $|\eta+\theta(\xi-\eta)| \leqslant |\eta|$.

We denote by $\hat{p}_2(\tau, \xi)$ the Fourier transform of the function $p_2(x, \xi)$ with respect to the variable x, and by $\hat{p}_2^{(\alpha)}(\tau, \xi)$ the derivative $\partial^{|\alpha|}\hat{p}_2/\partial\xi_1^{\alpha_1}\cdots\partial\xi_m^{\alpha_m}$.

LEMMA 2.2.3. *For any multi-index* α *and* $M \geqslant 0$ *there exists a constant* $C_{\alpha,M}$ *such that*

$$|\hat{p}_2^{(\alpha)}(\tau, \xi)| \leqslant C_{\alpha,M}(1+|\xi|^2)^{\frac{\sigma-|\alpha|}{2}}(1+|\tau|^2)^{-M}. \tag{2.2.9}$$

PROOF. This inequality is easily obtained by transforming the integral for $\hat{p}_2^{(\alpha)}$ by integration by parts, bearing in mind estimate (2.2.4) as well as property a) of the function $p_2(x, \xi)$ according to which $\operatorname{supp} p_2(x, \xi) \subset K$ for any ξ.

LEMMA 2.2.4. *Let* $\int|A(\zeta)|\,d\zeta < \infty$. *Then*

$$I \equiv \left|\iint A(\xi-\eta)\,a(\xi)\,b(\eta)\,d\xi\,d\eta\right| \leqslant \int|A(\zeta)|\,d\zeta\,\|a\|_0\,\|b\|_0. \tag{2.2.10}$$

PROOF. By the Schwarz inequality,

$$I \leqslant \int \Big\{\Big(\int |A(\xi-\eta)|\,d\xi\Big)^{1/2}\Big(\int |A(\xi-\eta)|\,|a(\xi)|^2\,d\xi\Big)^{1/2}|b(\eta)|\Big\}\,d\eta$$

$$\leqslant \Big(\int |A(\zeta)|\,d\zeta\Big)^{1/2}\Big(\int |b(\eta)|^2\,d\eta\Big)^{1/2}\Big(\int\int |a(\xi)|^2|A(\xi-\eta)|\,d\xi\,d\eta\Big)^{1/2}$$

$$\leqslant \int |A(\zeta)|\,d\zeta\,\|a\|_0\|b\|_0.$$

The lemma is proved.

THEOREM 2.2.1. (*On the boundedness of pseudodifferential operators.*) *Let* Pu *be an operator with symbol* $p(x, \xi)$, *satisfying conditions* a) *and* b). *For every* $s \in R^1$ *there exists a constant* C_s *depending on* s *such that*

$$\|Pu\|_s \leqslant C_s\|u\|_{s+\sigma} \tag{2.2.11}$$

for any function u *in* S.

PROOF. Clearly $\|Pu\|_s \leqslant \|P_1u\|_s + \|P_2u\|_s$, where P_1 is the pseudodifferential operator with symbol $p_1(\xi)$ and P_2 is the pseudodifferential operator with symbol $p_2(x, \xi)$. According to the definition of the norm $\mathcal{H}_s$ and condition (2.2.3) with $|\alpha| = 0$,

$$\|P_1u\|_s^2 = (2\pi)^{-m}\int (1+|\xi|^2)^s|\widehat{P_1u}|^2 d\xi$$
$$\leqslant C_1\int (1+|\xi|^2)^{s+\sigma}|\hat{u}(\xi)|^2\,d\xi \leqslant C_1\|u\|_{s+\sigma}^2. \tag{2.2.12}$$

We now estimate P_2u. We consider the Fourier transform of the function P_2u, where $u \in S$. It is easy to see that

$$\widehat{P_2u}(\eta) = (2\pi)^{-m}\int \hat{p}_2(\eta-\xi, \xi)\,\hat{u}(\xi)\,d\xi, \tag{2.2.13}$$

where $\hat{p}_2(\tau, \xi)$ is the Fourier transform of the function $p_2(x, \xi)$ with respect to the variable x. For any function $w \in S$, Parseval's equation implies

$$\int P_2(u)\,\overline{w}\,dx = (2\pi)^{-2m}\int\int \hat{p}_2(\eta-\xi, \xi)\,\overline{\hat{w}(\eta)}\,\hat{u}(\xi)\,d\xi\,d\eta.$$

Using (2.2.6) and (2.2.9) with $|\alpha| = 0$, we obtain

$$\Big|\int P_2(u)\,\overline{w}\,dx\Big| \leqslant (2\pi)^{-2m}\int\int |\hat{p}_2(\eta-\xi, \xi)|\,|\hat{w}(\eta)|\,|\hat{u}(\xi)|\,d\xi\,d\eta$$

$$\leqslant C_M\int\int (1+|\eta-\xi|^2)^{-M/2}(1+|\xi|^2)^{\frac{\sigma+s}{2}}|\hat{u}(\xi)|(1+|\xi|^2)^{-\frac{s}{2}}|\hat{w}(\eta)|\,d\xi\,d\eta$$

$$\leqslant C'_M\int\int (1+|\eta-\xi|^2)^{\frac{|s|-M}{2}}(1+|\xi|^2)^{\frac{\sigma+s}{2}}|\hat{u}(\xi)|(1+|\eta|^2)^{-s/2}|\hat{w}(\eta)|\,d\xi\,d\eta.$$

Here M is chosen so that $M - |s| \geqslant m + 1$. Therefore, applying (2.2.10) to the estimation of the last integral, we obtain

$$\left|\int P_2(u)\bar{w}dx\right| \leqslant C_1\left(\int |\hat{u}(\xi)|^2(1+|\xi|^2)^{\sigma+s}d\xi\right)^{1/2}\left(\int |\hat{w}(\eta)|^2(1+|\eta|^2)^{-s}d\eta\right)^{1/2}$$
$$\leqslant C_2\|u\|_{\sigma+s}\|w\|_{-s}.$$

From the last inequality and Theorem 2.1.9 it follows that

$$\|P_2u\|_s \leqslant C_s\|u\|_{\sigma+s}. \tag{2.2.14}$$

The assertion of the theorem follows from (2.2.12) and (2.2.14).

Since the set of functions u in S is dense in $\mathcal{H}_s$, we see that the operator P may be closed in $\mathcal{H}_s$ with retention of (2.2.11).

We note that it follows from Theorem 2.2.1, in particular, that if $\phi(x) \in C_0^\infty(R^m)$, then for any $s \in R^1$

$$\|\varphi u\|_s \leqslant C_s\|u\|_s, \tag{2.2.15}$$

where $C_s = \text{const} > 0$ and $u \in \mathcal{H}_s$.

For any operator P, the lower limit of numbers σ for which inequalities of the form (2.2.11) are fulfilled is called the order of P.

THEOREM 2.2.2. *(On the product of pseudodifferential operators.) Let P and Q be pseudodifferential operators with symbols $p(x, \xi)$ and $q(x, \xi)$ satisfying conditions* a) *and* b) *with $\sigma = \sigma_1$ and $\sigma = \sigma_2$ respectively. Let $(PQ)_N$ be the pseudodifferential operator with symbol*

$$\sum_{|\alpha|\leqslant N-1}\frac{1}{\alpha!}\frac{\partial^{|\alpha|}}{\partial\xi^\alpha}p(x,\xi)\cdot D_x^\alpha q(x,\xi).$$

Then for all u in S

$$P\cdot Qu = (PQ)_N u + T_N u,$$

where the operator T_N has order at most $\sigma_1 + \sigma_2 - N$, i. e.

$$\|T_N u\|_s \leqslant C_{N,s}\|u\|_{s+\sigma_1+\sigma_2-N}. \tag{2.2.16}$$

PROOF. Since $Pu = P_1u + P_2u$ and $Qu = Q_1u + Q_2u$, where P_1, P_2, Q_1, Q_2 are pseudodifferential operators with symbols $p_1(\xi)$, $p_2(x, \xi)$, $q_1(\xi)$, $q_2(x, \xi)$ respectively, we have

$$PQu = P_1Q_1u + P_2Q_1u + P_1Q_2u + P_2Q_2u.$$

The symbol of the operator $(PQ)_N$ may clearly be represented in the form

$$\sum_{|\alpha|\leqslant N-1}\frac{1}{\alpha!}\frac{\partial^{|\alpha|}}{\partial\xi^\alpha}p(x,\xi)\cdot D^\alpha q(x,\xi) = p_1(\xi)\,q_1(\xi)$$
$$+\,p_2(x,\xi)q_1(\xi) + \sum_{|\alpha|\leqslant N-1}\frac{1}{\alpha!}\frac{\partial^{|\alpha|}p_1(\xi)}{\partial\xi^\alpha}D^\alpha q_2(x,\xi) + \sum_{|\alpha|\leqslant N-1}\frac{1}{\alpha!}\frac{\partial^{|\alpha|}p_2}{\partial\xi^\alpha}D^\alpha q_2.$$

Since the symbol of the operator P_1Q_1 is equal to $p_1(\xi)q_1(\xi)$ and the symbol of the operator P_2Q_1 is equal to $p_2(x, \xi)q_1(\xi)$ it follows that $T_Nu \equiv PQu - (PQ)_Nu = P_1Q_2u - (P_1Q_2)_Nu + P_2Q_2u - (P_2Q_2)_Nu$, where $(P_1Q_2)_N$ has the symbol

$$\sum_{|\alpha|\leqslant N-1} \frac{1}{\alpha!} \frac{\partial^{|\alpha|}}{\partial \xi^\alpha} p_1(\xi)\cdot D^\alpha q_2(x, \xi),$$

and $(P_2Q_2)_N$ has the symbol

$$\sum_{|\alpha|\leqslant N-1} \frac{1}{\alpha!} \frac{\partial^{|\alpha|}}{\partial \xi^\alpha} p_2(x, \xi) D^\alpha q_2(x, \xi).$$

Therefore to prove Theorem 2.2.2 it is sufficient to prove that

$$\| T'_Nu \|_s^2 = \| (P_1Q_2 - (P_1Q_2)_N)\, u \|_s^2 \leqslant C'_{N,s} \| u \|^2_{s+\sigma_1+\sigma_2-N}$$

and

$$\| T''_Nu \|_s^2 = \| (P_2Q_2 - (P_2Q_2)_N)\, u \|_s^2 \leqslant C''_{N,s} \| u \|^2_{s+\sigma_1+\sigma_2-N}.$$

Thus it is sufficient to prove Theorem 2.2.2 for the case of operators P and Q for which $p_1(\xi) \equiv 0$ and $q_1(\xi) \equiv 0$ and also for the case when $p_2(x, \xi) \equiv 0$ and $q_1(\xi) \equiv 0$. We first prove the theorem for the case $p_1(\xi) \equiv 0$ and $q_1(\xi) \equiv 0$.

Consider the Fourier transform of the function $T''_Nu = P_2Q_2u - (P_2Q_2)_Nu$. According to (2.2.13) we have

$$\widehat{T''_Nu}(\eta) = (2\pi)^{-2m} \iint [\hat{p}_2(\eta - \tau, \tau)\, \hat{q}_2(\tau - \xi, \xi)\, \hat{u}(\xi)]\, d\xi d\tau$$
$$- (2\pi)^{-m} \iint z(\eta - \xi, \xi)\, \hat{u}(\xi)\, d\xi, \qquad (2.2.17)$$

where $z(\zeta, \xi)$ is the Fourier transform of the symbol of the operator $(P_2Q_2)_N$ with respect to the variable x. Using the known formula

$$\widehat{\varphi\cdot\psi} = (2\pi)^{-m} \hat{\varphi} * \hat{\psi}, \qquad (2.2.18)$$

where $*$, as usual, signifies the convolution operation, we obtain for $z(\zeta, \xi)$ an expression of the form

$$z(\zeta, \xi) = (2\pi)^{-m} \int \sum_{|\alpha|\leqslant N-1} \frac{1}{\alpha!} \hat{p}_2^{(\alpha)}(\zeta - \rho, \xi)\, \hat{q}_2(\rho, \xi)\, \rho^\alpha d\rho.$$

We change the variables of integration here to $\rho = \tau - \xi$ and substitute this expression for z into (2.2.17) to obtain

$$\widehat{T''_Nu}(\eta) = (2\pi)^{-2m} \iint \hat{q}_2(\tau - \xi, \xi)[\hat{p}_2(\eta - \tau, \tau)$$
$$- \sum_{|\alpha|\leqslant N-1} \frac{1}{\alpha!} \hat{p}_2^{(\alpha)}(\eta - \tau, \xi)(\tau - \xi)^\alpha]\, \hat{u}(\xi)\, d\xi d\tau. \qquad (2.2.19)$$

According to Lemma 2.2.3 for $|\alpha| = 0$ we have

$$|\hat{q}_2(\tau-\xi,\xi)| \leqslant C_M(1+|\tau-\xi|^2)^{-M/2}(1+|\xi|^2)^{\sigma_2/2}, \tag{2.2.20}$$

where $M = \text{const} > 0$.

According to Taylor's formula,

$$H(\tau,\xi,\eta) \equiv \hat{p}(\eta-\tau,\tau) - \sum_{|\alpha|\leqslant N-1} \frac{1}{\alpha!}\hat{p}^{(\alpha)}(\eta-\tau,\xi)(\tau-\xi)^\alpha$$
$$= \sum_{|\alpha|=N} \frac{1}{\alpha!}\hat{p}^{(\alpha)}(\eta-\tau,\xi+\theta(\tau-\xi))(\tau-\xi)^\alpha, \tag{2.2.21}$$

where $0 \leqslant \theta \leqslant 1$. Using (2.2.19) and Parseval's equation, we obtain

$$\int T''_N(u)\bar{w}\,dx = (2\pi)^{-m}\int \widehat{T''_N u}(\eta)\,\overline{\hat{w}(\eta)}\,d\eta$$
$$=(2\pi)^{-3m}\iiint \hat{q}_2(\tau-\xi,\xi)\,H(\tau,\xi,\eta)\,\hat{u}(\xi)\,\bar{\hat{w}}(\eta)\,d\tau\,d\xi\,d\eta, \tag{2.2.22}$$

where $u, w \in S$. From Lemma 2.2.3 and formula (2.2.21) we have

$$|H(\tau,\xi,\eta)| \leqslant C_1(1+|\eta-\tau|^2)^{-M_1/2}$$
$$\times(1+|\xi+\theta(\tau-\xi)|^2)^{\frac{\sigma_1-N}{2}}(1+|\tau-\xi|^2)^{N/2},$$

where $M_1 > 0$ is an arbitrary integer and C_1 depends on M_1. Applying the inequalities proved in Lemma 2.2.2, we obtain

$$|H(\tau,\xi,\eta)| \leqslant C_2(1+|\eta-\tau|^2)^{-M_1/2}(1+|\tau-\xi|^2)^{\frac{|\sigma_1-N|+N}{2}}$$
$$\times\left[(1+|\tau|^2)^{\frac{\sigma_1-N}{2}}+(1+|\xi|^2)^{\frac{\sigma_1-N}{2}}\right]. \tag{2.2.23}$$

Since according to Lemma 2.2.1

$$(1+|\tau|^2)^{\frac{\sigma_1-N}{2}} \leqslant 2^{\frac{|\sigma_1-N|}{2}}(1+|\tau-\xi|^2)^{\frac{|\sigma_1-N|}{2}}(1+|\xi|^2)^{\frac{\sigma_1-N}{2}},$$

$$(1+|\eta-\tau|^2)^{-M_1/2} \leqslant 2^{M_1/2}(1+|\eta-\xi|^2)^{-\frac{M_1}{2}}(1+|\tau-\xi|^2)^{M_1/2}$$

and since, moreover, for any s

$$(1+|\xi|^2)^{-s/2} \leqslant 2^{|s|/2}(1+|\eta|^2)^{-s/2}(1+|\eta-\xi|^2)^{|s|/2},$$

it follows from (2.2.23) that

$$|H(\tau,\xi,\eta)| \leqslant C_3(1+|\tau-\xi|^2)^{\frac{1}{2}[2|\sigma_1-N|+M_1+N]}(1+|\xi|^2)^{\frac{\sigma_1-N+s}{2}}$$
$$\times(1+|\eta-\xi|^2)^{\frac{|s|-M_1}{2}}(1+|\eta|^2)^{-s/2}. \tag{2.2.24}$$

Considering estimates (2.2.24), (2.2.20) and equation (2.2.22), we obtain

$$\left|\int T''_N(u)\bar{w}dx\right| \leqslant C_4 \iiint (1+|\xi|^2)^{\frac{1}{2}(\sigma_1+\sigma_2-N+s)} (1+|\eta|^2)^{-s/2}$$
$$\times (1+|\tau-\xi|^2)^{\frac{1}{2}(2|\sigma_1-N|+M_1-M+N)}$$
$$\times (1+|\eta-\xi|^2)^{\frac{|s|-M_1}{2}} |\hat{u}(\xi)||\hat{w}(\eta)|\,d\xi d\eta d\tau. \qquad (2.2.25)$$

Choosing M_1 so that $-M_1+|s| \leqslant -(m+1)$, we then choose M so that the inequality

$$-M+N+2|\sigma_1-N|+M_1 \leqslant -(m+1)$$

is fulfilled.

It is easy to see that after integration with respect to τ in the right side of (2.2.25) we obtain

$$\left|\int T''_N(u)\bar{w}\,dx\right| \leqslant C_5 \iint (1+|\eta-\xi|^2)^{\frac{|s|-M_1}{2}}$$
$$\times (1+|\xi|^2)^{\frac{1}{2}(\sigma_1+\sigma_2-N+s)} |\hat{u}(\xi)| (1+|\eta|^2)^{-s/2} |\hat{w}(\eta)|\,d\xi\,d\eta.$$

Applying Lemma 2.2.4 to the last integral, we find

$$\left|\int T''_N(u)\bar{w}\,dx\right| \leqslant C_6 \|w\|_{-s} \|u\|_{\sigma_1+\sigma_2-N+s}.$$

From this and Theorem 2.1.9 it follows that for all u in S,

$$\|T''_N u\|_s \leqslant C_7 \|u\|_{\sigma_1+\sigma_2-N+s},$$

which we wished to prove. In an analogous manner one may prove Theorem 2.2.2 in the case when $p_2(x,\xi) \equiv 0$ and $q_1(\xi) \equiv 0$. In this case the calculations simplify. We have

$$p(x,\xi) = p(\xi) \quad \text{and} \quad \widehat{Pu} = p(\xi)\hat{u}(\xi).$$

It is easy to see that

$$\widehat{T'_N u}(\eta) = (2\pi)^{-m} p(\eta) \int \hat{q}(\eta-\xi,\xi)\hat{u}(\xi)\,d\xi$$
$$-(2\pi)^{-m} \int \sum_{|\alpha|\leqslant N-1} \frac{1}{\alpha!}\frac{\partial^{|\alpha|}p(\xi)}{\partial\xi^\alpha}(\eta-\xi)^\alpha \hat{q}(\eta-\xi,\xi)\hat{u}(\xi)\,d\xi,$$

$$(T'_N u, w)_0 = (2\pi)^{-2m} \iint \hat{q}(\eta-\xi,\xi)$$
$$\times \Big[p(\eta) - \sum_{|\alpha|\leqslant N-1} \frac{1}{\alpha!} p^{(\alpha)}(\xi)(\eta-\xi)^\alpha\Big] \hat{u}(\xi)\overline{\hat{w}(\eta)}\,d\xi\,d\eta. \qquad (2.2.26)$$

Furthermore, integral (2.2.26) is estimated in the same way as (2.2.22). Using Taylor's formula, estimate (2.2.9), and Lemmas 2.2.1 and 2.2.2, we obtain

$$|(T'_N u, w)|_0 = \Big| (2\pi)^{-2m} \iint \hat{q}(\eta-\xi,\xi) \sum_{|\alpha|=N} \frac{1}{\alpha!} p^{(\alpha)}(\xi+\theta(\eta-\xi))$$
$$\times (\eta-\xi)^\alpha \hat{u}(\xi) \overline{\hat{w}(\eta)}\, d\xi\, d\eta \Big| \leqslant C_8 \iint (1+|\xi|^2)^{\frac{1}{2}(\sigma_1+\sigma_2+s-N)}$$
$$\times (1+|\eta|^2)^{-s/2} (1+|\eta-\xi|^2)^{-M+\frac{1}{2}(N+2|\sigma_1-N|+|s|)}$$
$$\times |\hat{u}(\xi)|\,|\hat{w}(\eta)|\, d\xi\, d\eta.$$

Choosing $M>0$ sufficiently large and applying Lemma 2.2.4, we find that $|(T'_N u, w)_0| \leqslant C_9 \|u\|_{\sigma_1+\sigma_2-N+s} \|w\|_{-s}$. This proves the theorem.

We recall the notation for commutators: $[A, B] \equiv AB - BA$ for any operators A and B.

THEOREM 2.2.3 (ON COMMUTATORS). *Let P and Q be pseudodifferential operators with symbols satisfying conditions* a) *and* b) *for $\sigma=\sigma_1$ and $\sigma=\sigma_2$ respectively. Then*

$$[P, Q]u = (PQ)_N u - (QP)_N u + T_N u, \tag{2.2.27}$$

where the operator T_N has order at most $\sigma_1+\sigma_2-N$:

$$\|T_N u\|_s \leqslant C_{s,N} \|u\|_{s+\sigma_1+\sigma_2-N} \quad \text{for } u\in S \text{ and } s\in R^1.$$

Theorem 2.2.3 follows immediately from Theorem 2.2.2.

If $N=1$, it follows from (2.2.27) that

$$\|[P, Q]u\|_s \leqslant C_s \|u\|_{s+\sigma_1+\sigma_2-1}. \tag{2.2.28}$$

By $\bar{G}$ we denote the closure of the bounded set G in R^m.

THEOREM 2.2.4 (PSEUDOLOCALNESS). *Let G_1 and G_2 be two bounded domains in R^m such that $\bar{G}_1 \cap \bar{G}_2 = \varnothing$. Let the function $\phi_1(x) \in C_0^\infty(R^m)$ be such that* supp $\phi_1 \subset G_1$, *and let u be any function in $C_0^\infty(R^m)$ such that* supp $u \subset G_2$. *Then for each $N>0$ and $s\in R^1$ there exists a constant $C(N, s, G_1, G_2)$ such that*

$$\|\varphi_1 P u\|_{N+s} \leqslant C(N, s, G_1, G_2)\|u\|_s. \tag{2.2.29}$$

PROOF. Let $\phi_2(x)$ be a function in $C_0^\infty(R^m)$ such that $\phi_2 \equiv 1$ on $\bar{G}_2$ and supp $\phi_2 \cap \bar{G}_1 = \varnothing$. Then $\phi_1 Pu = \phi_1 P\phi_2 u$. We estimate $\phi_1 P \phi_2 u$. The operator $\phi_1 P$ may be considered as a pseudodifferential operator P' with symbol $\phi_1(x)p(x,\xi)$, i.e. $\phi_1 P \phi_2 u \equiv P'\phi_2 u$. Furthermore, by Theorem 2.2.2 on the product of pseudodifferential operators we have

$$P'\varphi_2 u = P''u + T_{N_2} u,$$

where the operator P'' has the symbol

$$\sum_{|\alpha|\leqslant N_2} \frac{1}{\alpha!} p^{(\alpha)}(x, \xi)\varphi_1(x) D^{\alpha}\varphi_2(x),$$

which is identically zero, since $\operatorname{supp}\phi_1 \cap \operatorname{supp}\phi_2 = \varnothing$. If $N_2 > N + \sigma$, where σ is the order of the operator P, then T_{N_2} has order at most $-N$. Therefore $\phi_1 P\phi_2 u = T_{N_2} u$ and consequently

$$\| \varphi_1 Pu \|_{N+s} = \| \varphi_1 P\varphi_2 u \|_{N+s} \leqslant C \| u \|_s; \; u \in C_0^{\infty}(G_2),$$

which proves the theorem.

We denote by P^* the operator formally adjoint to the operator P in $\mathcal{H}_0$, i.e. for any u, v in S

$$(Pu, v)_0 = (u, P^*v)_0.$$

The following theorem is valid.

THEOREM 2.2.5 (ON THE ADJOINT OPERATOR). *Let P be a pseudodifferential operator with symbol $p(x, \xi)$ satisfying conditions* a) *and* b) *for some σ. Let $(P^*)_N$ be the pseudodifferential operator with symbol*

$$\sum_{|\alpha|\leqslant N-1} \frac{1}{\alpha!} D_x^{\alpha} \frac{\partial^{|\alpha|}}{\partial\xi^{\alpha}} \overline{p}(x, \xi). \tag{2.2.30}$$

Then for any $N \geqslant 1$

$$P^* = (P^*)_N + T_N, \tag{2.2.31}$$

where the operator T_N has order at most $\sigma - N$, i.e.

$$\| T_N u \|_s \leqslant C(N, s) \| u \|_{\sigma+s-N}; \; u \in S. \tag{2.2.32}$$

PROOF. Since $P = P_1 + P_2$, where P_1 and P_2 have symbols $p_1(\xi)$ and $p_2(x, \xi)$, respectively, and since

$$(P_1 u, v)_0 = (2\pi)^{-m} \int p_1(\xi)\hat{u}(\xi)\overline{\hat{v}}(\xi)\, d\xi$$
$$= (2\pi)^{-m} \int \hat{u}(\xi)\overline{\overline{p_1}(\xi)\hat{v}(\xi)}\, d\xi = (u, P_1^* v),$$

where P_1^* is the pseudodifferential operator with symbol $\overline{p_1(\xi)}$, it follows that it is sufficient to prove Theorem 2.2.5 for the case when $p_1 \equiv 0$. According to the definition of the norm in $\mathcal{H}_s$,

$$\| T_N u \|_s = \sup_v \frac{(T_N u, v)_0}{\| v \|_{-s}}, \quad v \in S.$$

We estimate

$$(T_N u, v)_0 = ((P^* - (P^*)_N) u, v)_0 = (u, Pv)_0 - ((P^*)_N u, v)_0.$$

Applying Parseval's equation, we obtain

$$(T_N u, v)_0 = (2\pi)^{-2m} \iint \hat{u}(\xi) \overline{\hat{\bar{p}}(\xi - \eta, \eta) \hat{v}(\eta)}\, d\xi d\eta$$
$$- (2\pi)^{-m} \int \widehat{(P^*)_N u}(\eta) \overline{\hat{v}(\eta)}\, d\eta.$$

Clearly the Fourier transform with respect to x of the symbol of the operator $(P^*)_N$ is equal to

$$\sum_{|\alpha| \leqslant N-1} \frac{1}{\alpha!} \eta^\alpha \frac{\partial^{|\alpha|}}{\partial \xi^\alpha} \hat{\bar{p}}(-\eta, \xi) = \sum_{|\alpha| \leqslant N-1} \frac{1}{\alpha!} \eta^\alpha \hat{\bar{p}}^{(\alpha)}(-\eta, \xi).$$

Therefore

$$(T_N u, v)_0 = (2\pi)^{-2m} \iint \hat{u}(\xi) \overline{\hat{v}(\eta)} [\hat{\bar{p}}(\xi - \eta, \eta)$$
$$- \sum_{|\alpha| \leqslant N-1} \frac{1}{\alpha!} \hat{\bar{p}}^{(\alpha)}(\xi - \eta, \xi)(\eta - \xi)^\alpha]\, d\xi d\eta.$$

For the estimate

$$H(\xi, \eta) = \hat{\bar{p}}(\xi - \eta, \eta) - \sum_{|\alpha| \leqslant N-1} \frac{1}{\alpha!} \hat{\bar{p}}^{(\alpha)}(\xi - \eta, \xi)(\eta - \xi)^\alpha$$

we apply Taylor's formula. According to Lemma 2.2.3 we have the inequality

$$|H(\xi, \eta)| = \left| \sum_{|\alpha| = N} \frac{1}{\alpha!} \hat{\bar{p}}^{(\alpha)}(\xi - \eta, \eta + \Theta(\xi - \eta))(\eta - \xi)^\alpha \right|$$
$$\leqslant C_1 (1 + |\eta - \xi|^2)^{-M/2} (1 + |\eta + \Theta(\xi - \eta)|^2)^{\frac{\sigma - N}{2}} (1 + |\eta - \xi|^2)^{\frac{N}{2}},$$

where $0 \leqslant \Theta \leqslant 1$. Using inequality (2.2.7) or (2.2.8) of Lemma 2.2.2 and inequality (2.2.6), we obtain

$$|H(\xi, \eta)| \leqslant C_2 (1 + |\xi|^2)^{\frac{\sigma - N}{2}} (1 + |\eta - \xi|^2)^{\frac{1}{2}(2|\sigma - N| + N - M)}$$

Therefore

$$|(T_N u, v)_0| \leqslant (2\pi)^{-2m} 2^{2|s|} \iint |\hat{u}(\xi)| |\hat{v}(\eta)|$$
$$\times (1 + |\xi|^2)^{s/2} (1 + |\eta|^2)^{-s/2} (1 + |\xi - \eta|^2)^{|s|/2} |H(\xi, \eta)|\, d\xi d\eta$$
$$\leqslant C_3 \iint |\hat{u}(\xi)| (1 + |\xi|^2)^{\frac{\sigma - N + s}{2}} |\hat{v}(\eta)| (1 + |\eta|^2)^{-\frac{s}{2}}$$
$$\times (1 + |\xi - \eta|^2)^{\frac{1}{2}(N + 2|\sigma - N| + |s| - M)}\, d\xi d\eta. \tag{2.2.33}$$

We choose M such that

$$(N + 2|\sigma - N| + |s| - M) \leqslant -(m + 1).$$

Applying Lemma 2.2.4 to the last integral in (2.2.33), we obtain

$$|(T_N u, v)_0| \leqslant C_4 \|u\|_{\sigma+s-N} \|v\|_{-s}$$

and consequently

$$\|T_N u\|_s \leqslant C_4 \|u\|_{\sigma-N+s}.$$

The theorem is proved.

It is easy to see that a linear differential operator $P(x, D)$ of order m_1, whose coefficients are infinitely differentiable and constant outside some bounded domain, is a pseudodifferential operator with symbol of the form $p(x, \xi)$, satisfying conditions a) and b) with $\sigma = m_1$.

We denote by $E_s u$ a pseudodifferential operator with symbol

$$\varphi(x)(1 + |\xi|^2)^{s/2}, \tag{2.2.34}$$

where $s \in R^1$, and the function $\phi(x)$ belongs to $C_0^\infty(R^m)$. It is clear that the symbol of the operator $E_s u$ satisfies conditions a) and b) with $\sigma = s$.

THEOREM 2.2.6. *For any bounded domain G in R^m and $s_1, s \in R^1$, there exists a constant C, depending on s, s_1, N and G, such that for any function u in $C_0^\infty(G)$ the inequality*

$$\|u\|_{s+s_1} \leqslant \|E_s u\|_{s_1} + C \|u\|_{s+s_1-N} \tag{2.2.35}$$

holds if $\phi(x) \equiv 1$ in $\bar{G}_1$, where $G_1 \subset \bar{G}$.

PROOF. Let the function $\psi(x) \in C_0^\infty(R^m)$, $\psi \equiv 1$ on $\bar{G}$ and $\phi \equiv 1$ on supp ψ. We consider the operator

$$\pi_s = K^s - E_s, \quad \text{where} \quad K^s u \equiv (2\pi)^{-m} \int (1 + |\xi|^2)^{\frac{s}{2}} \hat{u} e^{i(x,\xi)} d\xi.$$

Applying Theorem 2.2.2 on the product of pseudodifferential operators, for any function u in $C_0^\infty(G)$ we obtain

$$\pi_s u = \pi_s \psi u = \sum_{|\alpha| \leqslant N-1} \frac{1}{\alpha!} \psi_{(\alpha)} \left(K^{s(\alpha)} - E_s^{(\alpha)}\right) u + T_N u,$$

where $E_s^{(\alpha)} u$ is the pseudodifferential operator with symbol

$$\varphi(x) \frac{\partial^{|\alpha|}}{\partial \xi^\alpha} (1 + |\xi|^2)^{s/2},$$

$K^{s(\alpha)} u$ is the pseudodifferential operator with symbol $(\partial^{|\alpha|} / \partial \xi^\alpha)(1 + |\xi|^2)^{s/2}$, and the operator $T_N u$ has order at most $s - N$, i.e.

$$\|T_N u\|_{s'} \leqslant C_N \|u\|_{s'+s-N} \quad \text{for} \quad u \in S;\ s' \in R^1.$$

By $\psi_{(\alpha)}$, as always, we denote $D^\alpha \psi(x)$. Since $\phi(x) \equiv 1$ on the support of $\psi(x)$, we know that the symbol of the operator $\psi_{(\alpha)}(K^{s(\alpha)} - E_s^{(\alpha)})u$ is equal to zero for any multi-index α. Consequently $K^s u = E_s u + T_N u$ for $u \in C_0^\infty(G)$, and therefore

$$\|u\|_{s'+s} \leqslant \|E_s u\|_{s'} + C\|u\|_{s+s'-N} \quad \text{for } u \in C_0^\infty(G).$$

THEOREM 2.2.7. *Let P and Q be pseudodifferential operators whose symbols satisfy conditions* a) *and* b) *for $\sigma = 1$. Then for any $s \in R^1$ and $s_1 \leqslant 0$ there exists a constant $C(s, s_1)$ such that*

$$\|[P, Q]u\|^2_{\frac{s_1-1}{2}+s} \leqslant C(s, s_1)\{\|Pu\|_s^2 + \|P^*u\|_s^2 + \|Qu\|^2_{s+s_1} + \|Q^*u\|^2_{s+s_1} + \|u\|_s^2\} \tag{2.2.36}$$

for any function u in S.

PROOF. Let $2t = 2s + s_1 - 1$. The identity

$$\|[P, Q]u\|_t^2 = ((PQ - QP)u, K^{2t}[P, Q]u)_0 \tag{2.2.37}$$

is evident. According to (2.1.19) and the definition of adjoint operator,

$$\begin{aligned} &((PQ - QP)u, K^{2t}[P, Q]u)_0 \\ &\leqslant |(Qu, P^*K^{2t}[P, Q]u)_0| + |(Pu, Q^*K^{2t}[P, Q]u)_0| \\ &\leqslant \|Qu\|^2_{s+s_1} + \|P^*K^{2t}[P, Q]u\|^2_{-s-s_1} + \|Pu\|_s^2 + \|Q^*K^{2t}[P, Q]u\|^2_{-s}. \end{aligned} \tag{2.2.38}$$

By virtue of Theorems 2.2.1 and 2.2.3 we obtain the estimate

$$\begin{aligned} \|Q^*K^{2t}[P, Q]u\|^2_{-s} &\leqslant 2\|K^{2t}[P, Q]Q^*u\|^2_{-s} + 2\|[Q^*, K^{2t}[P, Q]]u\|^2_{-s} \\ &\leqslant C_1\{\|Q^*u\|^2_{s+s_1} + \|[(Q^*)_N, K^{2t}[P, Q]]u\|^2_{-s} + \|u\|^2_{s+s_1}\} \\ &\leqslant C_2\{\|Q^*u\|^2_{s+s_1} + \|u\|_s^2\}, \end{aligned} \tag{2.2.39}$$

since $s_1 \leqslant 0$. In exactly the same way we obtain

$$\|P^*K^{2t}[P, Q]u\|^2_{-s-s_1} \leqslant 2\|K^{2t}[P, Q]P^*u\|^2_{-s-s_1} + 2\|[P^*, K^{2t}[P, Q]]u\|^2_{-s-s_1} \leqslant C_3\{\|P^*u\|_s^2 + \|u\|_s^2\}. \tag{2.2.40}$$

From (2.2.37)–(2.2.40) follows (2.2.36).

The following theorem is a corollary of Theorems 2.2.7 and 2.2.5.

THEOREM 2.2.8. *Let P and Q be pseudodifferential operators with real symbols satisfying conditions* a) *and* b) *with $\sigma = 1$. Then for any $s_1 \leqslant 0$ and any $s \in R^1$ there exists a constant $C(s, s_1)$ such that*

$$|||[P, Q]\, u|||^2_{s+\frac{s_1-1}{2}} \leqslant C(s, s_1)\, \{|||Pu|||^2_s + |||Qu|||^2_{s+s_1} + |||u|||^2_s\}$$

for any function u *in* S.

THEOREM 2.2.9. *Let* $p_j(x, \xi), j = 1, \cdots, N_1$, *be symbols of pseudodifferential operators* P_j *satisfying conditions* a) *and* b) *for* $\sigma = 1$, *and let*

$$\sum_{j=1}^{N_1} |p_j(x, \xi)|^2 + 1 \geqslant c_0(1 + |\xi|^2), \quad c_0 = \text{const} > 0, \tag{2.2.41}$$

for all x *belonging to a closed set* $\bar{G}_1$. *Then for any function* u *in* $C_0^\infty(G)$, *where* $\bar{G} \subset G_1$,

$$\|u\|^2_{1+s} \leqslant C_s \left\{ \sum_{j=1}^{N_1} \|P_j u\|^2_s + \|u\|^2_s \right\}; \quad s \in R^1. \tag{2.2.42}$$

PROOF. We consider the pseudodifferential operator P_0 with symbol

$$p_0(x, \xi) = \psi(x) \left(\sum_{j=1}^{N_1} |p_j(x, \xi)|^2 + 1 \right)^{1/2},$$

where $\psi(x) \in C_0^\infty(G_1)$ and $\psi \equiv 1$ on $\bar{G}$. We denote by P_0^{-1} the pseudodifferential operator with symbol

$$\psi(x) \left(\sum_{j=1}^{N_1} |p_j(x, \xi)|^2 + 1 \right)^{-1/2}.$$

Then, according to Theorem 2.2.2,

$$P_0^{-1} P_0 u = \psi^2(x)\, u + Tu, \tag{2.2.43}$$

where Tu is an operator of order at most -1.

Since $\psi(x) \equiv 1$ on $\bar{G}$ and $u \in C_0^\infty(G)$, it follows from (2.2.43) that

$$\|u\|^2_{1+s} \leqslant 2\|P_0^{-1} P_0 u\|^2_{1+s} + 2\|Tu\|^2_{1+s} \leqslant C_1 \{\|P_0 u\|^2_s + \|u\|^2_s\}. \tag{2.2.44}$$

It is easy to see that

$$\|P_0 u\|^2_s \leqslant C_2 \{\Sigma \|P_j u\|^2_s + \|u\|^2_s\}. \tag{2.2.45}$$

In fact,

$$p_0(x, \xi) = \psi(x) \left\{ \sum_{j=1}^{N_1} \frac{\bar{p}_j}{\sqrt{1 + \sum_k |p_k|^2}}\, p_j + \frac{1}{\sqrt{1 + \sum_k |p_k|^2}} \right\}.$$

Since the symbol

$$\psi(x)\, \bar{p}_j \left(1 + \sum_{k=1}^{N_1} |p_k|^2 \right)^{-1/2}$$

corresponds to a pseudodifferential operator of order at most zero and P_0^{-1} has order at most -1, we have from Theorem 2.2.2 that

$$\|P_0 u\|_s^2 \leqslant C_3 \left\{ \sum_{j=1}^{N_1} \|P_j u\|_s^2 + \|T_1 u\|_s^2 + \|P_0^{-1} u\|_s^2 \right\},$$

where $T_1 u$ is an operator of order at most zero. Therefore inequality (2.2.45) is valid. The assertion of the theorem follows from (2.2.44) and (2.2.45).

§3. A necessary condition for hypoellipticity

A linear differential operator P with infinitely differentiable coefficients defined in a domain $\Omega \subset R^m$ is called hypoelliptic in Ω if for any distribution u in $D'(\Omega)$ and any domain $\Omega_1 \subset \Omega$ the condition that $Pu \in C^\infty(\Omega_1)$ implies that u is infinitely differentiable in Ω_1. The concept of hypoellipticity for differential operators was introduced in the book by Schwartz [120]. Hörmander studied hypoelliptic equations with constant coefficients (see [50, 51]) and hypoelliptic equations with variable coefficients [52, 53, 55]. Hypoelliptic equations were also the subject of the papers [21, 76, 114, 122, 130, 138].

The following theorem yields a necessary condition for hypoellipticity for a linear equation of second order. An analogous theorem is valid also for a differential operator of arbitrary order (see [55]).

THEOREM 2.3.1. *If the second order operator*

$$L(u) \equiv a^{kj} u_{x_k x_j} + b^k u_{x_k} + cu \tag{2.3.1}$$

with real coefficients $a^{kj}(x)$, $b^k(x)$, $c(x)$ in the class $C^\infty(\Omega)$ is hypoelliptic in the domain Ω, then for any point x in Ω

$$\text{either } a^{kj}(x)\xi_k\xi_j \geqslant 0, \quad \text{or} \quad a^{kj}(x)\xi_k\xi_j \leqslant 0 \tag{2.3.2}$$

for all $\xi \in R^m$.

PROOF. We denote $a^{kj}(x)\xi_k\xi_j$ by $a(x, \xi)$. We assume that at some point $x_0 \in \Omega$ condition (2.3.2) is not fulfilled. This means that there exist vectors ξ' and ξ'' such that $a(x_0, \xi') > 0$ and $a(x_0, \xi'') < 0$. Then it is easy to see that there exists a vector $\xi^0 \neq 0$ such that

$$a(x_0, \xi^0) = 0, \text{ and } \operatorname{grad}_\xi a(x_0, \xi^0) \neq 0. \tag{2.3.3}$$

Let Ω_1 be a neighborhood of the point x_0 such that $\overline{\Omega}_1 \subset \Omega$, and let $\{K_j, j = 1, \cdots\}$ be a system of closed regions such that $K_j \subset K_{j+1}$, $\Sigma_j K_j = \Omega_1$. We denote by M the linear space of functions u defined in Ω_1 and such that

$$u \in C^0(\Omega_1), \quad Lu \in C^\infty(\Omega_1).$$

In M we introduce a countable set of seminorms (see [144])

$$p_{j,l}(u) = \sup_{K_l} |u| + \sum_{|\alpha| \leq j} \sup_{K_l} |D^\alpha L u|.$$

Since the operator $L(u)$ is by assumption hypoelliptic, each function u in M is infinitely differentiable in Ω_1. Let $T(u)$ be the operator of imbedding of the space M in the space $C^\infty(\Omega_1)$, i.e. each function u in M corresponds with the same function, considered now in $C^\infty(\Omega_1)$, in which space we introduce the system of seminorms

$$q_{j,l}(u) = \sum_{|\alpha| \leq j} \sup_{K_l} |D^\alpha u|.$$

It is easy to see that the graph of the operator T is closed. Therefore by a known theorem for F-spaces (see [144], Chapter II, §6), the operator T is continuous. This means that for any seminorm $q_{j,l}$ of the space $C^\infty(\Omega_1)$ there is a seminorm $p_{t,s}$ and a constant C such that

$$q_{j,l}(Tu) \leq C p_{t,s}(u) \tag{2.3.4}$$

for any function u in M.

In particular, it follows from (2.3.4) that any function u in M satisfies the estimate

$$|\operatorname{grad} u(x_0)| \leq C \left\{ \sup_{K_\rho} |u| + \sum_{|\alpha| \leq N} \sup_{K_\rho} |D^\alpha L(u)| \right\}, \tag{2.3.5}$$

where N is some positive integer, and K_ρ is an element of the system $\{K_j\}$, $C =$ const.

We show that (2.3.5) is impossible if there exists a vector satisfying (2.3.3). For this we consider the function

$$v = \sum_{j=0}^{N} u_j(x) t^{-j} e^{it\phi}, \quad t = \text{const}, \tag{2.3.6}$$

defined in $\bar{\Omega}_1$. For the function ϕ, defined in Ω_1, we take a solution of the equation $a^{kj}\phi_{x_k}\phi_{x_j} = 0$ such that $\operatorname{grad} \phi(x_0) = \xi^0$. Since condition (2.3.3) is fulfilled for the vector ξ^0, it is known (see [101]) that such a function $\phi(x)$ exists and belongs to the class $C^\infty(\bar{\Omega}_1)$ if the neighborhood Ω_1 of the point x_0 is sufficiently small. The infinitely differentiable functions u_j ($j = 0, \cdots, N$) are defined from the condition

$$L(v) = O(t^{-N}), \quad D^\alpha L(v) = O(t^{-N+|\alpha|}) \quad \text{as } t \to \infty. \tag{2.3.7}$$

Then, substituting v into inequality (2.3.5), we obtain that the right side of this inequality is bounded for all sufficiently large t, whereas the left side grows unboundedly as $t \to \infty$. This contradiction proves the theorem.

Thus it only remains to show that functions u_j with the required properties exist. For this we examine $L(v)$. We have

$$L(v) = e^{it\varphi} \sum_{j=0}^{N} t^{-j} \Bigg[- t^2 u_j a(x, \operatorname{grad} \varphi) + t\left(- \sum_{k=1}^{m} a^{(k)}(x, \operatorname{grad} \varphi) D_k u_j + A_j u_j \right) + B_j \Bigg],$$

where the A_j do not depend on u_j, the B_j depend on u_j and their derivatives, and $a^{(k)}(x, \xi) = \partial a(x, \xi)/\partial \xi_k$.

We define u_0 as a solution of the equation

$$\sum_{k=1}^{m} a^{(k)}(x, \operatorname{grad} \varphi) D_k u_0 - A_0 u_0 = 0,$$

such that $u_0(x_0) = 1$. Since $\operatorname{grad}_\xi a(x_0, \xi^0) \neq 0$ for $\xi^0 = \operatorname{grad} \phi(x_0)$, it follows that the function u_0 exists if the neighborhood Ω_1 of the point x_0 is sufficiently small. For the function u_j we take a solution of the equation

$$- \sum_{k=1}^{m} a^{(k)}(x, \operatorname{grad} \varphi) D_k u_j + A_j u_j + B_{j-1} = 0.$$

This equation clearly also has an infinitely differentiable solution in the neighborhood of the point x_0. Thus

$$L(v) = e^{it\varphi} t^{-N} B_N,$$

where the functions B_N do not depend on t, and therefore $L(v)$ satisfies the required condition (2.3.7). The theorem is proved.

An analogous theorem (see [55]) also holds in the case of a differential operator of arbitrary order m_1. Consider a differential operator $P(x, D)$ of order m_1 with infinitely differentiable coefficients in the domain Ω. We denote by $p^0(x, \xi)$ the principal part of the symbol of the operator $P(x, D)$, i.e. the sum of all terms of order m_1 of the polynomial $p(x, \xi)$. Suppose the principal part $p^0(x, \xi)$ of the symbol of the operator $P(x, D)$ is real for $x \in \Omega$ and $\xi \in R^m$. If at some point $x_0 \in \Omega$ there exists a real vector $\xi^0 \neq 0$ such that $p^0(x_0, \xi^0) = 0$ but $\partial p^0(x_0, \xi^0)/\partial \xi_j \neq 0$ for some j, then the operator $P(x, D)$ cannot be hypoelliptic in the domain Ω.

§4. Sufficient conditions for local smoothness of weak solutions and hypoellipticity of differential operators

In this section we show that the satisfaction of certain a priori estimates for the operator $P(x, D)$ applied to functions of class $C_0^\infty(\Omega)$ is a sufficient condition for hypoellipticity of the operator P in the domain Ω. It is also a condition for the smoothness of weak solutions, if the right side of the equation has a definite degree of smoothness.

We first prove certain auxiliary propositions.

Let $\chi(x) \in C_0^\infty(R^m)$ and let $k \geqslant 0$. We assume that $\chi(x)$ is such that

$$\hat{\chi}(\xi) = O(|\xi|^k) \quad \text{as} \quad \xi \to 0, \tag{2.4.1}$$

and

$$\text{if } \hat{\chi}(t\xi) = 0 \text{ for some } \xi \in R^m \text{ for all real } t, \text{ then } \xi = 0 \tag{2.4.2}$$

We define $\chi_\epsilon(x) = \epsilon^{-m}\chi(x/\epsilon)$. It is easy to see that $\hat{\chi}_\epsilon(\xi) = \hat{\chi}(\epsilon\xi)$.

We introduce the following notation for polynomials and the differential operators corresponding to them: for any multi-indices α and β,

$$p^{(\alpha)}(x, \xi) = \frac{\partial^{|\alpha|} p(x, \xi)}{\partial \xi^\alpha}, \quad p_{(\alpha)}(x, \xi) = D_x^\alpha p(x, \xi),$$

$$p_{(\beta)}^{(\alpha)}(x, \xi) = \frac{\partial^{|\alpha|} D_x^\beta p(x, \xi)}{\partial \xi^\alpha},$$

$$p^{(j)}(x, \xi) = \frac{\partial}{\partial \xi_j} p(x, \xi), \quad p_{(j)}(x, \xi) = D_j p(x, \xi) \quad \text{for } j = 1, \cdots, m.$$

The operators $P^{(\alpha)}, P_{(\alpha)}, P_{(\beta)}^{(\alpha)}, P^{(j)}, P_{(j)}$ are obtained from the corresponding polynomials by replacing the vector ξ by the vector $(D_1, \cdots, D_m)$.

We consider the following two-parameter family of norms:

$$\|u\|_{s,\gamma}^2 = (2\pi)^{-m} \int |\hat{u}(\xi)|^2 (1 + |\xi|^2)^s (1 + |\delta\xi|^2)^{-\gamma} d\xi,$$

where $\gamma > 0$, $s \in R^1$, $\delta = \text{const} > 0$. It is clear that this norm is equivalent to the norm in the space $\mathcal{H}_{s-\gamma}$. It is also seen that

$$\|u\|_{s,\gamma} \nearrow \|u\|_s \quad \text{as} \quad \delta \searrow 0, \quad \text{if} \quad u \in \mathcal{H}_s;$$

in fact,

$$\|u\|_s = \sup_\delta \|u\|_{s,\gamma}. \tag{2.4.3}$$

THEOREM 2.4.1 (SEE [56]). *If the function* χ *satisfies condition* (2.4.1), (2.4.2) *for* $k > s$, *then for each* $N > \gamma$ *there exist positive constants* C_1, C_2 *and*

C_N, independent of u and δ, such that for $0<\delta\leqslant 1$ the inequality

$$C_1\|u\|^2_{s_1+s,\gamma}\leqslant\int_0^1\|u*\chi_\varepsilon\|^2_{s_1}\left(1+\frac{\delta^2}{\varepsilon^2}\right)^{-\gamma}\varepsilon^{-2s}\frac{d\varepsilon}{\varepsilon}+C_N\|u\|^2_{s+s_1-N}\leqslant C_2\|u\|^2_{s+s_1,\gamma} \tag{2.4.4}$$

holds for any function u in $\mathcal{H}_{s+s_1-\gamma}$.

Moreover, for any function u in $\mathcal{H}_{s+s_1}$ the inequality

$$C_1\|u\|^2_{s+s_1}\leqslant\int_0^1\|u*\chi_\varepsilon\|^2_{s_1}\varepsilon^{-2s}\frac{d\varepsilon}{\varepsilon}+C_N\|u\|^2_{s+s_1-N}\leqslant C_2\|u\|^2_{s+s_1} \tag{2.4.5}$$

holds.

Proof. According to the definition of the norm $\mathcal{H}_{s_1}$ we have

$$\int_0^1\|u*\chi_\varepsilon\|^2_{s_1}\left(1+\frac{\delta^2}{\varepsilon^2}\right)^{-\gamma}\varepsilon^{-2s}\frac{d\varepsilon}{\varepsilon}=(2\pi)^{-m}\int|\hat{u}(\xi)|^2F(\xi,\delta)\,d\xi,$$

where

$$F(\xi,\delta)=\int_0^1(1+|\xi|^2)^{s_1}|\hat{\chi}(\varepsilon\xi)|^2\varepsilon^{-2s}\left(1+\frac{\delta^2}{\varepsilon^2}\right)^{-\gamma}\frac{d\varepsilon}{\varepsilon}. \tag{2.4.6}$$

Here we have used the formula for the Fourier transform of a convolution:

$$\widehat{\varphi*\psi}=\hat{\varphi}\cdot\hat{\psi}.$$

Suppose $|\xi|\geqslant 1$. In the integral (2.4.6) we introduce a new variable of integration

$$t=\epsilon|\xi|. \tag{2.4.7}$$

We thus obtain

$$\begin{aligned}F(\xi,\delta)&=(1+|\xi|^2)^{s_1}|\xi|^{2s}\int_0^{|\xi|}t^{-2s}\left|\hat{\chi}\left(t\frac{\xi}{|\xi|}\right)\right|^2\left(1+\frac{\delta^2|\xi|^2}{t^2}\right)^{-\gamma}\frac{dt}{t}\\&\leqslant(1+|\xi|^2)^{s_1}|\xi|^{2s}(1+|\delta\xi|^2)^{-\gamma}\left\{\int_0^1\left|\hat{\chi}\left(t\frac{\xi}{|\xi|}\right)\right|^2t^{-2s}\frac{dt}{t}+\int_1^\infty\left|\hat{\chi}\left(t\frac{\xi}{|\xi|}\right)\right|^2t^{2(\gamma-s)}\frac{dt}{t}\right\}\\&\leqslant C_3(1+|\xi|^2)^{s+s_1}(1+|\delta\xi|^2)^{-\gamma},\end{aligned} \tag{2.4.8}$$

since $\hat{\chi}$ lies in class S and since, according to (2.4.1),

$$\left|\hat{\chi}\left(t\frac{\xi}{|\xi|}\right)\right|\leqslant C_4t^k$$

for small t, where $k>s$. For $|\xi|\leqslant 1$ the integral (2.4.6) is bounded uniformly in ξ and δ by virtue of condition (2.4.1); therefore we may be sure that there exists a constant C_5 such that for $|\xi| \leqslant 1$

$$F(\xi, \delta) \leqslant C_5 (1 + |\xi|^2)^{s+s_1}(1 + |\delta\xi|^2)^{-\gamma}. \tag{2.4.9}$$

It follows from (2.4.8) and (2.4.9) that

$$\int_0^1 \|u * \chi_\varepsilon\|_{s_1}^2 \varepsilon^{-2s} \left(1 + \frac{\delta^2}{\varepsilon^2}\right)^{-\gamma} \frac{d\varepsilon}{\varepsilon} \leqslant C_6 \|u\|_{s+s_1\gamma}^2.$$

In order to estimate the integral (2.4.6) from below for $|\xi| \geqslant 1$ we again introduce the new variable $t = \epsilon|\xi|$. We find

$$F(\xi, \delta) \geqslant (1 + |\xi|^2)^{s_1} |\xi|^{2s} 4^{-\gamma} (1 + |\delta\xi|^2)^{-\gamma} \int_{1/2}^1 \left|\hat{\chi}\left(t \frac{\xi}{|\xi|}\right)\right|^2 t^{-2s} \frac{dt}{t}$$
$$\geqslant C_7' (1 + |\xi|^2)^{s+s_1}(1 + |\delta\xi|^2)^{-\gamma}.$$

This last inequality follows from the fact that $\hat{\chi}(t\xi/|\xi|)$ is an analytic function of t and the integral

$$\int_{1/2}^1 \left|\hat{\chi}\left(t \frac{\xi}{|\xi|}\right)\right|^2 t^{-2s} \frac{dt}{t},$$

by virtue of (2.4.2), is a continuous function of $\eta = \xi/|\xi|$, positive for all points η on the unit sphere.

Furthermore, for $|\xi| \leqslant 1$ and any $N > 0$,

$$F(\xi, \delta) + (1 + |\xi|^2)^{s+s_1-N} \geqslant 2^{-N} (1 + |\xi|^2)^{s+s_1} (1 + |\delta\xi|^2)^{-\gamma},$$

since $F(\xi, \delta) \geqslant 0$. From this follows the first inequality of (2.4.4).

Inequality (2.4.5) follows from (2.4.4) if in these inequalities we let δ tend to zero and use relation (2.4.3).

The following lemma (see [53]) is similar to Theorem 2.2.2 and is proved by analogous methods.

LEMMA 2.4.1. *Let $P(x, D)$ be a differential operator of order m_1 with coefficients in the space $C_0^\infty(R^m)$. Let $\chi_\epsilon^\alpha \equiv \epsilon^{-|\alpha|} x^\alpha \chi_\epsilon(x)$ and let N be a positive integer. Then for any function u in $\mathcal{H}_{s+s_1+m_1-N+1/2}$*

$$\int_0^1 \Big\|P(x, D)(u * \chi_\varepsilon) - \sum_{|\alpha| \leqslant N-1} \frac{i^{|\alpha|}\varepsilon^{|\alpha|}}{\alpha!} (P_{(\alpha)}(x, D)\, u) * \chi_\varepsilon^\alpha \Big\|_{s_1}^2 \varepsilon^{-2s} \frac{d\varepsilon}{\varepsilon}$$
$$\leqslant C_N \|u\|^2_{s+s_1+m_1-N+\frac{1}{2}}, \tag{2.4.10}$$

if k *in condition* (2.4.1) *is sufficiently large and* $s \leqslant N - ½$.

PROOF. Let

$$v(x, \varepsilon) = P(x, D)(u * \chi_\varepsilon) - \sum_{|\alpha| \leqslant N-1} \frac{i^{|\alpha|}\varepsilon^{|\alpha|}}{\alpha!} (P_{(\alpha)}(x, D)\, u) * \chi^\alpha_{\cdot \varepsilon}.$$

We compute the Fourier transform of the function v with respect to x. We have

$$\hat{v}(\eta, \varepsilon) = (2\pi)^{-m} \int [\hat{p}(\eta - \xi, \xi)\, \hat{u}(\xi)\, \hat{\chi}(\varepsilon\xi)$$

$$- \sum_{|\alpha| \leqslant N-1} \frac{1}{\alpha!} \hat{p}(\eta - \xi, \xi)(\xi - \eta)^\alpha\, \hat{u}(\xi)\, \varepsilon^{|\alpha|} \hat{\chi}^{(\alpha)}(\varepsilon\eta)]\, d\xi.$$

Using Taylor's formula, we obtain

$$\hat{v}(\eta, \varepsilon) = (2\pi)^{-m} \int \hat{p}(\eta - \xi, \xi)\, \hat{u}(\xi)\, R_N(\varepsilon, \xi, \eta)\, d\xi,$$

where

$$R_N(\varepsilon, \xi, \eta) = \sum_{|\alpha| = N} \frac{1}{\alpha!} \varepsilon^{|\alpha|} (\xi - \eta)^\alpha \hat{\chi}^{(\alpha)}(\varepsilon\eta + \varepsilon\theta(\xi - \eta)),$$

and $\hat{\chi}^{(\alpha)}$, as usual, denotes the derivative $\partial^{|\alpha|}\hat{\chi}(\xi)/\partial\xi^\alpha$.

According to Theorem 2.1.9,

$$\|v\|_{s_1} = \frac{\sup\, (v, w)_0}{\|w\|_{-s_1}}.$$

We therefore estimate $(v, w)_0$. By the Parseval equation,

$$(v, w)_0 = (2\pi)^{-2m} \iint \hat{p}(\eta - \xi, \xi)\, \hat{u}(\xi)\, R_N(\varepsilon, \xi, \eta)\, \overline{\hat{w}(\eta)}\, d\xi\, d\eta$$

$$= (2\pi)^{-2m} \iint \hat{u}(\xi)(1 + |\xi|^2)^{\frac{1}{2}\left(s + s_1 + m_1 - N + \frac{1}{2}\right)} \overline{\hat{w}(\eta)}(1 + |\eta|^2)^{-\frac{s_1}{2}}$$

$$\times H_\varepsilon(\xi, \eta)\, d\xi\, d\eta, \tag{2.4.11}$$

where

$$H_\varepsilon(\xi, \eta) = \hat{p}(\eta - \xi, \xi)(1 + |\eta|^2)^{\frac{s_1}{2}} (1 + |\xi|^2)^{-\frac{1}{2}\left(s + s_1 + m_1 - N + \frac{1}{2}\right)} R_N(\varepsilon, \xi, \eta).$$

It is easy to see that for any M

$$|R_N(\varepsilon, \xi, \eta)| \leqslant C_1 \varepsilon^N (1 + |\xi - \eta|^2)^{\frac{N}{2}} (1 + |\varepsilon\eta|^2)^{-M}, \tag{2.4.12}$$

if $|\xi - \eta| \leqslant |\eta|/2$, and

$$|R_N(\varepsilon, \xi, \eta)| \leqslant C_2 \varepsilon^N (1 + |\xi - \eta|^2)^{N/2} \tag{2.4.13}$$

for any ξ and η.

Since the coefficients of the operator $P(x, D)$ belong to the class $C_0^\infty(R^m)$, it follows from Lemma 2.2.3 that

$$|\hat{p}(\eta-\xi,\xi)| \leqslant C_3(1+|\xi|^2)^{\frac{m_1}{2}}(1+|\eta-\xi|^2)^{-N_1/2},$$

where N_1 is an arbitrary positive number. Let $|\xi-\eta| \leqslant |\eta|/2$. Since $s-N+\frac{1}{2} \leqslant 0$, according to (2.4.12) we have

$$|R_N(\varepsilon,\xi,\eta)| \leqslant C_4\varepsilon^N(1+|\xi-\eta|^2)^{N/2}(\varepsilon^2+|\varepsilon\eta|^2)^{\frac{s-N+\frac{1}{2}}{2}},$$

$$|H_\varepsilon(\xi,\eta)| \leqslant C_5(1+|\xi|^2)^{-\frac{1}{2}\left(s+s_1-N+\frac{1}{2}\right)}(1+|\eta|^2)^{\frac{1}{2}\left(s_1+s-N+\frac{1}{2}\right)}$$
$$\times\,\varepsilon^{\frac{2s+1}{2}}(1+|\eta-\xi|^2)^{\frac{N-N_1}{2}}.$$

Applying Lemma 2.2.1, we find that for $|\xi-\eta| \leqslant |\eta|/2$

$$|H_\varepsilon(\xi,\eta)| \leqslant C_6\varepsilon^{s+\frac{1}{2}}(1+|\eta-\xi|^2)^{\frac{1}{2}\left(N-N_1+|s_1|+\left|s-N+\frac{1}{2}\right|\right)}.$$

If $|\xi-\eta| > |\eta|/2$, then, using the fact that

$$(1+|\eta|^2) \leqslant (1+4|\xi-\eta|^2) \leqslant 4(1+|\xi-\eta|^2),$$

as well as inequality (2.4.13) and the inequality proved in Lemma 2.2.1, we have

$$|H_\varepsilon(\xi,\eta)| \leqslant C_7(1+|\eta-\xi|^2)^{\frac{N-N_1}{2}}(1+|\xi|^2)^{-\frac{1}{2}\left(s+s_1-N+\frac{1}{2}\right)}\varepsilon^N(1+|\eta|^2)^{\frac{s_1}{2}}$$
$$\leqslant C_8(1+|\eta-\xi|^2)^{\frac{1}{2}\left(N-N_1+\left|s+s_1-N+\frac{1}{2}\right|+N-s-\frac{1}{2}\right)}\varepsilon^{s+\frac{1}{2}},$$

since $\frac{1}{2}+s<N$ and $0 \leqslant \epsilon \leqslant 1$.

We choose N_1 so large that

$$-N_1+N+|s_1|+\left|s-N+\tfrac{1}{2}\right| \leqslant -(m+1)$$

and

$$2N-N_1+\left|s+s_1-N+\tfrac{1}{2}\right|-s \leqslant -m-\tfrac{1}{2}.$$

Then we may apply Lemma 2.2.4 to the integral (2.4.11). We have

$$|(v,w)_0| \leqslant C_9\|w\|_{-s_1}\|u\|_{s+s_1+m_1-N+\frac{1}{2}}\cdot\varepsilon^{s+\frac{1}{2}}.$$

From this it follows that

$$\|v\|^2_{s_1} \leqslant C_9^2\|u\|^2_{s+s_1+m_1-N+\frac{1}{2}}\cdot\varepsilon^{2s+1}. \tag{2.4.14}$$

Multiplying (2.4.14) by $\epsilon^{-(2s+1)}$ and integrating with respect to ϵ from 0 to 1, we obtain the required inequality (2.4.10).

LEMMA 2.4.2. *Let $P(x, D)$ be a linear differential operator of order m_1 with* *coefficients in the space* $C_0^\infty(R^m)$. *Then for any integer* $N>0$ *and any function* u *in* $\mathcal{H}_{s+s_1+m_1-\gamma}$ *the inequality*

$$\sum_{1\leqslant|\alpha|\leqslant N-1}\int_0^1 \|(P_{(\alpha)}u)*\chi_\varepsilon^\alpha\|_{s_1}^2\varepsilon^{-2(s-|\alpha|)}\left(1+\frac{\delta^2}{\varepsilon^2}\right)^{-\gamma}\frac{d\varepsilon}{\varepsilon}$$

$$\leqslant \sum_{1\leqslant|\alpha|\leqslant N-1} C_\alpha\int_0^1 \|(P_{(\alpha)}u)*\chi_\varepsilon\|_{s_1-|\alpha|}^2\varepsilon^{-2s}\left(1+\frac{\delta^2}{\varepsilon^2}\right)^{-\gamma}\frac{d\varepsilon}{\varepsilon}$$

$$+C_{N_1}\|u\|_{-N_1}^2 \tag{2.4.15}$$

holds, where $s, s_1\in R^1$ *and* $N_1>0$ *is any sufficiently large number*

PROOF. According to (2.4.4) we have

$$\sum_{1\leqslant|\alpha|\leqslant N-1}\int_0^1 \|(P_{(\alpha)}u)*\chi_\varepsilon^\alpha\|_{s_1}^2\varepsilon^{-2(s-|\alpha|)}\left(1+\frac{\delta^2}{\varepsilon^2}\right)^{-\gamma}\frac{d\varepsilon}{\varepsilon}$$

$$\leqslant C_1\sum_{1\leqslant|\alpha|\leqslant N-1}\|P_{(\alpha)}u\|_{s_1+s-|\alpha|,\gamma}^2. \tag{2.4.16}$$

Here we have applied Theorem 2.4.1 to the function χ_ε^α instead of χ_ε; this is possible since the Fourier transform of the function χ^α is equal to $i^{|\alpha|}\partial^{|\alpha|}\hat{\chi}/\partial\xi^\alpha$ and therefore it also satisfies properties (2.4.1) and (2.4.2) for some k. Again applying Theorem 2.4.1, we obtain

$$\|P_{(\alpha)}u\|_{s_1+s-|\alpha|,\gamma}^2\leqslant C_2\int_0^1\|(P_{(\alpha)}u)*\chi_\varepsilon\|_{s_1-|\alpha|}^2\varepsilon^{-2s}\left(1+\frac{\delta^2}{\varepsilon^2}\right)^{-\gamma}\frac{d\varepsilon}{\varepsilon}$$

$$+C_3\|P_{(\alpha)}u\|_{s_1+s-|\alpha|-N_0}^2. \tag{2.4.17}$$

Since $N_0>0$ is an arbitrary number, $\|P_{(\alpha)}u\|_{s+s_1-|\alpha|-N_0}\leqslant C_4\|u\|_{-N_1}$ and (2.4.15) follows from (2.4.16) and (2.4.17). The lemma is proved.

We now establish a basic theorem on the smoothness of generalized solutions and hypoellipticity of the operator $P(x, D)$ in the domain Ω.

Here K will denote any closed bounded set in R^m.

THEOREM 2.4.2. *Let* P *be a linear differential operator of order* m_1 *with coefficients in the class* $C_0^\infty(R^m)$ *for which the following conditions are fulfilled:*

I. *For each compact set* $K\subset\Omega$ *there exists a constant* $s_0=s_0(K)$ *such that for sufficiently large* $N>0$ *the inequality*

$$\|u\|_{s_0}^2\leqslant C(K,N)\|u\|_{-N}^2+C(K)\|Pu\|_0^2 \tag{2.4.18}$$

holds, where $-N<s_0$ *and* u *is any function in* $C_0^\infty(K)$.

II. *For each compact set* $K\subset\Omega$, *each* $s\in R^1$, *and each* $\delta_1>0$, *there exists a constant* $C(K, s, \delta_1, N)$ *such that for sufficiently large* $N>0$ *the inequality*

$$\sum_{j=1}^{m} \| P_{(j)}u \|_s^2 \leqslant \delta_1 \| Pu \|_{s+1}^2 + C(K, s, \delta_1, N) \| u \|_{-N}^2 \tag{2.4.19}$$

holds, where $-N < s + s_0$, $u \in C_0^\infty(K)$.

IIIa. *For each compact set* $K \subset \Omega \backslash M$, *where* M *is some bounded closed set in* Ω, *for each* $s \in R^1$, *and for each sufficiently large* N, *the inequality*

$$\sum_{j=1}^{m} \| P^{(j)}u \|_s^2 \leqslant C(K, s) \{ \| Pu \|_{s-\mu}^2 + C(N) \| u \|_{-N}^2 \} \tag{2.4.20}$$

holds, where $\mu = \mu(K) > 0$, $-N < s + s_0$, $u \in C_0^\infty(K)$.

IIIb. *For each compact set* $K \subset \Omega$, *for each* $s \in R^1$, *for each* $\delta_1 > 0$, *and for each sufficiently large* $N > 0$, *the inequality*

$$\sum_{j=1}^{m} \| P^{(j)}u \|_s^2 \leqslant \delta_1 \| Pu \|_s^2 + C(\delta_1, N, s, K) \| u \|_{-N}^2 \tag{2.4.21}$$

holds, where $-N < s + s_0$, $u \in C_0^\infty(K)$.

Then for any distribution u *in* $D'(\Omega)$ *such that* $Pu \in \mathcal{H}_s^{\text{loc}}(\Omega)$ *we have the estimates*

$$\| \varphi u \|_{s+s_0}^2 \leqslant C(\varphi_1) \{ \| \varphi_1 Pu \|_s^2 + \| \varphi_1 u \|_{\gamma_0}^2 \}, \tag{2.4.22}$$

where the functions $\phi, \phi_1 \in C_0^\infty(\Omega)$, $\phi_1 \equiv 1$ *on* $\operatorname{supp} \phi$ *and, moreover, either* $\phi \equiv 1$ *on* M *or* $\operatorname{supp} \phi \cap M = \varnothing$; $\gamma = \text{const} < s + s_0$.

Differential operators P *satisfying conditions* I–III *are globally hypoelliptic, i. e. if* $u \in D'(\Omega)$ *and* $Pu \in C^\infty(\Omega)$, *then* $u \in C^\infty(\Omega)$. *(If the set* M *is empty, then the operators* Pu *are clearly hypoelliptic in the usual sense (see* §3).)

For the proof of Theorem 2.4.2 we shall use the following auxiliary results.

LEMMA 2.4.3. *Suppose the operator* P *satisfies the conditions* II *and* IIIb *of Theorem* 2.4.2. *Then for each compact set* $K \subset \Omega$, *each* $s \in R^1$, *any* $\delta_1 > 0$, *and for each sufficiently large* $N > 0$, *the estimate*

$$\| P_{(\beta)}^{(\alpha)}u \|_s^2 \leqslant \delta_1 \| Pu \|_{s+|\beta|}^2 + C(\delta_1, s, K, N, \alpha, \beta) \| u \|_{-N}^2 \tag{2.4.23}$$

holds, where the constant $-N < s + s_0$, *the function* u *belongs to the space* $C_0^\infty(K)$, *and* α *and* β *are any multi-indices.*

PROOF. We prove (2.4.23) by induction. For $|\alpha| = 1$, $|\beta| = 0$ and $|\alpha| = 0$, $|\beta| = 1$ the estimate (2.4.23) is satisfied by virtue of conditions II and IIIb. Let $\alpha = (\alpha_1, \cdots, \alpha_m)$ and $\alpha' = (\alpha_1, \cdots, \alpha_j + 1, \cdots, \alpha_m)$. We show that if (2.4.23) is valid for the multi-index α, then it is valid also for α'. Actually, Leibniz' formula (2.1.1) implies the equation $[P_{(\beta)}^{(\alpha)}, Z_j]\, u = P_{(\beta)}^{(\alpha')} u$, where Z_j is the operator of multiplication by a function equal to x_j in K and belonging to the set $C_0^\infty(R^m)$. Hence

$$\| P_{(\beta)}^{(\alpha')} u \|_s^2 = \| [P_{(\beta)}^{(\alpha)}, Z_j]\, u \|_s^2 \leqslant C_1 \{ \| Z_j P_{(\beta)}^{(\alpha)} u \|_s^2 + \| P_{(\beta)}^{(\alpha)} Z_j u \|_s^2 \}. \quad (2.4.24)$$

According to (2.2.15), $\| Z_j P_{(\beta)}^{(\alpha)} u \|_s^2 \leqslant C_2 \| P_{(\beta)}^{(\alpha)} u \|_s^2$. By the induction hypothesis

$$\| P_{(\beta)}^{(\alpha)} Z_j u \|_s^2 + \| P_{(\beta)}^{(\alpha)} u \|_s^2 \leqslant \delta_2 \| Pu \|_{s+|\beta|}^2$$
$$+ C(\delta_2, s, K, N) \| u \|_{-N}^2 + \delta_2 \| P Z_j u \|_{s+|\beta|}^2$$

Using the identity $PZ_j u = Z_j Pu + P^{(j)} u$, we obtain

$$\| P Z_j u \|_{s+|\beta|}^2 \leqslant C_3 \{ \| Pu \|_{s+|\beta|}^2 + \| P^{(j)} u \|_{s+|\beta|}^2 \}.$$

To estimate the last term we use inequality (2.4.21) of condition IIIb. By choosing δ_2 in a suitable manner, we obtain from (2.4.24) that

$$\| P_{(\beta)}^{(\alpha')} u \|_s^2 \leqslant \delta_1 \| Pu \|_{s+|\beta|}^2 + C(\delta_1, s, K, N, \alpha, \beta) \| u \|_{-N}^2.$$

Now suppose that $\beta = (\beta_1, \cdots, \beta_m)$ and $\beta' = (\beta_1, \cdots, \beta_j + 1, \cdots, \beta_m)$. We assume that (2.4.23) is satisfied for β. We show that it is also valid for β'. Clearly

$$\| P_{(\beta')}^{(\alpha)} u \|_s^2 = \| [P_{(\beta)}^{(\alpha)}, D_j]\, u \|_s^2 \leqslant C_4 (\| P_{(\beta)}^{(\alpha)} u \|_{s+1}^2 + \| P_{(\beta)}^{(\alpha)} D_j u \|_s^2). \quad (2.4.25)$$

According to the induction hypothesis

$$\| P_{(\beta)}^{(\alpha)} u \|_{s+1}^2 + \| P_{(\beta)}^{(\alpha)} D_j u \|_s^2$$
$$\leqslant \delta_2 (\| Pu \|_{s+|\beta|+1}^2 + \| P D_j u \|_{s+|\beta|}^2) + C_N \| u \|_{-N}^2. \quad (2.4.26)$$

Since $PD_j u = D_j Pu - P_{(j)} u$, we obtain from condition II that

$$\| P D_j u \|_{s+|\beta|}^2 \leqslant 2 (\| Pu \|_{s+|\beta|+1}^2 + \| P_{(j)} u \|_{s+|\beta|}^2)$$
$$\leqslant 2 \| Pu \|_{s+|\beta|+1}^2 + \delta_2 \| Pu \|_{s+|\beta'|}^2 + C_N \| u \|_{-N}^2. \quad (2.4.27)$$

From estimates (2.4.25)–(2.4.27) it follows that

$$\| P_{(\beta')}^{(\alpha)} u \|_s^2 \leqslant \delta_1 \| Pu \|_{s+|\beta'|}^2 + C(\delta_1, N, s, K, \alpha, \beta') \| u \|_{-N}^2,$$

which is the desired conclusion.

LEMMA 2.4.4. *Suppose the differential operator* P *satisfies conditions* II *and* IIIa *of Theorem* 2.4.2. *Then for each compact set* $K \subset \Omega \backslash M$, *each* $s \in R^1$ *and each sufficiently large* $N > 0$ *the inequality*

$$\| P_{(\beta)}^{(\alpha)} u \|_s^2 \leqslant C(\alpha, \beta, s, K) \{ \| Pu \|_{s+|\beta|-\mu}^2 + C(N) \| u \|_{-N}^2 \} \quad (2.4.28)$$

holds if $|\alpha| \geqslant 1$; $u \in C_0^\infty(K)$.

PROOF. Estimate (2.4.28) is established in the same way as Lemma 2.4.3. For

$|\alpha| + |\beta| = 1$, (2.4.28) is valid by virtue of condition IIIa. If $\alpha = (\alpha_1, \cdots, \alpha_m)$, $|\alpha| \geq 1$, and $\alpha' = (\alpha_1, \cdots, \alpha_j + 1, \cdots, \alpha_m)$, then, as in the proof of Lemma 2.4.3, we have

$$\|P_{(\beta)}^{(\alpha')} u\|_s^2 = \|[P_{(\beta)}^{(\alpha)}, Z_j]\, u\|_s^2 \leq C_1 \{\|P_{(\beta)}^{(\alpha)} u\|_s^2 + \|P_{(\beta)}^{(\alpha)} Z_j u\|_s^2\}.$$

By the induction hypothesis (2.4.29)

$$\|P_{(\beta)}^{(\alpha)} u\|_s^2 + \|P_{(\beta)}^{(\alpha)} Z_j u\|_s^2 \leq C(\alpha, \beta, s, K) \{\|Pu\|_{s+|\beta|-\mu}^2 + \|PZ_j u\|_{s+|\beta|-\mu}^2 + C(N) \|u\|_{-N}^2\}.$$

Therefore, using condition IIIa and estimate (2.4.29), we obtain

$$\|P_{(\beta)}^{(\alpha')} u\|_s^2 \leq C(\alpha', \beta, s, K) \{\|Pu\|_{s+|\beta|-\mu}^2 + C(N) \|u\|_{-N}^2\}. \tag{2.4.30}$$

If $\beta = (\beta_1, \cdots, \beta_m)$ and $\beta' = (\beta_1, \cdots, \beta_j + 1, \cdots, \beta_m)$, then

$$\|P_{(\beta')}^{(\alpha)} u\|_s^2 = \|[P_{(\beta)}^{(\alpha)}, D_j]\, u\|_s^2 \leq C_2 \{\|P_{(\beta)}^{(\alpha)} u\|_{s+1}^2 + \|P_{(\beta)}^{(\alpha)} D_j u\|_s^2\}.$$

Using the induction hypothesis, the equation $PD_j u = D_j Pu - P_{(j)} u$ and condition II, we obtain

$$\|P_{(\beta')}^{(\alpha)} u\|_s^2 \leq C(\alpha, \beta', s, K) \{\|Pu\|_{s+|\beta'|-\mu}^2 + C(N) \|u\|_{-N}^2\}.$$

This last inequality and (2.4.30) imply the required inequality (2.4.28).

Proof of Theorem 2.4.2. We first show that inequality (2.4.22) is satisfied for functions $\phi \in C_0^\infty(\Omega \setminus M)$. Let ψ_j, $j = 0, 1, \cdots$, be a family of functions such that $\psi_j \in C_0^\infty(\Omega \setminus M)$, and moreover $\psi_{j+1} \equiv 1$ on $\operatorname{supp} \psi_j$, with $\psi_0 = \phi$.

Let $v_\epsilon = \psi_j u * \chi_\epsilon$. Since $u \in D'(\Omega)$ and $\psi_j \in C_0^\infty(\Omega)$, it follows by Theorem 2.1.10 that the functions $\psi_j u \in \mathcal{H}_t$ for some $t \in R^1$ and all j, if $\operatorname{supp} \psi_j$ lie in some fixed compact set in $\Omega \setminus M$.

The following identity is obvious:

$$P(\psi_j u * \chi_\varepsilon) \equiv (\psi_j Pu) * \chi_\varepsilon + ([P, \psi_j]\, \psi_{j+1} u) * \chi_\varepsilon + [P(\psi_j u * \chi_\varepsilon) - (P\psi_j u) * \chi_\varepsilon]. \tag{2.4.31}$$

We estimate the last two terms on the right of (2.4.31).

If $N > s_1 + s + m_1 - t + ½$, then for any positive $\gamma > s_1 + s + m_1 - t$ we apply Lemma 2.4.1 and the inequality $(1 + \delta^2/\epsilon^2)^{-\gamma} < 1$, $\gamma > 0$, to obtain

$$\int_0^1 \|P(\psi_j u * \chi_\varepsilon) - (P\psi_j u) * \chi_\varepsilon\|_{s_1}^2 \varepsilon^{-2s} \left(1 + \frac{\delta^2}{\varepsilon^2}\right)^{-\gamma} \frac{d\varepsilon}{\varepsilon}$$

$$\leq \sum_{1 \leq |\alpha| \leq N-1} C'_\alpha \int_0^1 \|(P_{(\alpha)} \psi_j u) * \chi_\varepsilon^\alpha\|_{s_1}^2 \varepsilon^{-2(s-|\alpha|)} \left(1 + \frac{\delta^2}{\varepsilon^2}\right)^{-\gamma} \frac{d\varepsilon}{\varepsilon} + C_N \|\psi_j u\|_t^2, \tag{2.4.32}$$

where all integrals in (2.4.32) are finite, since $\psi_j u \in \mathcal{H}_t$ and (2.4.4) is fulfilled. According to Lemma 2.4.2,

$$\sum_{1\leqslant|\alpha|\leqslant N-1}\int_0^1 \|(P_{(\alpha)}\psi_j u)*\chi_\varepsilon^\alpha\|_{s_1}^2\, \varepsilon^{-2(s-|\alpha|)}\left(1+\frac{\delta^2}{\varepsilon^2}\right)^{-\gamma}\frac{d\varepsilon}{\varepsilon}$$

$$\leqslant \sum_{1\leqslant|\alpha|\leqslant N-1} C_\alpha \int_0^1 \|(P_{(\alpha)}\psi_j u)*\chi_\varepsilon\|_{s_1-|\alpha|}^2\, \varepsilon^{-2s}\left(1+\frac{\delta^2}{\varepsilon^2}\right)^{-\gamma}\frac{d\varepsilon}{\varepsilon}+C_{N_1}\|\psi_j u\|_{-N_1}^2, \tag{2.4.33}$$

where $t>-N_1$, $N_1>0$ being any sufficiently large number, since $\psi_j u \in \mathcal{H}_t$ and $t>s_1+s+m_1-\gamma$. We apply Lemma 2.4.1 to the operator $P_{(\alpha)}$, replacing s_1 by $s_1-|\alpha|$. We thus obtain

$$\int_0^1 \|(P_{(\alpha)}\psi_j u)*\chi_\varepsilon\|_{s_1-|\alpha|}^2\, \varepsilon^{-2s}\left(1+\frac{\delta^2}{\varepsilon^2}\right)^{-\gamma}\frac{d\varepsilon}{\varepsilon}$$

$$\leqslant \int_0^1 \|P_{(\alpha)}(\psi_j u*\chi_\varepsilon)\|_{s_1-|\alpha|}^2\, \varepsilon^{-2s}\left(1+\frac{\delta^2}{\varepsilon^2}\right)^{-\gamma}\frac{d\varepsilon}{\varepsilon}$$

$$+\sum_{1\leqslant|\beta|\leqslant N-|\alpha|-1}\int_0^1 C_\beta\|(P_{(\alpha+\beta)}\psi_j u)*\chi_\varepsilon^\beta\|_{s_1-|\alpha|}^2\, \varepsilon^{-2(s-|\beta|)}\left(1+\frac{\delta^2}{\varepsilon^2}\right)^{-\gamma}\frac{d\varepsilon}{\varepsilon}$$

$$+\sum_{N-|\alpha|\leqslant|\beta|\leqslant N-1}\int_0^1 C_\beta\|(P_{(\alpha+\beta)}\psi_j u)*\chi_\varepsilon^\beta\|_{s_1-|\alpha|}^2\, \varepsilon^{-2(s-|\beta|)}\left(1+\frac{\delta^2}{\varepsilon^2}\right)^{-\gamma}\frac{d\varepsilon}{\varepsilon}+C_N^1\|\psi_j u\|_t^2. \tag{2.4.34}$$

From Lemma 2.4.2 follows

$$\int_0^1 \|(P_{(\alpha+\beta)}\psi_j u)*\chi_\varepsilon^\beta\|_{s_1-|\alpha|}^2\, \varepsilon^{-2(s-|\beta|)}\left(1+\frac{\delta^2}{\varepsilon^2}\right)^{-\gamma}\frac{d\varepsilon}{\varepsilon}$$

$$\leqslant C_1\int_0^1 \|(P_{(\alpha+\beta)}\psi_j u)*\chi_\varepsilon\|_{s_1-|\alpha|-|\beta|}^2\, \varepsilon^{-2s}\left(1+\frac{\delta^2}{\varepsilon^2}\right)^{-\gamma}\frac{d\varepsilon}{\varepsilon}+C_{N_2}\|\psi_j u\|_{-N_2}^2, \tag{2.4.35}$$

where $t>-N_2$, $N_2>0$ being any sufficiently large number, since $\psi_j u\in\mathcal{H}_t$ and $t>s+s_1+m_1-|\alpha|-\gamma$. The terms of the last summation on the right of (2.4.34) do not exceed $C_2\|\psi_j u\|_t^2$. In fact, according to Theorem 2.4.1,

$$\int_0^1 \|(P_{(\alpha+\beta)}\psi_j u)*\chi_\varepsilon^\beta\|_{s_1-|\alpha|}^2\, \varepsilon^{-2(s-|\beta|)}\left(1+\frac{\delta^2}{\varepsilon^2}\right)^{-\gamma}\frac{d\varepsilon}{\varepsilon}$$

$$\leqslant C_3\|P_{(\alpha+\beta)}\psi_j u\|_{s_1+s-|\alpha|-|\beta|,\,\gamma}^2.$$

Since the order of $P_{(\alpha+\beta)}$ is no larger than m_1 and $|\beta|\geqslant N-|\alpha|$, it follows that

$$\|P_{(\alpha+\beta)}\psi_j u\|^2_{s_1+s-|\alpha|-|\beta|,\,\gamma} \leqslant C_4\|\psi_j u\|^2_{s_1+s+m_1-N,\,\gamma} \leqslant C_5\|\psi_j u\|^2_t.$$

From these estimates and from the repeated application of (2.4.34) it follows that

$$\sum_{1\leqslant|\alpha|\leqslant N-1}\int_0^1 \|(P_{(\alpha)}\psi_j u)*\chi_\varepsilon\|^2_{s_1-|\alpha|}\,\varepsilon^{-2s}\left(1+\frac{\delta^2}{\varepsilon^2}\right)^{-\gamma}\frac{d\varepsilon}{\varepsilon}$$
$$\leqslant C_6\left[\sum_{1\leqslant|\alpha|\leqslant N-1}\int_0^1 \|P_{(\alpha)}(\psi_j u*\chi_\varepsilon)\|^2_{s_1-|\alpha|}\,\varepsilon^{-2s}\left(1+\frac{\delta^2}{\varepsilon^2}\right)^{-\gamma}\frac{d\varepsilon}{\varepsilon}+\|\psi_j u\|^2_t\right]. \tag{2.4.36}$$

Thus it follows from (2.4.32) that

$$\int_0^1 \|P(\psi_j u*\chi_\varepsilon)-(P\psi_j u)*\chi_\varepsilon\|^2_{s_1}\,\varepsilon^{-2s}\left(1+\frac{\delta^2}{\varepsilon^2}\right)^{-\gamma}\frac{d\varepsilon}{\varepsilon}$$
$$\leqslant C_7\left[\sum_{1\leqslant|\alpha|\leqslant N-1}\int_0^1 \|P_{(\alpha)}(\psi_j u*\chi_\varepsilon)\|^2_{s_1-|\alpha|}\,\varepsilon^{-2s}\left(1+\frac{\delta^2}{\varepsilon^2}\right)^{-\gamma}\frac{d\varepsilon}{\varepsilon}+\|\psi_j u\|^2_t\right]. \tag{2.4.37}$$

We estimate the right side of (2.4.37), using Lemma 2.4.3. We obtain

$$\sum_{1\leqslant|\alpha|\leqslant N-1}\int_0^1 \|P_{(\alpha)}(\psi_j u*\chi_\varepsilon)\|^2_{s_1-|\alpha|}\,\varepsilon^{-2s}\left(1+\frac{\delta^2}{\varepsilon^2}\right)^{-\gamma}\frac{d\varepsilon}{\varepsilon}$$
$$\leqslant \delta_1\int_0^1 \|P(\psi_j u*\chi_\varepsilon)\|^2_{s_1}\,\varepsilon^{-2s}\left(1+\frac{\delta^2}{\varepsilon^2}\right)^{-\gamma}\frac{d\varepsilon}{\varepsilon}+C(\delta_1)\|\psi_j u\|^2_t. \tag{2.4.38}$$

From inequalities (2.4.37) and (2.4.38) we obtain the first basic inequality:

$$\int_0^1 \|P(\psi_j u*\chi_\varepsilon)-(P\psi_j u)*\chi_\varepsilon\|^2_{s_1}\,\varepsilon^{-2s}\left(1+\frac{\delta^2}{\varepsilon^2}\right)^{-\gamma}\frac{d\varepsilon}{\varepsilon}$$
$$\leqslant \delta_2\int_0^1 \|P(\psi_j u*\chi_\varepsilon)\|^2_{s_1}\,\varepsilon^{-2s}\left(1+\frac{\delta^2}{\varepsilon^2}\right)^{-\gamma}\frac{d\varepsilon}{\varepsilon}+C(\delta_2)\|\psi_j u\|^2_t. \tag{2.4.39}$$

We now estimate the terms of the form $([P,\psi_j]\psi_{j+1}u)*\chi_\epsilon$ entering into (2.4.31).

Applying Leibniz' formula to calculate $[P,\psi_j]v$, we obtain

$$\int_0^1 \|([P,\psi_j]\psi_{j+1}u)*\chi_\varepsilon\|^2_{s_1}\varepsilon^{-2s}\left(1+\frac{\delta^2}{\varepsilon^2}\right)^{-\gamma}\frac{d\varepsilon}{\varepsilon}$$
$$\leqslant \sum_{1\leqslant|\alpha|\leqslant m_1} C_\alpha\int_0^1 \|(\psi_{j\,(\alpha)}P^{(\alpha)}\psi_{j+1}u)*\chi_\varepsilon\|^2_{s_1}\varepsilon^{-2s}\left(1+\frac{\delta^2}{\varepsilon^2}\right)^{-\gamma}\frac{d\varepsilon}{\varepsilon}. \tag{2.4.40}$$

To estimate the latter integrals inequality (2.4.37) is employed with the operators $\psi_{j(\alpha)}P^{(\alpha)}$ taking place of the operator P. We obtain

$$\int_0^1 \| (\psi_{j\,(\alpha)} P^{(\alpha)} \psi_{j+1} u)_* \chi_\varepsilon \|_{s_1}^2 \varepsilon^{-2s} \left(1 + \frac{\delta^2}{\varepsilon^2}\right)^{-\gamma} \frac{d\varepsilon}{\varepsilon}$$

$$\leqslant \int_0^1 \| \psi_{j\,(\alpha)} P^{(\alpha)} (\psi_{j+1} u_* \chi_\varepsilon) \|_{s_1}^2 \varepsilon^{-2s} \left(1 + \frac{\delta^2}{\varepsilon^2}\right)^{-\gamma} \frac{d\varepsilon}{\varepsilon}$$

$$+ \sum_{\substack{1 \leqslant |\beta| \leqslant N-1-|\alpha| \\ \beta=\beta'+\beta''}} C_\beta \int_0^1 \| \psi_{j(\alpha+\beta')} P^{(\alpha)}_{(\beta'')} (\psi_{j+1} u_* \chi_\varepsilon) \|_{s_1-|\beta|}^2 \varepsilon^{-2s} \left(1 + \frac{\delta^2}{\varepsilon^2}\right)^{-\gamma} \frac{d\varepsilon}{\varepsilon}$$

$$+ C_8 \| \psi_{j+1} u \|_t^2. \tag{2.4.41}$$

Here we have used the fact that

$$\sum_{\substack{N-|\alpha| \leqslant |\beta| \leqslant N-1 \\ \beta=\beta'+\beta''}} \int_0^1 \| \psi_{j(\alpha+\beta')} P^{(\alpha)}_{(\beta'')} (\psi_{j+1} u_* \chi_\varepsilon) \|_{s_1-|\beta|}^2 \varepsilon^{-2s} \left(1 + \frac{\delta^2}{\varepsilon^2}\right)^{-\gamma} \frac{d\varepsilon}{\varepsilon}$$

$$\leqslant C_9 \| \psi_{j+1} u \|_t^2.$$

We apply Lemma 2.4.4 for the estimation of the last integrals on the right side of (2.4.41). Considering Theorem 2.2.1, we obtain

$$\sum_{\substack{1 \leqslant |\beta| \leqslant N-1-|\alpha| \\ \beta=\beta'+\beta''}} \int_0^1 \| \psi_{j(\alpha+\beta')} P^{(\alpha)}_{(\beta'')} (\psi_{j+1} u_* \chi_\varepsilon) \|_{s_1-|\beta|}^2 \varepsilon^{-2s} \left(1 + \frac{\delta^2}{\varepsilon^2}\right)^{-\gamma} \frac{d\varepsilon}{\varepsilon}$$

$$\leqslant C_{10} \sum_{0 \leqslant |\beta| \leqslant N-1-|\alpha|} \int_0^1 \| P^{(\alpha)}_{(\beta)} (\psi_{j+1} u_* \chi_\varepsilon) \|_{s_1-|\beta|}^2 \varepsilon^{-2s} \left(1 + \frac{\delta^2}{\varepsilon^2}\right)^{-\gamma} \frac{d\varepsilon}{\varepsilon}$$

$$\leqslant C_{11} \left\{ \int_0^1 \| P (\psi_{j+1} u_* \chi_\varepsilon) \|_{s_1-\mu}^2 \varepsilon^{-2s} \left(1 + \frac{\delta^2}{\varepsilon^2}\right)^{-\gamma} \frac{d\varepsilon}{\varepsilon} + C_{N_3} \| \psi_{j+1} u \|_{-N_3}^2 \right\},$$

where $t > -N_3$, $N_3 > 0$ being any sufficiently large number. We thus obtain the second basic inequality:

$$\int_0^1 \| ([P, \psi_j] \psi_{j+1} u)_* \chi_\varepsilon \|_{s_1}^2 \varepsilon^{-2s} \left(1 + \frac{\delta^2}{\varepsilon^2}\right)^{-\gamma} \frac{d\varepsilon}{\varepsilon}$$

$$\leqslant C_{12} \left[\int_0^1 \| P (\psi_{j+1} u_* \chi_\varepsilon) \|_{s_1-\mu}^2 \varepsilon^{-2s} \left(1 + \frac{\delta^2}{\varepsilon^2}\right)^{-\gamma} \frac{d\varepsilon}{\varepsilon} + \| \psi_{j+1} u \|_t^2 \right]. \tag{2.4.42}$$

We now estimate the left side of (2.4.31). From the estimates (2.4.39) and (2.4.42) it follows that

$$\int_0^1 \| P (\psi_j u_* \chi_\varepsilon) \|_{s_1}^2 \varepsilon^{-2s} \left(1 + \frac{\delta^2}{\varepsilon^2}\right)^{-\gamma} \frac{d\varepsilon}{\varepsilon} \leqslant$$

$$\leqslant C_{13}\int_0^1 \| P(\psi_{j+1}u * \chi_\varepsilon)\|^2_{s_1-\mu}\varepsilon^{-2s}\left(1+\frac{\delta^2}{\varepsilon^2}\right)^{-\gamma}\frac{d\varepsilon}{\varepsilon}$$

$$+\delta_2\int_0^1 \| P(\psi_j u * \chi_\varepsilon)\|^2_{s_1}\varepsilon^{-2s}\left(1+\frac{\delta^2}{\varepsilon^2}\right)^{-\gamma}\frac{d\varepsilon}{\varepsilon}$$

$$+C_{14}\left[\|\psi_{j+1}u\|^2_t+\int_0^1 \|(\psi_j Pu)*\chi_\varepsilon\|^2_{s_1}\varepsilon^{-2s}\frac{d\varepsilon}{\varepsilon}\right] \tag{2.4.43}$$

for any positive $\gamma \geqslant s + s_1 + m_1 - t$ and $s_1 \leqslant 0$, since $\psi_j Pu \in \mathcal{H}_s$. Choosing $\delta_2 = ½$, we deduce from (2.4.43) that for any $s \in R^1$ and $s_1 \leqslant 0$

$$\int_0^1 \|P(\psi_j u * \chi_\varepsilon)\|^2_{s_1}\varepsilon^{-2s}\left(1+\frac{\delta^2}{\varepsilon^2}\right)^{-\gamma}\frac{d\varepsilon}{\varepsilon}$$

$$\leqslant C_{15}\left\{\int_0^1 \| P(\psi_{j+1}u * \chi_\varepsilon)\|^2_{s_1-\mu}\varepsilon^{-2s}\left(1+\frac{\delta^2}{\varepsilon^2}\right)^{-\gamma}\frac{d\varepsilon}{\varepsilon}\right.$$

$$\left.+\int_0^1 \|(\psi_j Pu)*\chi_\varepsilon\|^2_{s_1}\varepsilon^{-2s}\frac{d\varepsilon}{\varepsilon}+\|\psi_{j+1}u\|^2_t\right\}. \tag{2.4.44}$$

Since by Theorem 2.2.1

$$\|\psi_j u\|^2_t = \|\psi_j\psi_{j+1}u\|^2_t \leqslant C_{16}\|\psi_{j+1}u\|^2_t,$$

we may apply inequality (2.4.44) successively for $s_1 = 0, -\mu, -2\mu, \cdots, -l\mu$, with $j = 0, 1, \cdots, l$ respectively, and use each inequality to estimate the first term appearing on the right of the preceding inequality. We thus obtain

$$\int_0^1 \| P(\varphi u * \chi_\varepsilon)\|^2_0\varepsilon^{-2s}\left(1+\frac{\delta^2}{\varepsilon^2}\right)^{-\gamma}\frac{d\varepsilon}{\varepsilon}$$

$$\leqslant C_{17}\left\{\int_0^1 \| P(\psi_{l+1}u * \chi_\varepsilon)\|^2_{-\mu(l+1)}\varepsilon^{-2s}\left(1+\frac{\delta^2}{\varepsilon^2}\right)^{-\gamma}\frac{d\varepsilon}{\varepsilon}\right.$$

$$\left.+\sum_{j=0}^{l}\int_0^1 \|(\psi_j Pu)*\chi_\varepsilon\|^2_{-\mu j}\varepsilon^{-2s}\frac{d\varepsilon}{\varepsilon}+\|\psi_{l+1}u\|^2_t\right\}. \tag{2.4.45}$$

Since $Pu \in \mathcal{H}^{\text{loc}}_s$, all integrals on the right of (2.4.45) are finite. If l is so large that $-\mu(l+1) + m_1 + s < t$, it follows from (2.4.45) that

$$\int_0^1 \| P(\varphi u * \chi_\varepsilon)\|^2_0\varepsilon^{-2s}\left(1+\frac{\delta^2}{\varepsilon^2}\right)^{-\gamma}\frac{d\varepsilon}{\varepsilon} \leqslant C_{18}\left\{\sum_{j=0}^{l}\|\psi_j Pu\|^2_s+\|\psi_{l+1}u\|^2_t\right\}. \tag{2.4.46}$$

On the basis of this inequality it is easy to prove the required estimate (2.4.22).

In fact, from Theorem 2.4.1 and from estimate (2.4.18), which is fulfilled by virtue of Condition I of the theorem being proved, it follows that

$$\| \varphi u \|^2_{s_0+s,\gamma} \leqslant C_{19}\left[\| \varphi u \|^2_t + \int_0^1 \| \varphi u * \chi_\varepsilon \|^2_{s_0} \varepsilon^{-2s}\left(1+\frac{\delta^2}{\varepsilon^2}\right)^{-\gamma}\frac{d\varepsilon}{\varepsilon}\right]$$

$$\leqslant C_{20}\left\{\| \varphi u \|^2_t + \int_0^1 \| P(\varphi u * \chi_\varepsilon) \|^2_0 \varepsilon^{-2s}\left(1+\frac{\delta^2}{\varepsilon^2}\right)^{-\gamma}\frac{d\varepsilon}{\varepsilon}\right\}.$$

Applying inequality (2.4.46), we find

$$\| \varphi u \|^2_{s+s_0} \leqslant \sup_\delta \| \varphi u\|^2_{s+s_0,\gamma} \leqslant C_{21}\left\{\sum_{j=0}^{l} \| \psi_j P u \|^2_s + \| \psi_{l+1} u \|^2_t\right\}. \qquad (2.4.47)$$

We may assume that the system of functions $\{\psi_j\}$ is chosen so that $\psi_0 = \phi$, and $\psi_{l+1} = \phi_1$. The constant C_{21} will generally depend on t, which is determined by the function $u \in D'(\Omega)$. However, from inequalities of the type (2.4.47) it follows that for each compact set $K \subset \Omega \setminus M$ we may suppose that the constant t is the same for all ψu, where $u \in D'(\Omega)$, $Pu \in \mathcal{H}^{\mathrm{loc}}_s$ and $\psi \in C_0^\infty(K)$. We may choose $t = \gamma_0 < s + s_0$. Thus (2.4.47) yields the required inequality (2.4.22) if $\operatorname{supp} \phi \subset \Omega \setminus M$.

Inequality (2.4.22) yields the hypoellipticity of the operator Pu in $\Omega\setminus M$, since, if $\phi_1 Pu \in C^\infty(\Omega\setminus M)$, then according to (2.4.22) $\phi u \in \mathcal{H}_s$ for any s, and it follows from the Imbedding Theorem 2.1.8 that $\phi u \in C^\infty(\Omega \setminus M)$.

We now prove (2.4.22) in the case when $\phi \equiv 1$ on M.

We consider functions ϕ and ϕ_0 in $C_0^\infty(\Omega)$ such that $\phi_0 \equiv 1$ on $\operatorname{supp} \phi$, and $\phi \equiv 1$ in some neighborhood of M.

For the functions ϕ and ϕ_0 considered, estimates (2.4.39) and (2.4.41) also hold. Lemma 2.4.3 and Theorem 2.2.1 now yield

$$\sum_{\substack{1 \leqslant |\beta| \leqslant N-1-|\alpha| \\ \beta=\beta'+\beta''}} \int_0^1 \| \varphi_{(\alpha+\beta')} P^{(\alpha)}_{(\beta'')}(\varphi_0 u * \chi_\varepsilon)\|^2_{-|\beta|} \varepsilon^{-2s}\left(1+\frac{\delta^2}{\varepsilon^2}\right)^{-\gamma}\frac{d\varepsilon}{\varepsilon}$$

$$\leqslant \delta_1 \int_0^1 \| P(\varphi_0 u * \chi_\varepsilon) \|^2_0 \varepsilon^{-2s}\left(1+\frac{\delta^2}{\varepsilon^2}\right)^{-\gamma}\frac{d\varepsilon}{\varepsilon} + C_{22} \| \varphi_0 u \|^2_t$$

$$\leqslant 2\delta_1 \int_0^1 \| P(\varphi u * \chi_\varepsilon) \|^2_0 \varepsilon^{-2s}\left(1+\frac{\delta^2}{\varepsilon^2}\right)^{-\gamma}\frac{d\varepsilon}{\varepsilon}$$

$$+ 2\delta_1 \int_0^1 \| P([\varphi_0 - \varphi]\, u * \chi_\varepsilon) \|^2_0 \varepsilon^{-2s}\left(1+\frac{\delta^2}{\varepsilon^2}\right)^{-\gamma}\frac{d\varepsilon}{\varepsilon} + C_{23} \| \varphi_0 u \|^2_t. \qquad (2.4.48)$$

Since the support of $\phi_0 - \phi$ lies in $\Omega\setminus M$, according to estimate (2.4.46) we have

$$\int_0^1 \| P([\varphi_0 - \varphi] u * \chi_\varepsilon) \|_0^2 \varepsilon^{-2s} \left(1 + \frac{\delta^2}{\varepsilon^2}\right)^{-\gamma} \frac{d\varepsilon}{\varepsilon}$$

$$\leqslant C_{24} \left\{ \sum_{k=0}^{l} \| \psi_k P u \|_s^2 + \| \psi_{l+1} u \|_t^2 \right\}, \tag{2.4.49}$$

where $\{\psi_k\}$ is a certain system of functions such that $\psi_k \in C_0^\infty(\Omega \setminus M)$, $\psi_{k+1} \equiv 1$ on $\operatorname{supp} \psi_k$, and all $\psi_k \equiv 1$ on $\operatorname{supp}(\phi_0 - \phi)$. As in the preceding case, we again consider identity (2.4.31).

From this identity and from estimates (2.4.39), (2.4.41) and (2.4.49) it follows that

$$\int_0^1 \| P(\varphi u * \chi_\varepsilon) \|_0^2 \varepsilon^{-2s} \left(1 + \frac{\delta^2}{\varepsilon^2}\right)^{-\gamma} \frac{d\varepsilon}{\varepsilon}$$

$$\leqslant \int_0^1 \| (\varphi P u) * \chi_\varepsilon \|_0^2 \varepsilon^{-2s} \frac{d\varepsilon}{\varepsilon} + \delta_1 \int_0^1 \| P(\varphi u * \chi_\varepsilon) \|_0^2 \varepsilon^{-2s} \left(1 + \frac{\delta^2}{\varepsilon^2}\right)^{-\gamma} \frac{d\varepsilon}{\varepsilon}$$

$$+ C_{25} \left\{ \| \varphi_0 u \|_t^2 + \| \psi_{l+1} u \|_t^2 + \sum_{j=0}^{l} \| \psi_k P u \|_s^2 \right\}.$$

Setting $\delta_1 = ½$, we obtain

$$\int_0^1 \| P(\varphi u * \chi_\varepsilon) \|_0^2 \varepsilon^{-2s} \frac{d\varepsilon}{\varepsilon}$$

$$\leqslant C_{26} \left\{ \| \varphi P u \|_s^2 + \sum_{k=0}^{l} \| \psi_k P u \|_s^2 + \| \psi_{l+1} u \|_t^2 + \| \varphi_0 u \|_t^2 \right\}. \tag{2.4.50}$$

According to condition I of the theorem,

$$\int_0^1 \| \varphi u * \chi_\varepsilon \|_{s_0}^2 \varepsilon^{-2s} \left(1 + \frac{\delta^2}{\varepsilon^2}\right)^{-\gamma} \frac{d\varepsilon}{\varepsilon}$$

$$\leqslant C_{27} \left\{ \int_0^1 \| P(\varphi u * \chi_\varepsilon) \|_0^2 \varepsilon^{-2s} \left(1 + \frac{\delta^2}{\varepsilon^2}\right)^{-\gamma} \frac{d\varepsilon}{\varepsilon} + \| \varphi u \|_t^2 \right\}. \tag{2.4.51}$$

Applying inequality (2.4.50) to estimate the right side of (2.4.51), we obtain from this and from Theorem 2.4.1 that

$$\|\varphi u\|^2_{s+s_0} \leqslant \sup_\delta \|\varphi u\|^2_{s+s_0,\gamma}$$

$$\leqslant C_{28}\left\{\|\varphi P u\|_s^2 + \sum_{k=0}^{l}\|\psi_k P u\|_s^2 + \|\varphi_0 u\|_t^2 + \|\psi_{l+1} u\|_t^2\right\}$$

$$\leqslant C_{29}\{\|\varphi_1 P u\|_s^2 + \|\varphi_1 u\|_t^2\}, \tag{2.4.52}$$

where $\phi_1 \in C_0^\infty(\Omega)$ and $\phi_1 \equiv 1$ on the set $\operatorname{supp} \phi_0 \cup \operatorname{supp} \psi_{l+1}$.

Estimate (2.4.52) implies (2.4.22) and the global hypoellipticity of the operator P in a neighborhood of the set M, exactly as was done in the case when $\operatorname{supp} \phi_1 \subset \Omega \setminus M$.

The theorem is proved.

§5. A priori estimates and hypoellipticity theorems for the operators of Hörmander

Hörmander [55] proved a theorem on the hypoellipticity of second order differential operators of the type

$$Pu \equiv \sum_{j=1}^{r} X_j^2 u + iX_0 u + \gamma u, \tag{2.5.1}$$

where $X_j(x, D) \equiv \Sigma_{l=1}^m a_j^l(x) D_l$ $(j = 0, 1, \cdots, r)$ under certain conditions on the Lie algebra of the operators X_j $(j = 0, 1, \cdots, r)$. The first order differential operators X_j have real infinitely differentiable coefficients in the domain Ω; and γ is a real infinitely differentiable function in Ω.

In this section we give another proof of Hörmander's theorem, and we also prove some more general theorems on hypoellipticity of the operator Pu, as well as theorems on local smoothness of weak solutions of the equation $Pu = f$. For this purpose we establish a priori estimates for the operator (2.5.1) and find conditions on P under which Theorem 2.4.2 is applicable. These same methods will be used in the following section to investigate questions having to do with the smoothness of solutions and the hypoellipticity of general second order equations with nonnegative characteristic form. An analogous theorem holds also for certain classes of higher order equations. Since all considerations will be carried out on some compact set K_1 in Ω, we may suppose that the coefficients of (2.5.1) belong to the class $C_0^\infty(\Omega)$, if necessary multiplying them by a function $\phi(x) \in C_0^\infty(\Omega)$ such that $\phi \equiv 1$ on K_1. Clearly, $-Pu = f$ will be an equation with nonnegative characteristic form.

In §§5 and 6 we denote $(u, v)_0$ by (u, v) for simplicity. We first prove the following auxiliary result.

LEMMA 2.5.1. *Let A_s be a pseudodifferential operator whose symbol satisfies* *conditions* a) *and* b) *of* §2, *with* $\sigma = s$. *Then for each compact set* $K \subset \Omega$ *the inequality*

$$\| PA_s u \|_{-s}^2 \leqslant C(s, K)(\| Pu \|_0^2 + \| u \|_0^2) \tag{2.5.2}$$

is fulfilled for $u \in C_0^\infty(K)$; $C(s, K)$ *is a constant depending on* s *and* K.

PROOF. The following identity is easily verified:

$$[P, A_s]\, u \equiv \sum_{j=1}^{r} (2\,[X_j, A_s]\, X_j + [X_j, [X_j, A_s]])\, u + i\,[X_0, A_s]\, u + [\gamma, A_s]\, u. \tag{2.5.3}$$

Applying Theorem 2.2.3 on commutators, inequality (2.2.28), and Theorem 2.2.1 on the order of a pseudodifferential operator, we see from (2.5.3) that

$$\| [P, A_s]\, u \|_{-s}^2 \leqslant C_1 \left(\sum_{j=1}^{r} \| [X_j, [X_j, A_s]]\, u \|_{-s}^2 + \sum_{j=1}^{r} \| [X_j, A_s]\, X_j u \|_{-s}^2 + \| [X_0, A_s]\, u \|_{-s}^2 + \| [\gamma, A_s]\, u \|_{-s}^2 \right) \leqslant C_2 \left(\| u \|_0^2 + \sum_{j=1}^{r} \| X_j u \|_0^2 \right). \tag{2.5.4}$$

In the equation

$$\operatorname{Re}(Pu, u) = \operatorname{Re}\left[\sum_{j=1}^{r} (X_j^2 u, u) + i\,(X_0 u, u) + (\gamma u, u) \right] \tag{2.5.5}$$

we transform by integration by parts the following integrals:

$$(X_j^2 u, u) = \| X_j u \|_0^2 + (X_j u, (X_j^* - X_j)\, u),$$

$$\operatorname{Re}\,[i\,(X_0 u, u)] = \operatorname{Re}\left[\frac{i}{2}(X_0 u, u) + \frac{i}{2}(u, X_0 u) + \frac{i}{2}(u, (X_0^* - X_0)\, u) \right] = \operatorname{Re} \frac{i}{2}(u, (X_0^* - X_0)\, u).$$

Equation (2.5.5) yields

$$\sum_{j=1}^{r} \| X_j u \|_0^2 \leqslant |(Pu, u)| + \left| \frac{1}{2}(u, (X_0^* - X_0)\, u) \right| + \sum_{j=1}^{r} |(X_j u, (X_j^* - X_j)\, u)| + |(\gamma u, u)|. \tag{2.5.6}$$

From this we obtain

$$\sum_{j=1}^{r} \| X_j u \|_0^2 \leqslant C_3 (\| Pu \|_0^2 + \| u \|_0^2) + \frac{1}{2} \sum_{j=1}^{r} \| X_j u \|_0^2$$

which implies

$$\sum_{j=1}^{r} \| X_j u \|_0^2 \leqslant C_4 (\| Pu \|_0^2 + \| u \|_0^2). \tag{2.5.7}$$

Estimates (2.5.4) and (2.5.7) yield

$$\| PA_s u \|_{-s}^2 \leqslant 2 \| A_s Pu \|_{-s}^2 + 2 \| [P, A_s] u \|_{-s}^2 \leqslant C_5 (\| Pu \|_0^2 + \| u \|_0^2).$$

The lemma is proved.

REMARK. Lemma 2.5.1 is clearly also valid for operators A_s of the form $A_s = A_{s,N} + T_N$, where $A_{s,N}$ is a pseudodifferential operator of order at most s, and T_N is an operator of order at most $-N$, where $N > 0$ is a sufficiently large number.

THEOREM 2.5.1 (ENERGY ESTIMATE). *For any function* $u \in C_0^\infty(K)$ *and any* $s \geqslant 0$

$$\| X_0 u \|_{s-\frac{1}{2}}^2 + \sum_{j=1}^{r} \| X_j u \|_s^2 \leqslant C(K, s) \{ \| Pu \|_0^2 + \| u \|_{2s}^2 \}. \tag{2.5.8}$$

PROOF. We note that for $s \geqslant 1$ the inequality (2.5.8) is trivial. Let E_s be the pseudodifferential operator with symbol defined by formula (2.2.34), where the function $\phi \in C_0^\infty(K_1)$ and $\phi \equiv 1$ on K, $K \subset K_1$. We investigate Re $(PE_s u, E_s u)$, transforming this expression in the same way as we transformed (2.5.5). We obtain

$$\operatorname{Re} (PE_s u, E_s u) = \sum_{j=1}^{r} (X_j E_s u, X_j E_s u) + (\gamma E_s u, E_s u) + \operatorname{Re} \left\{ \sum_{j=1}^{r} (X_j E_s u, (X_j^* - X_j) E_s u) + \frac{i}{2} (E_s u, (X_0^* - X_0) E_s u) \right\}. \tag{2.5.9}$$

By Theorem 2.2.6 on operators E_s, for each function $u \in C_0^\infty(K)$ we have

$$\sum_{j=1}^{r} \| X_j u \|_s^2 \leqslant 2 \sum_{j=1}^{r} \| E_s X_j u \|_0^2 + C_1 \| u \|_0^2 \leqslant \sum_{j=1}^{r} 4 (\| X_j E_s u \|_0^2 + \| [E_s, X_j] u \|_0^2) + C_1 \| u \|_0^2.$$

From estimate (2.2.28) for the norm of the commutator $[E_s, X_j]$ we obtain

$$\| [E_s, X_j] u \|_0^2 \leqslant C_2 \| u \|_s^2.$$

Hence

$$\sum_{j=1}^{r} \| X_j u \|_s^2 \leqslant C_3 \left(\sum_{j=1}^{r} \| X_j E_s u \|_0^2 + \| u \|_s^2 \right). \tag{2.5.10}$$

We now estimate the norm of $X_j E_s u$, using (2.5.9). Applying Theorem 2.2.1 on the order of a pseudodifferential operator, and inequality (2.1.19), we obtain

$$\sum_{j=1}^{r} \| X_j E_s u \|_0^2 \leqslant |(PE_s u, E_s u)| + \sum_{j=1}^{r} |(X_j E_s u, (X_j^* - X_j) E_s u)|$$
$$+ \frac{1}{2} |(E_s u, (X_0^* - X_0) E_s u)| + C_4 \| u \|_s^2$$
$$\leqslant 2 \| PE_s u \|_{-s}^2 + C_5 \| u \|_{2s}^2 + \frac{1}{2} \sum_{j=1}^{r} \| X_j E_s u \|_0^2. \tag{2.5.11}$$

On the basis of Lemma 2.5.1 we have

$$\| PE_s u \|_{-s}^2 \leqslant C_6 (\| Pu \|_0^2 + \| u \|_0^2). \tag{2.5.12}$$

From inequalities (2.5.10)–(2.5.12) we conclude that

$$\sum_{j=1}^{r} \| X_j u \|_s^2 \leqslant C_7 (\| Pu \|_0^2 + \| u \|_{2s}^2). \tag{2.5.13}$$

In order to estimate $\| X_0 u \|_{s-1/2}^2$, we consider $(PE_s u, E_{-1/2}^* X_0 E_s u)$. Here E_s^* is the operator adjoint to E_s. It is clear that

$$(PE_s u, E_{-1/2}^* E_{-1/2} X_0 E_s u) = i \| E_{-1/2} X_0 E_s u \|_0^2$$
$$+ \left(\gamma E_s u, E_{-\frac{1}{2}}^* E_{-\frac{1}{2}} X_0 E_s u \right)$$
$$+ \sum_{j=1}^{r} \left(X_j E_s u, X_j^* E_{-\frac{1}{2}}^* E_{-\frac{1}{2}} X_0 E_s u \right). \tag{2.5.14}$$

Applying (2.1.19) and Theorem 2.2.1 on the order of a pseudodifferential operator, we obtain the estimate

$$\left| \left(PE_s u, E_{-\frac{1}{2}}^* E_{-\frac{1}{2}} X_0 E_s u \right) \right| \leqslant 2 \| PE_s u \|_{-s}^2 + C_8 \| u \|_{2s}^2$$
$$\leqslant C_9 (\| Pu \|_0^2 + \| u \|_{2s}^2). \tag{2.5.15}$$

Here we have also used (2.5.12). It is easy to see that

$$\left| \left(\gamma E_s u, \ E_{-\frac{1}{2}}^* E_{-\frac{1}{2}} X_0 E_s u \right) \right| \leqslant C_{10} \| u \|_s^2.$$

Again using inequality (2.1.19), Theorems 2.2.1 and 2.2.3, we obtain

$$\left| \left(X_j E_s u, \ X_j^* E_{-\frac{1}{2}}^* E_{-\frac{1}{2}} X_0 E_s u \right) \right| \leqslant \| X_j E_s u \|_0^2 + \left\| X_j^* E_{-\frac{1}{2}}^* E_{-\frac{1}{2}} X_0 E_s u \right\|_0^2,$$

as well as the inequality

$$\| X_j E_s u \|_0^2 \leqslant C_{11} \left(\| X_j u \|_s^2 + \| u \|_s^2 \right)$$

and the estimate

$$\left\|X_j^* E^*_{-\frac{1}{2}} E_{-\frac{1}{2}} X_0 E_s u\right\|_0^2 \leqslant C_{12}\left(\left\|E^*_{-\frac{1}{2}} E_{-\frac{1}{2}} X_0 E_s X_j u\right\|_0^2 + \|u\|_s^2\right)$$
$$\leqslant C_{13}\left(\|X_j u\|_s^2 + \|u\|_s^2\right).$$

From this it follows that

$$\left|\left(X_j E_s u,\ X_j^* E^*_{-\frac{1}{2}} E_{-\frac{1}{2}} X_0 E_s u\right)\right| \leqslant C_{14}\left(\|X_j u\|_s^2 + \|u\|_s^2\right).$$

Therefore from (2.5.14) we have

$$\left\|E_{-\frac{1}{2}} X_0 E_s u\right\|_0^2 \leqslant C_{15}\left\{\|Pu\|_0^2 + \sum_{j=1}^{r} \|X_j u\|_s^2 + \|u\|_{2s}^2\right\}. \tag{2.5.16}$$

By Theorem 2.2.6, for any function $u \in C_0^\infty(K)$ we obtain the inequality

$$\|X_0 u\|^2_{s-\frac{1}{2}} \leqslant 2\,\|E_s X_0 u\|^2_{-\frac{1}{2}} + C_{16}\|u\|_0^2 \leqslant 4\left\|E_{-\frac{1}{2}} E_s X_0 u\right\|_0^2 + C_{17}\|u\|_0^2$$
$$\leqslant C_{18}\left\{\left\|E_{-\frac{1}{2}} X_0 E_s u\right\|_0^2 + \left\|E_{-\frac{1}{2}} [X_0, E_s]\, u\right\|_0^2 + \|u\|_0^2\right\}.$$

In estimating the right side of this last inequality, we apply (2.5.16) and Theorem 2.2.3 on commutators, obtaining

$$\|X_0 u\|^2_{s-\frac{1}{2}} \leqslant C_{19}\left(\|Pu\|_0^2 + \|u\|_{2s}^2\right). \tag{2.5.17}$$

From the a priori estimates (2.5.13) and (2.5.17) follows the required result. The theorem is proved.

We consider the system of operators $\{X_0, \cdots, X_r\}$ defined by (2.5.1). For any multi-index $I = (\alpha_1, \cdots, \alpha_t)$ where $\alpha_l = 0, 1, \cdots, r$ $(l = 1, \cdots, t)$ we set $|I| = \sum_{l=1}^t \lambda_l$, where $\lambda_l = 1$ if $\alpha_l = 1, \cdots, r$ and $\lambda_l = 2$ if $\alpha_l = 0$. To each multi-index I we associate the operator

$$X_I = \operatorname{ad} X_{\alpha_1} \ldots \operatorname{ad} X_{\alpha_{t-1}} X_{\alpha_t},$$

where $\operatorname{ad} AB = [A, B] = AB - BA$ for any operators A and B. The following lemma yields an estimate for the norm of the operator X_I.

Lemma 2.5.2. *For each compact set K, each integer $k \geqslant 1$, and each $s \geqslant 0$, there exists a constant $C(K, s, k)$ such that for each function u in $C_0^\infty(K)$ the inequality*

$$\sum_{|I|=k} \|X_I u\|^2_{s+2^{1-k}-1} \leqslant C(K, s, k)\{\|Pu\|_0^2 + \|u\|^2_{2^k s}\} \tag{2.5.18}$$

is fulfilled.

Proof. We prove (2.5.18) by induction on k. For $k \leqslant 2$ this inequality follows from (2.5.17), (2.5.13) and Theorem 2.2.7, according to which

$$\sum_{j,l=1}^{r} \|[X_j, X_l] u\|^2_{s-\frac{1}{2}} \leqslant C_1 \left\{ \sum_{j=1}^{r} \|X_j u\|_s^2 + \|u\|_s^2 \right\} \leqslant C_2 \{\|Pu\|_0^2 + \|u\|_{2s}^2\}.$$

We now assume that (2.5.18) is satisfied for all $k \leqslant k_0$, and show that it must also be satisfied for $k = k_0 + 1$ Two cases are possible:

1) $X_I = [X_j, X_{I_1}]$, where $|I| = k_0 + 1$ and $|I_1| = k_0$; $j = 1, \ldots, r$,

2) $X_I = [X_0, X_{I_0}]$, where $|I| = k_0 + 1$ and $|I_0| = k_0 - 1$.

In order to estimate the norm of $X_I u$ in the first case, we use inequality (2.2.36) for $P = X_j$, $Q = X_{I_1}$ and $s_1 = 2^{1-k_0} - 1$. We obtain

$$\sum_{|I_1|=k_0} \sum_{j=1}^{r} \|[X_j, X_{I_1}] u\|^2_{s-\frac{1}{2}+\left(2^{-k_0}-\frac{1}{2}\right)}$$
$$\leqslant C_3 \left\{ \sum_{j=1}^{r} \|X_j u\|_s^2 + \sum_{|I_1|=k_0} \|X_{I_1} u\|^2_{s+2^{1-k_0}-1} + \|u\|_s^2 \right\}.$$

The estimate (2.5.13) and the induction hypothesis yield

$$\sum_{|I_1|=k_0} \sum_{j=1}^{r} \|[X_j, X_{I_1}] u\|^2_{s+2^{-k_0}-1} \leqslant C_4 \{\|Pu\|_0^2 + \|u\|^2_{2^{k_0}s}\}. \tag{2.5.19}$$

We now obtain estimate (2.5.18) for the case of the operator $X_I = [X_0, X_{I_0}]$, $|I_0| = k_0 - 1$. We denote $2^{-k_0} - 1$ by ρ. It is easy to see that

$$[P, X_{I_0}] u = \sum_{j=1}^{r} (2X_j [X_j, X_{I_0}] u - [X_j, [X_j, X_{I_0}]] u)$$
$$+ i [X_0, X_{I_0}] u + [\gamma, X_{I_0}] u.$$

In this equation we put $E_s u$ in place of u, and take the scalar product of the resulting equation with $E_\rho^* E_\rho [X_0, X_{I_0}] E_s u$ in the space $\mathcal{H}_0$. We thus have

$$([P, X_{I_0}] E_s u, E_\rho^* E_\rho [X_0, X_{I_0}] E_s u) = i \|E_\rho [X_0, X_{I_0}] E_s u\|_0^2$$
$$+ ([\gamma, X_{I_0}] E_s u, E_\rho^* E_\rho [X_0, X_{I_0}] E_s u)$$
$$+ \sum_{j=1}^{r} \{2 ([X_j, X_{I_0}] E_s u, X_j^* E_\rho^* E_\rho [X_0, X_{I_0}] E_s u)$$
$$- (E_\rho [X_j, [X_j, X_{I_0}]] E_s u, E_\rho [X_0, X_{I_0}] E_s u)\}. \tag{2.5.20}$$

This equation yields an estimate for $\|E_\rho [X_0, X_{I_0}] E_s u\|_0^2$. We estimate the remaining terms in (2.5.20). We have

$$V_1 \equiv |([\gamma, X_{I_0}] E_s u, E_\rho^* E_\rho [X_0, X_{I_0}] E_s u)|$$
$$\leqslant \|[\gamma, X_{I_0}] E_s u\|_0^2 + \|E_\rho^* E_\rho [X_0, X_{I_0}] E_s u\|_0^2.$$

Applying Theorem 2.2.1 on the order of a pseudodifferential operator, we obtain

$$V_1 \leqslant C_5 \{\|u\|_s^2 + \|u\|^2_{s+2^{1-k_0}-1}\} \leqslant C_6 \|u\|_s^2,$$

since $2^{1-k_0} - 1 \leqslant 0$. Furthermore, by the Schwarz inequality (2.1.19) we have

$$V_2 \equiv \sum_{j=1}^{r} |([X_j, X_{I_0}] E_s u, X_j^* E_\rho^* E_\rho [X_0, X_{I_0}] E_s u)|$$

$$\leqslant \sum_{j=1}^{r} \{\|[X_j, X_{I_0}] E_s u\|^2_{2^{1-k_0}-1} + \|X_j^* E_\rho^* E_\rho [X_0, X_{I_0}] E_s u\|^2_{1-2^{1-k_0}}\}$$

$$\leqslant C_7 \sum_{j=1}^{r} \{\|[X_j, X_{I_0}] u\|^2_{s+2^{1-k_0}-1} + \|[[X_j, X_{I_0}], E_s] u\|^2_{2^{1-k_0}-1}$$

$$+ \|X_j^* u\|_s^2 + \|[X_j^*, E_\rho^* E_\rho [X_0, X_{I_0}] E_s] u\|^2_{1-2^{1-k_0}}\}.$$

Applying Theorem 2.2.1 and estimate (2.5.13), we obtain

$$V_2 \leqslant C_8 \left\{\sum_{j=1}^{r} \|[X_j, X_{I_0}] u\|^2_{s+2^{1-k_0}-1} + \|Pu\|_0^2 + \|u\|_{2s}^2\right\}.$$

It is easy to see that

$$V_3 \equiv \sum_{j=1}^{r} |(E_\rho [X_j, [X_j, X_{I_0}]] E_s u, E_\rho [X_0, X_{I_0}] E_s u)|$$

$$\leqslant r\delta \|E_\rho [X_0, X_{I_0}] E_s u\|_0^2 + \frac{1}{\delta} \sum_{j=1}^{r} \|E_\rho [X_j, [X_j, X_{I_0}]] E_s u\|_0^2,$$

where $\delta > 0$ is an arbitrary constant. For the last term of this inequality the estimate

$$\sum_{j=1}^{r} \|E_\rho [X_j, [X_j, X_{I_0}]] E_s u\|_0^2$$

$$\leqslant C_9 \sum_{j=1}^{r} \{\|[X_j, [X_j, X_{I_0}]] u\|^2_{s+2^{-k_0}-1} + \|E_\rho [[X_j, [X_j, X_{I_0}]], E_s] u\|_0^2\}$$

$$\leqslant C_{10} \sum_{j=1}^{r} \{\|u\|^2_{s+2^{-k_0}-1} + \|[X_j, [X_j, X_{I_0}]] u\|^2_{s+2^{-k_0}-1}\}$$

is valid. Therefore

$$V_3 \leqslant r\delta \|E_\rho [X_0, X_{I_0}] E_s u\|_0^2 + C_\delta \{\|u\|_s^2 + \sum_{j=1}^{r} \|[X_j, [X_j, X_{I_0}]] u\|^2_{s+2^{-k_0}-1}\}.$$

In order to estimate the expression

$$V_4 \equiv ([P, X_{I_0}] E_s u, E_\rho^* E_\rho [X_0, X_{I_0}] E_s u),$$

we represent it in the form $V_4 = V_5 + V_6$, where

$$V_5 = (X_{l_0}E_s u,\ P^* E_\rho^* E_\rho [X_0, X_{l_0}] E_s u),$$
$$V_6 = -(PE_s u,\ X_{l_0}^* E_\rho^* E_\rho [X_0, X_{l_0}] E_s u).$$

Using the Schwarz inequality, we obtain

$$|V_6| \leqslant \|PE_s u\|_{-s}^2 + \|X_{l_0}^* E_\rho^* E_\rho [X_0, X_{l_0}] E_s u\|_s^2.$$

We estimate the first term on the right of this inequality using (2.5.12), and the second does not surpass

$$3\{\|E_\rho^* E_\rho [X_0, X_{l_0}] E_s X_{l_0} u\|_s^2 + \|E_\rho^* E_\rho [X_0, X_{l_0}] E_s (X_{l_0}^* - X_{l_0}) u\|_s^2 + \|[X_{l_0}^*, E_\rho^* E_\rho [X_0, X_{l_0}] E_s] u\|_s^2\}.$$

Bearing Theorems 2.2.1 and 2.2.3 in mind, we obtain

$$|V_6| \leqslant C_{11} (\|Pu\|_0^2 + \|X_{l_0} u\|_{2s+2^{1-k_0}-1}^2 + \|u\|_{2s}^2),$$

since $2^{1-k_0} - 1 \leqslant 0$ for $k_0 \geqslant 1$.

In order to estimate $|V_5|$ we use the following expression for the operator P^*:

$$P^* v = -Pv + \sum_{j=1}^{r} (2X_j^2 + \beta_j X_j) v + \beta_0 v,$$

where β_j and β_0 are certain functions. It is easy to see that

$$\begin{aligned}|V_5| \leqslant\ & |(X_{l_0} E_s u, -PE_\rho^* E_\rho [X_0, X_{l_0}] E_s u)| \\ & + \sum_{j=1}^{r} \{|(X_{l_0} E_s u, 2X_j^2 E_\rho^* E_\rho [X_0, X_{l_0}] E_s u)| \\ & + |(X_{l_0} E_s u, \beta_j X_j E_\rho^* E_\rho [X_0, X_{l_0}] E_s u)|\} \\ & + |(X_{l_0} E_s u, \beta E_\rho^* E_\rho [X_0, X_{l_0}] E_s u)|. \end{aligned} \tag{2.5.21}$$

We estimate the scalar product on the right of (2.5.21). According to (2.1.19) we have

$$\begin{aligned} W_1 &= |(X_{l_0} E_s u, PE_\rho^* E_\rho [X_0, X_{l_0}] E_s u)| \\ &\leqslant \|X_{l_0} E_s u\|_{s+2^{1-k_0}-1}^2 + \|PE_\rho^* E_\rho [X_0, X_{l_0}] E_s u\|_{1-s-2^{1-k_0}}^2. \end{aligned}$$

Since the operator $E_\rho^* E_\rho [X_0, X_{l_0}] E_s$ is of order at most $s - 1 + 2^{1-k_0}$, it follows from Lemma 2.5.1 and the Remark to it, and from Theorems 2.2.1 and 2.2.3, that

$$W_1 \leqslant C_{12} \{\|Pu\|_0^2 + \|u\|_{2s}^2 + \|X_{l_0} u\|_{2s+2^{1-k_0}-1}^2\}.$$

Here we have used the fact that $2^{1-k_0} - 1 \leqslant 0$ for $k_0 \geqslant 1$.

Further applying inequality (2.1.19) as well as Theorems 2.2.1 and 2.2.3, we obtain

$$W_2 = \sum_{j=1}^{r} |(X_{l_0} E_s u, \beta_j X_j E_\rho^* E_\rho [X_0, X_{l_0}] E_s u)| \leqslant$$

$$\leqslant \sum_{j=1}^{r} \{\|X_{I_0} E_s u\|^2_{2^{1-k_0}-1} + \|\beta_j X_j E^*_\rho E_\rho [X_0, X_{I_0}] E_s u\|^2_{1-2^{1-k_0}}\}$$

$$\leqslant C_{13} \{\|X_{I_0} u\|^2_{s+2^{1-k_0}-1} + \sum_{j=1}^{r} \|X_j u\|^2_s$$

$$+ \|[X_{I_0}, E_s] u\|^2_{2^{1-k_0}-1} + \sum_{j=1}^{r} \|\beta_j [X_j, E^*_\rho E_\rho [X_0, X_{I_0}] E_s] u\|^2_{1-2^{1-k_0}}\}.$$

From this and estimate (2.5.13) follows

$$W_2 \leqslant C_{14} \{\|X_{I_0} u\|^2_{s+2^{1-k_0}-1} + \|Pu\|^2_0 + \|u\|^2_{2s}\}.$$

In the same way we obtain

$$W_3 = |(X_{I_0} E_s u, \beta E^*_\rho E_\rho [X_0, X_{I_0}] E_s u)| \leqslant C_{15} \{\|X_{I_0} u\|^2_{s+2^{1-k_0}-1} + \|u\|^2_s\}.$$

We further examine

$$W_4 = \sum_{j=1}^{r} |(X^*_j X_{I_0} E_s u, X_j E^*_\rho E_\rho [X_0, X_{I_0}] E_s u)|.$$

By the Schwarz inequality (2.1.19),

$$W_4 \leqslant \sum_{j=1}^{r} \{\|X^*_j X_{I_0} E_s u\|^2_{2^{1-k_0}-1} + \|X_j E^*_\rho E_\rho [X_0, X_{I_0}] E_s u\|^2_{1-2^{1-k_0}}\}$$

$$\leqslant C_{16} \sum_{j=1}^{r} \{\|X_j u\|^2_s + \|u\|^2_s + \|X^*_j X_{I_0} E_s u\|^2_{2^{1-k_0}-1}\}. \tag{2.5.22}$$

In order to estimate

$$W_5 = \sum_{j=1}^{r} \|X^*_j X_{I_0} E_s u\|^2_{2^{1-k_0}-1},$$

we first obtain an estimate for $\Sigma_1^r \|X_j X_{I_0} E_{2^{1-k_0}-1} E_s u\|^2_0$. Setting the function $v = X_{I_0} E_{2^{1-k_0}-1} E_s u$ in place of $E_s u$ in the equation (2.5.9), we obtain

$$\sum_{j=1}^{r} \|X_j v\|^2_0 \leqslant C_{17} \{|(Pv, v)| + \|v\|^2_0\}. \tag{2.5.23}$$

From this it follows that

$$\sum_{j=1}^{r} \|X_j v\|^2_0 \leqslant C_{18} \{\|X_{I_0} u\|^2_{s+2^{1-k_0}-1} + \|Pv\|^2_{-s-2^{1-k_0}} + \|u\|^2_s + \|v\|^2_{s+2^{1-k_0}}\}. \tag{2.5.24}$$

To estimate $\|Pv\|^2_{-s-2^{1-k_0}}$ we apply Lemma 2.5.1. It is easy to see that

$$\|v\|^2_{s+2^{1-k_0}} \leqslant C_{19} \{\|X_{I_0} u\|^2_{2s+2^{2-k_0}-1} + \|u\|^2_{2s+2^{2-k_0}-1}\}. \tag{2.5.25}$$

Since $2^{2-k_0}-1 \leqslant 0$ for $k_0 \geqslant 2$ and $s \geqslant 0$, we obtain from (2.5.24) and (2.5.25) the estimate

$$\sum_{j=1}^{r} \|X_j v\|_0^2 \leqslant C_{20}\{\|X_{l_0}u\|^2_{2s+2^{2-k_0}-1}+\|Pu\|_0^2+\|u\|_{2s}^2\}. \tag{2.5.26}$$

Here we also have used the inequality

$$\|u\|_\mu^2 \leqslant \|u\|_l^2 \quad \text{for } \mu \leqslant l. \tag{2.5.27}$$

Thus, applying Theorem 2.2.6 and Theorem 2.2.3, we obtain the inequality

$$W_5 \leqslant 2\sum_{j=1}^{r} \|E_{2^{1-k_0}-1}X_j^* X_{l_0} E_s u\|_0^2 + C_{21}\|u\|_0^2$$

$$\leqslant 4\sum_{j=1}^{r} \|X_j X_{l_0} E_{2^{1-k_0}-1} E_s u\|_0^2 + C_{22}\{\|X_{l_0}u\|^2_{s+2^{1-k_0}-1}$$

$$+\sum_{j=1}^{r} \|X_j u\|^2_{s+2^{1-k_0}-1}+\|u\|^2_{s+2^{1-k_0}-1}+\|u\|_0^2\} \tag{2.5.28}$$

for any function $u \in C_0^\infty(K)$, $K \subset K_1$.

Taking into account that $2^{1-k_0}-1 \leqslant 0$ for $k_0 \geqslant 1$ and $s \geqslant 0$, we obtain

$$W_5 \leqslant C_{23}\left\{\sum_{j=1}^{r} \|X_j v\|_0^2 + \|X_{l_0}u\|^2_{s+2^{1-k_0}-1}+\|u\|_s^2+\sum_{j=1}^{r}\|X_j u\|_s^2\right\}. \tag{2.5.29}$$

Bearing in mind (2.5.22), (2.5.29), (2.5.13) and (2.5.26), we finally obtain the following relation for W_4:

$$W_4 \leqslant C_{24}\{\|Pu\|_0^2+\|u\|_{2s}^2+\|X_{l_0}u\|^2_{2s+2^{2-k_0}-1}\} \tag{2.5.30}$$

and hence

$$|V_5| \leqslant W_1+W_2+W_3+W_4 \leqslant C_{25}\{\|Pu\|_0^2+\|X_{l_0}u\|^2_{2s+2^{2-k_0}-1}+\|u\|_{2s}^2\}. \tag{2.5.31}$$

Furthermore, from (2.5.20) we obtain the relation

$$\|E_\rho[X_0, X_{l_0}]\,E_s u\|_0^2 \leqslant V_1+V_2+V_3+|V_5|+|V_6|. \tag{2.5.32}$$

From Theorem 2.2.6 we have

$$\|[X_0, X_{l_0}]\,u\|^2_{s+2^{-k_0}-1} \leqslant 2\,\|E_\rho E_s[X_0, X_{l_0}]\,u\|_0^2+C_{26}\|u\|_0^2$$

$$\leqslant 4\,\|E_\rho[X_0, X_{l_0}]\,E_s u\|_0^2+C_{27}\{\|E_\rho[E_s, [X_0, X_{l_0}]]\,u\|_0^2+\|u\|_0^2\}$$

$$\leqslant 4\,\|E_\rho[X_0, X_{l_0}]\,E_s u\|_0^2+C_{28}\|u\|_s^2. \tag{2.5.33}$$

Therefore, considering (2.5.33) and (2.5.32), we obtain

$$\|[X_0, X_{I_0}]\,u\|^2_{s+2^{-k_0-1}}$$

$$\leqslant C_{29}\{\|Pu\|_0^2+C(\delta)\,\|u\|^2_{2s}+\sum_{j=1}^{r}\|[X_j, X_{I_0}]\,u\|^2_{s+2^{1-k_0-1}}+\|X_{I_0}u\|^2_{2s+2^{2-k_0-1}}$$

$$+\sum_{j=1}^{r}C(\delta)\,\|[X_j, [X_j, X_{I_0}]]\,u\|^2_{s+2^{-k_0-1}}+\delta\|\,E_\rho\,[X_0, X_{I_0}]E_s u\|_0^2\},$$

(2.5.34)

where $\delta > 0$ is an arbitrary positive constant and C_{29} does not depend on δ.

Since

$$\delta\|E_\rho\,[X_0, X_{I_0}]\,E_s u\|_0^2\leqslant 2\delta\,(\|\,[X_0, X_{I_0}]\,u\|^2_{s+2^{-k_0-1}}+C_{30}\|u\|_s^2),$$

we may use (2.5.27) and choose δ sufficiently small to obtain

$$\|[X_0, X_{I_0}]\,u\|^2_{s+2^{-k_0-1}}\leqslant C_{31}\Big\{\|Pu\|_0^2+\|u\|^2_{2s}+\sum_{j=1}^{r}\|[X_j, X_{I_0}]\,u\|^2_{s+2^{1-k_0-1}}$$

$$+\sum_{j=1}^{r}\|[X_j, [X_j, X_{I_0}]]\,u\|^2_{s+2^{-k_0-1}}+\|X_{I_0}u\|^2_{2s+2^{2-k_0-1}}\Big\}.$$

(2.5.35)

Since for $X_I = [X_j, X_{I_0}]$ with $j \neq 0$ we have $|I| = k_0$, the induction hypothesis implies

$$\sum_{j=1}^{r}\|\,[X_j, X_{I_0}]\,u\,\|^2_{s+2^{1-k_0-1}}\leqslant C_{32}\,\{\|\,Pu\,\|_0^2+\|\,u\,\|^2_{2^{k_0}s}\}.$$

For $X_I = [X_j, [X_j, X_{I_0}]],\, j \neq 0$, we have $|I| = k_0 + 1$, and for such X_I the inequality (2.5.19) proved above is valid.

Since $|I_0| = k_0 - 1$, from the induction hypothesis we have

$$\|\,X_{I_0}u\,\|^2_{2s+2^{1-(k_0-1)-1}}\leqslant C_{33}\,\{\|\,Pu\,\|_0^2+\|\,u\,\|^2_{2^{k_0-1}\cdot 2s}\}.$$

Therefore from (2.5.35) we finally obtain

$$\|\,[X_0, X_{I_0}]\,u\,\|^2_{s+2^{-k_0-1}}\leqslant C_{34}\,\{\,\|\,Pu\,\|_0^2+\|\,u\,\|^2_{2^{k_0+1}s}\},$$

which is the desired conclusion.

Definition. The system of differential operators $\{X_0, \cdots, X_r\}$ is called a system of rank m at the point x_0 if there exists a number $R(x_0)$ such that

$$\sum_{|I|\leqslant R(x_0)}|X_I\,(x_0, \xi)\,|>0 \quad \text{for} \quad \xi\neq 0, \tag{2.5.36}$$

where $X_I(x, \xi)$ is the symbol of the differential operator $X_I = \operatorname{ad} X_{\alpha_1} \cdots \operatorname{ad} X_{\alpha_{t-1}} X_{\alpha_t}$ corresponding to the multi-index $I = (\alpha_1, \cdots, \alpha_t)$.

It is clear that condition (2.5.36) is equivalent to the fact that for $x = x_0$, among the operators $\{X_I\}$, where $|I| \leqslant R(x_0)$, there exist m linearly independent operators, i.e. the rank of the Lie algebra generated by the operators $\{X_0, \cdots, X_r\}$ is equal to m.

LEMMA 2.5.3. *If at each point* $x \in K$, *where* K *is a compact set in* R^m, *the system of operators* $\{X_0, \cdots, X_r\}$ *is of rank* m, *then for any* $s \in R^1$ *there exists a constant* $C(K, s)$ *such that*

$$\| u \|_{1+s}^2 \leqslant C(K, s) \Big\{ \sum_{|I| \leqslant R(K)} \| X_I u \|_s^2 + \| u \|_s^2 \Big\}, \tag{2.5.37}$$

where $R(K) = \sup_{x \in K} R(x)$, $u \in C_0^\infty(K)$.

PROOF. Since $X_I(x, \xi)$ is a homogeneous function of ξ of first degree, continuous with respect to x and ξ, it follows from condition (2.5.36) that for all x in some sufficiently small neighborhood $O(x_0)$ of the point x_0 the inequality

$$1 + \sum_{|I| \leqslant R(x_0)} | X_I(x, \xi) |^2 \geqslant C_0 (1 + |\xi|^2); \ \xi \in R^m \tag{2.5.38}$$

holds, where $C_0 = \text{const} > 0$ and x_0 is any point belonging to K.

We select a finite subcovering from the infinite covering of K by such neighborhoods $O(x)$, and thus obtain that for all x belonging to some closed domain $K_0 \supset K$ the inequality

$$\sum_{|I| \leqslant R(K)} | X_I(x, \xi) |^2 + 1 \geqslant C (1 + |\xi|^2) \tag{2.5.39}$$

is valid, where $C = \text{const} > 0$ and K lies strictly inside K_0.

On the basis of Theorem 2.2.9 it follows from condition (2.5.39) that inequality (2.5.37) is valid for functions $u \in C_0^\infty(K)$.

THEOREM 2.5.2. *Suppose the system of operators* $\{X_0, \cdots, X_r\}$ *has rank* m *at every point* $x \in \Omega$. *Then for any function* $u \in D'(\Omega)$ *such that* $Pu \in \mathcal{H}_s^{\text{loc}}(\Omega)$ *an estimate of the form*

$$\| \varphi u \|_{s+\varepsilon(K)}^2 \leqslant C(K, s) \{ \| \varphi_1 P u \|_s^2 + \| \varphi_1 u \|_\gamma^2 \} \tag{2.5.40}$$

holds, where $\gamma = \text{const} < s + \epsilon(K)$, *the functions* $\phi, \phi_1 \in C_0^\infty(K)$, K *is a compact set in* Ω, $\phi_1 \equiv 1$ *in a neighborhood of the support of* ϕ, *and* $\epsilon(K) > 0$ *is some number. Moreover, the operator* P *given by* (2.5.1) *is hypoelliptic in the domain* Ω.

PROOF. It suffices to prove that if the system $\{X_0, \cdots, X_r\}$ of operators has rank m at each point $x \in \Omega$, then the operator P from (2.5.1) satisfies the conditions of Theorem 2.4.2 with the set M understood to be empty. Suppose that the domain $G \subset \Omega$ and the closure $\bar{G}$ also belongs to Ω. Suppose that $\bar{G}$ is such that $G \supset K$ and $R(\bar{G}) = R(K)$. On the basis of Lemmas 2.5.3 and 2.5.2 we conclude that for any $t \in R^1$ the inequality

$$\|u\|_{t+1}^2 \leqslant C_1\Big(\sum_{|I|\leqslant R(K)} \|X_I u\|_t^2 + \|u\|_t^2\Big)$$

$$\leqslant C_2\{\|Pu\|_0^2 + \|u\|^2_{2^{R(K)}(t+1-2^{1-R(K)})}\} \tag{2.5.41}$$

is valid for any function u in $C_0^\infty(G)$, if $s = t + 1 - 2^{1-R(K)} \geqslant 0$. We choose t so that

$$t + 1 > 2^{R(K)}(t + 1 - 2^{1-R(K)}),$$
$$t + 1 > 2^{1-R(K)}. \tag{2.5.42}$$

Then it follows from (2.5.41) and Theorem 2.1.12 that for $u \in C_0^\infty(G)$ we have

$$\|u\|_{t+1}^2 \leqslant C_3\{\|Pu\|_0^2 + \|u\|_{-N}^2\} \tag{2.5.43}$$

for any $N > 0$ and for some t satisfying the inequality $-1 + 2^{1-R(K)} \leqslant t < -1 + 2(2^{R(K)} - 1)^{-1}$. We set $t + 1 = \epsilon(K)$. Clearly $\epsilon(K) > 0$. Then from (2.5.43) we obtain that for any $N > 0$ and $u \in C_0^\infty(G)$

$$\|u\|_\varepsilon^2 \leqslant C_4\{\|Pu\|_0^2 + \|u\|_{-N}^2\}. \tag{2.5.44}$$

This means that condition I of Theorem 2.4.2 is fulfilled for $u \in C_0^\infty(G)$. It follows from Theorem 2.5.1 and inequality (2.5.44) that if we set $2s = \epsilon$, then

$$\|X_0 u\|^2_{\frac{\varepsilon-1}{2}} + \sum_{j=1}^r \|X_j u\|^2_{\frac{\varepsilon}{2}} \leqslant C_5\{\|Pu\|_0^2 + \|u\|_{-N}^2\}.$$

It is clear that

$$P^{(j)}(x, \xi) = \sum_{k=1}^r 2a_k^j(x) X_k(x, \xi) + i a_0^j(x),$$

$$P_{(j)}(x, \xi) = \sum_{k=1}^r 2X_{k(j)}(x, \xi) X_k(x, \xi) + iX_{0(j)}(x, \xi) + \gamma_{(j)}(x),$$

where $X_k(x, \xi) = \Sigma_{l=1}^m a_k^l(x)\xi_l$; $k = 0, 1, \cdots, r$. Hence on the basis of Theorem 2.2.1 we have the inequality

$$\sum_{j=1}^r \{\|P^{(j)}u\|_s^2 + \|P_{(j)}u\|_{s-1}^2\} \leqslant C_6\left\{\sum_{j=1}^r \|X_j u\|_s^2 + \|u\|_s^2\right\}, \quad u \in C_0^\infty(G). \tag{2.5.45}$$

We take the operator $E_s u$ such that $E_s u \in C_0^\infty(G)$ and set $\phi(x) \equiv 1$ at points of the domain Ω_1 such that $K \subset \Omega_1$, $\bar{\Omega}_1 \subset G$, where $\bar{\Omega}_1$ is the closure of Ω_1.

In the inequality

$$\|u\|_\varepsilon^2 + \sum_{j=1}^r \|X_j u\|_{\varepsilon/2}^2 \leqslant C_7\{\|Pu\|_0^2 + \|u\|_{-N}^2\}$$

we replace u by $E_s u$ and thus obtain, for $u \in C_0^\infty(K)$,

$$\|E_s u\|_\varepsilon^2 + \sum_{j=1}^{r} \|E_s X_j u\|_{\varepsilon/2}^2$$
$$\leqslant C_8 \left\{ \|PE_s u\|_0^2 + \sum_{j=1}^{r} \|[X_j, E_s]\, u\|_{\varepsilon/2}^2 + \|u\|_{-N_1}^2 \right\}. \tag{2.5.46}$$

From Theorem 2.2.3 on commutators it follows that

$$\|[P, E_s]\, u\|_0^2 \leqslant C_9 \left\{ \sum_{j=1}^{m} (\|E_{s(j)} P^{(j)} u\|_0^2 + \|E_s^{(j)} P_{(j)} u\|_0^2 + \|u\|_s^2 \right\}$$
$$\leqslant C_{10} \left\{ \|u\|_s^2 + \sum_{j=1}^{m} (\|P_{(j)} u\|_{s-1}^2 + \|P^{(j)} u\|_s^2) \right\}, \tag{2.5.47}$$

where $E_{s(j)}$ and $E_s^{(j)}$ are the pseudodifferential operators with symbols

$$D_j \varphi(x)(1 + |\xi|^2)^{s/2} \quad \text{and} \quad \frac{\partial}{\partial \xi_j} (\varphi(x)(1 + |\xi|^2)^{s/2})$$

respectively. Using inequality (2.5.45), we obtain

$$\|[P, E_s]\, u\|_0^2 \leqslant C_{11} \left\{ \|u\|_s^2 + \sum_{j=1}^{r} \|X_j u\|_s^2 \right\}.$$

From (2.5.46) and Theorem 2.2.6 follows

$$\|u\|_{s+\varepsilon}^2 + \sum_{j=1}^{r} \|X_j u\|_{s+\varepsilon/2}^2$$
$$\leqslant C_{12} \left\{ \|Pu\|_s^2 + \|u\|_{s+\varepsilon/2}^2 + \sum_{j=1}^{r} \|X_j u\|_s^2 + \|u\|_{-N_2}^2 \right\}. \tag{2.5.48}$$

Bearing Theorem 2.1.12 in mind, we conclude that for $u \in C_0^\infty(K)$

$$\|u\|_{s+\varepsilon}^2 + \sum_{j=1}^{r} \|X_j u\|_{s+\varepsilon/2}^2 \leqslant C_{13} \{\|Pu\|_s^2 + \|u\|_{-N_3}^2\}. \tag{2.5.49}$$

Thus from (2.5.45) and (2.5.49) it follows that for $u \in C_0^\infty(K)$

$$\sum_{j=1}^{m} \|P^{(j)} u\|_s^2 \leqslant C_{14} \{\|Pu\|_{s-\frac{\varepsilon}{2}}^2 + \|u\|_{-N_4}^2\}, \tag{2.5.50}$$

where N_4 is an arbitrary positive number, and C_{14} depends on N_4.

Bearing Theorem 2.1.12 in mind, we deduce that

$$\sum_{j=1}^{m} \| P_{(j)} u \|_{s-1}^2 \leqslant C_{15} \{ \| Pu \|_{s-\varepsilon/2}^2 + \| u \|_{-N}^2 \}_4$$

$$\leqslant \delta \| Pu \|_s^2 + C_{16} \| u \|_{-N_5}^2 \tag{2.5.51}$$

for any $\delta > 0$ and $N_5 > 0$, where C_{16} depends on δ and N_5, and $u \in C_0^\infty(K)$. This means that the operator P satisfies conditions I, II and IIIa of Theorem 2.4.2 for the domain Ω, and therefore the assertions of Theorem 2.5.2 are valid.

We now prove a theorem on the smoothness of weak solutions and on hypoellipticity of P in the case when on some set $M \in \Omega$ the system of operators $\{X_0, \cdots, X_r\}$ does not satisfy the conditions of Theorem 2.5.2.

The proof of Theorem 2.5.3 given below does not depend on the special form (2.5.1) of the operator P.

THEOREM 2.5.3. *Suppose the following conditions are fulfilled relative to the operator P of the form* (2.5.1) *in a domain Ω:*

1) *At points $x \in \Omega \backslash M$ the system of operators $\{X_0, \cdots, X_r\}$ has rank m, where M is a bounded set of points lying on a finite number of $(m-1)$-dimensional smooth manifolds $\mathfrak{M}$ such that the closure of M belongs to Ω.*

2) *At each point $x \in M$ the following conditions hold: either for some $k = 1, \cdots, r$*

$$a_k^j(x) \Phi_{x_j} \neq 0, \tag{2.5.52}$$

or, if $\sum_{k=1}^r |a_k^j(x)\Phi_{x_j}| = 0$, then

$$\sum_{j=1}^{r} X_j^2 \Phi + i X_0 \Phi \neq 0, \tag{2.5.53}$$

where $\Phi(x_1, \cdots, x_m) = 0$ is the equation for $\mathfrak{M}$ in a neighborhood of the point x, $\operatorname{grad} \Phi(x) \neq 0$, $X_k = a_k^j D_j$, $k = 0, 1, \cdots, r$.

Then for each distribution u in $D'(\Omega)$ such that $Pu \in \mathcal{H}_s^{\mathrm{loc}}(\Omega)$ the estimate

$$\|\varphi u\|_s^2 \leqslant C(\varphi, \gamma) \{ \|\varphi_1 Pu\|_s^2 + \|\varphi_1 u\|_\gamma^2 \} \tag{2.5.54}$$

holds, where the functions $\phi, \phi_1 \in C_0^\infty(\Omega)$, $\phi_1 \equiv 1$ on $\operatorname{supp} \phi$ *and, moreover, either* $\operatorname{supp} \phi_1 \cap M = \varnothing$, *or $\phi \equiv 1$ on M, and* $\gamma = \mathrm{const} < s$.

A differential operator P satisfying conditions 1) *and* 2) *is globally hypoelliptic in Ω, i. e. a distribution u in $D'(\Omega)$ such that $Pu \in C^\infty$ also belongs to $C^\infty(\Omega)$*

PROOF. We cover the set M by a finite number of domains Ω_j, $j = 1, \cdots, N_1$, such that in each of the domains Ω_j we may choose a system of local coordinates $y_1, \cdots, y_m$ such that the set $\mathfrak{M} \cap \Omega_j$ lies in the plane $y_m = 0$. Moreover, we assume that the operator P in the new coordinates takes the form

$$-Pu = \alpha^{kj}u_{y_ky_j} + \beta^j u_{y_j} + cu, \alpha^{kj}\xi_k\xi_j \geqslant 0, \tag{2.5.55}$$

where at points of Ω_j

$$\text{either } \alpha^{mm} \neq 0, \text{ or } \beta^m \neq 0. \tag{2.5.56}$$

It is easy to see from the hypotheses of the theorem that such a covering of M by domains Ω_j is possible. Suppose $u \in C_0^\infty(G_{j,\beta})$, where $G_{j,\beta} = \Omega_j \cap \{|y_m| \leqslant \beta\}$, and the constant β will be chosen below.

In order to estimate $\|u\|_0^2$, we consider in $G_{j,\beta}$ the equation for a function v such that $u = v(T - e^{\mu y_m})$, where μ, $T = \text{const}$ and $\text{sign}\,\mu = \text{sign}\,\beta^m$ in Ω_j if $\beta^m \neq 0$ in Ω_j; and $T > 0$. We have

$$\begin{aligned} -(T - e^{\mu y_m})^{-1} Pu &= \alpha^{kj} v_{y_k y_j} + \beta^j v_{y_j} + cv \\ &- (\alpha^{mm}\mu^2 + \beta^m\mu) e^{\mu y_m} (T - e^{\mu y_m})^{-1} v - 2\alpha^{mj}\mu e^{\mu y_m} (T - e^{\mu y_m})^{-1} v_{y_j}. \end{aligned} \tag{2.5.57}$$

We multiply equation (2.5.57) by v and integrate over the domain $G_{j,\beta}$. Suppose v is a real function. Then after integration by parts we obtain

$$\begin{aligned} -((T - e^{\mu y_m})^{-1} Pu, v) &= -(\alpha^{kj} v_{y_k}, v_{y_j}) \\ &+ \frac{1}{2}(\alpha^{kj}_{y_k y_j} v, v) - \frac{1}{2}(\beta^j_{y_j} v, v) \\ &+ ((c - [\alpha^{mm}\mu^2 + \beta^m\mu] e^{\mu y_m} (T - e^{\mu y_m})^{-1}) v, v) \\ &- 2(\alpha^{mj}\mu e^{\mu y_m} (T - e^{\mu y_m})^{-1} v_{y_j}, v). \end{aligned} \tag{2.5.58}$$

From (2.5.58) it follows that

$$\begin{aligned} (\alpha^{kj} v_{y_k}, v_{y_j}) &- \left(\left[c + \frac{1}{2}\alpha^{kj}_{y_k y_j} - \frac{1}{2}\beta^j_{y_j} - (\alpha^{mm}\mu^2 + \beta^m\mu) e^{\mu y_m} (T - e^{\mu y_m})^{-1}\right] v, v\right) \\ &\leqslant |((T - e^{\mu y_m})^{-1} Pu, v)| + 2|(\alpha^{mj}\mu e^{\mu y_m} (T - e^{\mu y_m})^{-1} v_{y_j}, v)| \\ &\leqslant \|(T - e^{\mu y_m})^{-1} Pu\|_0 \|v\|_0 + \frac{1}{2}\left\|\frac{\alpha^{mj}}{\sqrt{\alpha^{mm}}} v_{y_j}\right\|_0^2 \\ &+ 2(\alpha^{mm}\mu^2 e^{2\mu y_m} (T - e^{\mu y_m})^{-2} v, v). \end{aligned} \tag{2.5.59}$$

According to Lemma 1.7.1 we have the inequality

$$(\alpha^{mj}(x)\xi_j)^2 \leqslant 2\alpha^{mm}(x)\alpha^{kj}(x)\xi_k\xi_j \quad \text{for any } \xi \in R^m$$

Therefore

$$\frac{1}{2}\left\|\frac{\alpha^{mj}}{\sqrt{\alpha^{mm}}} v_{y_j}\right\|_0^2 \leqslant (\alpha^{kj} v_{y_k}, v_{y_j}).$$

From (2.5.59) it follows that

$$\begin{aligned} -\Big(\Big(c + \frac{1}{2}\alpha^{kj}_{y_k y_j} - \frac{1}{2}\beta^j_{y_j} &- (\alpha^{mm}\mu^2 + \beta^m\mu) e^{\mu y_m} (T - e^{\mu y_m})^{-1} \\ &+ 2\alpha^{mm}\mu^2 e^{2\mu y_m} (T - e^{\mu y_m})^{-2}\Big) v, v\Big) \leqslant \|v\|_0 \|(T - e^{\mu y_m})^{-1} Pu\|_0. \end{aligned} \tag{2.5.60}$$

We choose the constant $\beta(\mu)$ so that for $|y_m| \leqslant \beta$ the inequality $\frac{1}{2} \leqslant e^{\mu y_m} \leqslant 2$ holds, and we choose the constant $T > 0$ from the condition that for $|y_m| \leqslant \beta$

$$2e^{\mu y_m}(T - e^{\mu y_m})^{-1} \leqslant 4(T-2)^{-1} < \frac{1}{2}.$$

Then by virtue of condition (2.5.56) and inequality (2.5.60) it follows that for sufficiently large $|\mu|$

$$a_0 |\mu| \, \|v\|_0^2 \leqslant (T-2)^{-1} \|Pu\|_0 \|v\|_0,$$

where $a_0 = \text{const} > 0$ does not depend on μ and T, with $v \in C_0^\infty(G_{j,\beta})$.

From (2.5.60) it follows that

$$\|u\|_0^2 \leqslant \frac{C_1}{\mu^2} \|Pu\|_0^2 \quad \text{for} \quad u \in C_0^\infty(G_{j,\beta}), \tag{2.5.61}$$

where the constant C_1 does not depend on μ.

For $u \in C_0^\infty(G_{j,\beta})$ we now estimate $\|P^{(j)}u\|_0^2$ and $\|P_{(j)}u\|_{-1}^2$. In the proof of Theorem 2.5.2 we have shown that

$$\|P_{(j)}u\|_{-1}^2 + \|P^{(j)}u\|_0^2 \leqslant C_2 \left\{ \sum_{j=1}^{r} \|X_j u\|_0^2 + \|u\|_0^2 \right\}. \tag{2.5.62}$$

Multiplying the equation $Pu = f$ by u and integrating over R^m, we obtain

$$\sum_{j=1}^{r} \|X_j u\|_0^2 \leqslant C_3 \{\|u\|_0^2 + \|Pu\|_0 \|u\|_0\}. \tag{2.5.63}$$

Therefore, bearing in mind (2.5.61), we conclude from (2.5.62) and (2.5.63) that for $u \in C_0^\infty(G_{j,\beta})$

$$\|P_{(j)}u\|_{-1}^2 + \|P^{(j)}u\|_0^2 \leqslant \frac{C_4}{|\mu|} \|Pu\|_0^2, \tag{2.5.64}$$

where C_4 does not depend on μ.

Suppose $\psi_j \in C_0^\infty(\Omega_j)$, $0 \leqslant \psi_j \leqslant 1$, $\Sigma_j \psi_j \equiv 1$ on M, and $G_\beta = \Sigma_j G_{j,\beta}$ for all $\beta \leqslant \beta_0$, where β_0 is some sufficiently small number. Suppose $u \in C_0^\infty(G_\beta)$. Then according to (2.5.61) we have

$$\|u\|_0 = \left\| \sum_j \psi_j u \right\|_0 \leqslant \sum_j \|\psi_j u\|_0 \leqslant \frac{C_5}{|\mu|} \sum_j \|P\psi_j u\|_0.$$

Considering inequality (2.5.64), we obtain

$$\begin{aligned} \|u\|_0^2 &\leqslant \frac{C_6}{\mu^2} \sum_j \|P\psi_j u\|_0^2 \leqslant \frac{C_7}{\mu^2} \sum_j \{\|\psi_j Pu\|_0^2 + \|D_l \psi_j P^{(l)} u\|_0^2 + \|u\|_0^2\} \\ &\leqslant \frac{C_8}{\mu^2} \sum_j \{\|Pu\|_0^2 + \|P^{(l)}(D_l \psi_j) \cdot u\|_0^2 + \|u\|_0^2\} \\ &\leqslant \frac{C_9}{\mu^2} \{\|Pu\|_0^2 + \|u\|_0^2\}. \end{aligned}$$

If $1/|\mu|$ is sufficiently small, then

$$\|u\|_0^2 \leqslant \frac{C_{10}}{|\mu|^2}\|Pu\|_0^2 \quad \text{for} \quad u \in C_0^\infty(G_\beta), \tag{2.5.65}$$

where C_{10} does not depend on μ. Since the coefficients of the operator P are real, inequality (2.5.65) is valid for any complex valued functions in $C_0^\infty(G_\beta)$. We note that β is of order $1/\mu$.

In order to prove Theorem 2.5.3, we shall show that conditions I, II and III of Theorem 2.4.2 are fulfilled.

First we verify condition I of Theorem 2.4.2 for $s_0 = 0$. In the proof of Theorem 2.5.2 we showed that for each compact set $K \subset \Omega$, each $s \in R^1$ and each $\beta > 0$ there exist constants $C(K, s, \beta)$ and $\epsilon(K, \beta) > 0$ such that

$$\sum_{j=1}^{m} \{\|P_{(j)}u\|_{s-1}^2 + \|P^{(j)}u\|_s^2\} + \|u\|_s^2$$
$$\leqslant C(K, s, \beta)\{\|Pu\|_{s-\varepsilon}^2 + C_N\|u\|_{s-N}^2\} \tag{2.5.66}$$

for each function $u \in C_0^\infty(K \backslash G_\beta)$. We now show that (2.5.65) and (2.5.66) yield the estimate

$$\|u\|_0^2 \leqslant \frac{C_{11}}{|\mu|}\|Pu\|_0^2 + C(\mu, N)\|u\|_{-N}^2 \tag{2.5.67}$$

for any function $u \in C_0^\infty(K)$ where C_{11} does not depend on μ.

Let $\{\psi_1, \psi_2\}$ be a partition of unity on K such that $\psi_1 \in C_0^\infty(G_{2\beta})$ and $\psi_1 \equiv 1$ on G_β. Then for any $u \in C_0^\infty(K)$ according to (2.5.64), (2.5.65) and (2.5.66) we have

$$I \equiv \|u\|_0^2 + \sum_{j=1}^{m} \{\|P^{(j)}u\|_0^2 + \|P_{(j)}u\|_{-1}^2\}$$
$$\leqslant C_{12}\sum_{l=1}^{2}\left\{\|\psi_l u\|_0^2 + \sum_{j=1}^{m}(\|P^{(j)}\psi_l u\|_0^2 + \|P_{(j)}\psi_l u\|_{-1}^2)\right\}$$
$$\leqslant \frac{C_{13}}{|\mu|}\|P\psi_1 u\|_0^2 + C_{14}(\mu)\{\|P\psi_2 u\|_{-\varepsilon}^2 + \|\psi_2 u\|_{-\varepsilon}^2\}.$$

Applying Theorem 2.2.3, we obtain

$$I \leqslant \frac{C_{15}}{|\mu|}\Big\{\|\psi_1 Pu\|_0^2 + \|D_j\psi_1 P^{(j)}u\|_0^2 + \sum_{|\alpha|=2}\|D^\alpha\psi_1 \cdot u\|_0^2\Big\}$$
$$+ C_{16}(\mu)\Big\{\|\psi_2 Pu\|_{-\varepsilon}^2 + \|D_j\psi_2 P^{(j)}u\|_{-\varepsilon}^2 + \sum_{|\alpha|=2}\|D^\alpha\psi_2 \cdot u\|_{-\varepsilon}^2 + \|\psi_2 u\|_{-\varepsilon}^2\Big\}. \tag{2.5.68}$$

Since the functions $D^\alpha\psi_1$ for $1 \leqslant |\alpha| \leqslant 2$ have support lying in $K\backslash G_\beta$, we obtain from (2.5.66) the estimate

$$\|D_j\psi_1 P^{(j)}u\|_0^2+\sum_{1\leqslant|\alpha|\leqslant 2}\|D^\alpha\psi_1 u\|_0^2$$
$$\leqslant C_{17}\Big\{\|P^{(j)}(D_j\psi_1\cdot u)\|_0^2+\sum_{1\leqslant|\alpha|\leqslant 2}\|D^\alpha\psi_1\cdot u\|_0^2\Big\}$$
$$\leqslant C_{18}(\mu)\Big\{\sum_{1\leqslant|\alpha|\leqslant 2}\|P(D^\alpha\psi_1\cdot u)\|_{-\varepsilon}^2+\|u\|_{-\varepsilon}^2\Big\}$$
$$\leqslant C_{19}(\mu)\Big\{\|Pu\|_{-\varepsilon}^2+\|u\|_{-\varepsilon}^2+\sum_{j=1}^m\|P^{(j)}u\|_{-\varepsilon}^2\Big\}. \tag{2.5.69}$$

From (2.5.68), and bearing (2.5.69) in mind, we deduce that

$$I\leqslant\frac{C_{20}}{|\mu|}\|Pu\|_0^2+C_{19}(\mu)\Big\{\|Pu\|_{-\varepsilon}^2+\sum_{j=1}^m\|P^{(j)}u\|_{-\varepsilon}^2+\|u\|_{-\varepsilon}^2\Big\}.$$

Applying inequality (2.1.30) proved in Theorem 2.1.12, we find that for $u\in C_0^\infty(K)$

$$\|u\|_0^2+\sum_{j=1}^m\{\|P_{(j)}u\|_{-1}^2+\|P^{(j)}u\|_0^2\}$$
$$\leqslant\frac{C_{21}}{|\mu|}\|Pu\|_0^2+C(\mu,N)\|u\|_{-N}^2 \tag{2.5.70}$$

if $1/|\mu|$ is sufficiently small, where $N>0$ is arbitrary and C_{21} does not depend on μ.

Condition I of Theorem 2.4.2 follows from (2.5.70).

Since inequality (2.5.70) is valid for every compact set K belonging to Ω and containing M, it follows from (2.5.70) and Theorem 2.2.6 that

$$\|u\|_s^2\leqslant 2\|E_s u\|_0^2+C_{22}\|u\|_{-N}^2$$
$$\leqslant\frac{C_{23}}{|\mu|}\|PE_s u\|_0^2+C_{24}(\mu,N_1)\|u\|_{-N_1}^2. \tag{2.5.71}$$

With the help of Theorem 2.2.3 on commutators and Theorem 2.2.1 on the order of a pseudodifferential operator, we deduce from (2.5.71) that for $u\in C_0^\infty(K)$

$$\|u\|_s^2\leqslant\frac{C_{25}}{|\mu|}\Big\{\|Pu\|_s^2+\sum_{j=1}^m(\|P_{(j)}u\|_{s-1}^2+\|P^{(j)}u\|_s^2+\|u\|_s^2\Big\}+C_{26}\|u\|_{-N_1}^2,$$

where C_{25} does not depend on μ. Thus for sufficiently small $1/|\mu|$ we have

$$\|u\|_s^2\leqslant\frac{C_{27}}{|\mu|}\Big\{\|Pu\|_s^2+\sum_{j=1}^m(\|P_{(j)}u\|_{s-1}^2+\|P^{(j)}u\|_s^2)\Big\}+C_{28}\|u\|_{-N_1}^2. \tag{2.5.72}$$

Furthermore, it follows from Theorem 2.2.6 and inequality (2.5.70) that

$$\sum_{j=1}^{m} \{\| P_{(j)} u \|_{s-1}^2 + \| P^{(j)} u \|_s^2\}$$

$$\leqslant 2 \sum_{j=1}^{m} (\| E_s P_{(j)} u \|_{-1}^2 + \| E_s P^{(j)} u \|_0^2) + C_{29} \| u \|_{-N}^2$$

$$\leqslant C_{30} \left\{ \sum_{j=1}^{m} (\| P^{(j)} E_s u \|_0^2 + \| P_{(j)} E_s u \|_{-1}^2) + \| u \|_s^2 \right\}$$

$$\leqslant \frac{C_{31}}{|\mu|} \| P E_s u \|_0^2 + C_{32} (\| u \|_s^2 + C(\mu) \| u \|_{-N}^2),$$

where C_{31} and C_{32} do not depend on μ. Applying Theorem 2.2.3 on commutators and estimate (2.5.72), we obtain

$$\sum_{j=1}^{m} \{\| P_{(j)} u \|_{s-1}^2 + \| P^{(j)} u \|_s^2\}$$

$$\leqslant \frac{C_{33}}{|\mu|} \left\{ \| P u \|_s^2 + \sum_{j=1}^{m} (\| P_{(j)} u \|_{s-1}^2 + \| P^{(j)} u \|_s^2) \right\} + C_{34}(\mu) \| u \|_{-N}^2. \tag{2.5.73}$$

Since C_{33} does not depend on μ and $|\mu|$ may be chosen as large as desired, it follows from (2.5.73) that conditions II and III of Theorem 2.4.2 are fulfilled. With this, Theorem 2.5.3 is proved.

The following result may easily be obtained from Theorem 2.5.3.

THEOREM 2.5.4. *Assume that at each point of Ω except the point x_0, the system of operators $\{X_0, \cdots, X_r\}$ has rank m, and assume that at least one of the coefficients a_k^j of the operators $X_k \equiv \Sigma_{j=1}^m a_k^j D_j$, $k = 0, 1, \cdots, r$, is different from zero at the point x_0. Then for each distribution $u \in D'(\Omega)$ such that $Pu \in \mathcal{H}_s^{\text{loc}}(\Omega)$, estimate* (2.5.54) *is valid and the operator P is hypoelliptic in the domain Ω.*

PROOF. We choose a vector $\mathbf{n}$ such that either $a_k^j n_j \neq 0$ for some $k \geqslant 1$, or $a_0^j n_j \neq 0$ at the point x_0. As the manifold M we choose a surface through the point x_0 and orthogonal to the vector $\mathbf{n}$. With such a choice of $\mathfrak{M}$ the assertion of Theorem 2.5.4 follows from Theorem 2.5.3.

Results analogous to Lemmas 2.5.2 and 2.5.3 in the case $X_0 \equiv 0$ were obtained independently by Kohn [66].

REMARK 1. The condition of Theorem 2.5.2 with regard to the rank of the system of operators $\{X_j, j = 0, 1, \cdots, r\}$ is also a necessary condition for hypoellipticity in the class of operators Pu satisfying the following condition: in the domain Ω all operators X_I are generated by m_1 linearly independent operators X_{I_s}, $s = 1, \cdots, m_1$. If $m_1 < m$, then the operator Pu is not hypoelliptic in Ω under the condition that the equation $Pu = 0$ has at least one nontrivial solution.

In fact, by Frobenius' theorem [25] in the case $m_1 < m$ the operator Pu may be transformed by a local change of independent variables to an operator acting only with respect to the variables $y_1, \cdots, y_{m_1}$. Thus if u is a nontrivial solution of the equation $Pu = 0$ in a neighborhood of some point x_0, we may replace u by zero on one side of the hyperplane y_m = const passing through x_0 to obtain a nonsmooth solution of $Pv = 0$, where

$$v = \begin{cases} u & \text{for } y_m > y_m^0 = \text{const}, \\ 0 & \text{for } y_m < y_m^0 = \text{const}. \end{cases}$$

§6. A priori estimates and hypoellipticity theorems for general second order differential equations

In this section, sufficient conditions for hypoellipticity in a domain Ω will be given, along with conditions for local smoothness of weak solutions of general second order equations with nonnegative characteristic form

$$L(u) = a^{kj}(x) u_{x_k x_j} + b^k u_{x_k} + cu = f \tag{2.6.1}$$

with

$$a^{kj}(x)\xi_k \xi_j \geqslant 0 \quad \text{for } \xi \in R^m,$$

where a^{kj}, b^k, c and f are real functions defined in a domain Ω belonging to $C^\infty(\Omega)$. As in §5, we may assume without loss of generality that the coefficients of (2.6.1) are functions in $C_0^\infty(\Omega)$ with support in a compact set K_1.

Our treatment of the operator $L(u)$ will follow the same pattern as that of the operators (2.5.1). The question of conditions under which the general operator (2.6.1) may be reduced to one of the form (2.5.1) has still not been completely investigated. (In connection with this see [102] and also the remark at the end of this section.)

We let $L^0(x, \xi)$ denote the expression $a^{kj}(x)\xi_k \xi_j$. By L^0 we shall mean the differential operator with symbol $L^0(x, \xi)$. By a compact set K we shall mean a closed region contained in Ω. If A is any pseudodifferential operator, we shall denote by $A^{(j)}$ and $A_{(j)}$ the pseudodifferential operators with symbols $\partial A(x, \xi)/\partial \xi_j$ and $D_j A(x, \xi)$ respectively, where $A(x, \xi)$ is the symbol of the operator A and (as elsewhere in Chapter II) $D_j = -i\partial/\partial x_j$.

LEMMA 2.6.1. *Let A be an operator representable in the form $A = A_N + T_N$, where A_N is a pseudodifferential operator with symbol satisfying conditions* a) *and*

b) *of* §2 *for* $\sigma = t$, *and* T_N *is an operator of order at most* $t-N$, *where* N *is a positive integer. Then for any function* u *in* $C_0^\infty(K)$

$$\|LAu\|_{-t}^2 \leqslant C(K)\{\|Lu\|_0^2 + \sum_{j=1}^m (\|L_{(j)}^0 u\|_{-1}^2 + \|L^{0(j)}u\|_0^2) + \|u\|_0^2\}. \tag{2.6.2}$$

PROOF. According to Theorem 2.2.3 we have

$$\|[L, A_N]\,u\|_{-t}^2 \leqslant C_1 \left\{\sum_{j=1}^m (\|A_N^{(j)} L_{(j)} u\|_{-t}^2 + \|A_{N(j)} L^{(j)} u\|_{-t}^2) + \|u\|_0^2\right\}.$$

Since the operators $L_{(j)} - L_{(j)}^0$ are of order at most 1 and $L^{(j)} - L^{0(j)}$ have order at most 0, we obtain from Theorem 2.2.1 that

$$\|[L, A_N]\,u\|_{-t}^2 \leqslant C_2 \left\{\sum_{j=1}^m (\|L_{(j)}^0 u\|_{-1}^2 + \|L^{0(j)} u\|_0^2) + \|u\|_0^2\right\}. \tag{2.6.3}$$

It is easy to see that

$$\begin{aligned}\|LAu\|_{-t}^2 &\leqslant 2\|LA_N u\|_{-t}^2 + 2\|LT_N u\|_{-t}^2 \\ &\leqslant 4\{\|A_N L u\|_{-t}^2 + \|[L, A_N]\,u\|_{-t}^2 + \|LT_N u\|_{-t}^2\}.\end{aligned}$$

Choosing $N > 2$ and using (2.6.3), we obtain

$$\|LAu\|_{-t}^2 \leqslant C_3 \{\|Lu\|_0^2 + \sum_{j=1}^m (\|L_{(j)}^0 u\|_{-1}^2 + \|L^{0(j)} u\|_0^2) + \|u\|_0^2\}$$

which is the desired result.

LEMMA 2.6.2. *If* L *is a second order operator* (2.6.1) *with nonnegative characteristic form, then for any* $u \in C_0^\infty(K)$

$$\sum_{j=1}^m (\|L^{0(j)} u\|_0^2 + \|L_{(j)}^0 u\|_{-1}^2) \leqslant C_1 \{\|Lu\|_0^2 + \|u\|_0^2\}. \tag{2.6.4}$$

PROOF. Since $a^{kj}(x)\xi_k\xi_j \geqslant 0$ for $x \in \Omega$ and $\xi \in R^m$, it follows from inequality (1.7.10) that

$$\sum_{j=1}^m |a^{kj}(x)\,D_k u|^2 \leqslant C_2 a^{ks} D_k u \overline{D_s u}, \tag{2.6.5}$$

where the constant C_2 depends on the maxima of the absolute values of the functions $a^{kj}(x)$ on K. Integrating inequality (2.6.5) over the space R^m, we obtain

$$\sum_{j=1}^m \|L^{0(j)} u\|_0^2 \leqslant 4C_2 (a^{kj} D_k u, D_j u). \tag{2.6.6}$$

We consider the integral (Lu, u) and transform it by integration by parts (here, as in §5, we denote $(u, v)_0$ by (u, v)). We represent Lu in the form

$$L(u) = -D_j(a^{kj}D_ku) + (D_ja^{kj} + ib^k)D_ku + cu, \tag{2.6.7}$$

and set $il^k \equiv D_ja^{kj} + ib^k$, so that l^k are real functions. Then $(Lu, u) = -(a^{kj}D_ku, D_ju) + (il^kD_ku, u) + (cu, u)$ and

$$\operatorname{Re}(Lu, u) = -(a^{kj}D_ku, D_ju) + \operatorname{Re}(il^kD_ku, u) + (cu, u). \tag{2.6.8}$$

It is easy to see that

$$(l^kD_ku, u) = \frac{1}{2}(l^kD_ku, u) + \frac{1}{2}(u, l^kD_ku) + \frac{1}{2}(u, (D_kl^k)u).$$

Therefore

$$\operatorname{Re}(il^kD_ku, u) = \frac{1}{2}\operatorname{Re}[(iu, (D_kl^k)u)].$$

From the last equation and (2.6.8) it follows that

$$(a^{kj}D_ku, D_ju) \leqslant |(Lu, u)| + C_3\|u\|_0^2. \tag{2.6.9}$$

It follows from (2.6.6) and (2.6.9) that

$$\sum_{j=1}^m \|L^0{}_{(j)}v\|_0^2 \leqslant C_4\{\|Lu\|_0^2 + \|u\|_0^2\}. \tag{2.6.10}$$

To estimate $L^0_{(j)}u$ we use Lemma 1.7.1, according to which

$$\sum_{l=1}^m |a^{kj}_{x_l}(x)D_kD_jv|^2 \leqslant C_5a^{kj}(x)D_jD_lv\overline{D_kD_lv} \tag{2.6.11}$$

for functions $v \in C_0^\infty(K)$, where the constant C_5 depends on the maximum of the absolute values of the second derivatives of the functions $a^{kj}(x)$ on the compact set $K_1 \supset K$.

Integrating (2.6.11) over the space R^m, we obtain

$$\sum_{l=1}^m \|L^0_{(l)}v\|_0^2 \leqslant C_6(a^{kj}D_lD_kv, D_lD_jv). \tag{2.6.12}$$

We replace u by D_lv in (2.6.9) and use (2.6.12) to obtain

$$\sum_{l=1}^m \|L^0_{(l)}v\|_0^2 \leqslant C_7\{|(L(D_lv), D_lv)| + \|v\|_1^2\}. \tag{2.6.13}$$

We set $v = E_{-1}u$, where the pseudodifferential operator E_{-1} has the symbol $\phi(x)(1 + |\xi|^2)^{-1/2}$, with $\phi(x) \in C_0^\infty(K_1)$ and $\phi(x) \equiv 1$ on a compact set K contained in K_1. From (2.6.13) it follows that for any function $u \in C_0^\infty(K)$

$$\sum_{l=1}^m \|L^0_{(l)}E_{-1}u\|_0^2 \leqslant C_8\{|(LD_lE_{-1}u, D_lE_{-1}u)| + \|u\|_0^2\} \leqslant$$

$$\leqslant C_9\left\{\sum_{t=1}^{m} \mu \|LD_t E_{-1}u\|_0^2 + \frac{1}{\mu}\|u\|_0^2\right\}. \tag{2.6.14}$$

By Theorems 2.2.6 and 2.2.3 we have, for any function $u \in C_0^\infty(K)$,

$$\sum_{l=1}^{m} \|L^0_{(l)}u\|_{-1}^2 \leqslant 2\sum_{l=1}^{m} \|E_{-1}L^0_{(l)}u\|_0^2 + C_{10}\|u\|_0^2$$

$$\leqslant 4\sum_{l=1}^{m} \|L^0_{(l)}E_{-1}u\|_0^2 + C_{11}\|u\|_0^2. \tag{2.6.15}$$

Hence, applying Lemma 2.6.1, we deduce from (2.6.15) and (2.6.14) that

$$\sum_{l=1}^{m} \|L^0_{(l)}u\|_{-1}^2 \leqslant C_{12}\left\{\frac{1}{\mu}\|u\|_0^2 + \mu\|Lu\|_0^2 + \mu\sum_{l=1}^{m}(\|L^0_{(l)}u\|_{-1}^2 + \|L^{0(l)}u\|_0^2)\right\}.$$

Choosing μ sufficiently small and using (2.6.10), we obtain

$$\sum_{j=1}^{m} \|L^0_{(j)}u\|_{-1}^2 \leqslant C_{13}\{\|Lu\|_0^2 + \|u\|_0^2\}. \tag{2.6.16}$$

From (2.6.10) and (2.6.16) we obtain the required inequality (2.6.4).

LEMMA 2.6.3. *If A is any operator satisfying the conditions of Lemma 2.6.1, then the inequality*

$$\|LAu\|_{-t}^2 \leqslant C(K)\{\|Lu\|_0^2 + \|u\|_0^2\} \tag{2.6.17}$$

holds for any function $u \in C_0^\infty(K)$.

This is an immediate consequence of Lemmas 2.6.1 and 2.6.2.

We write the operator Lu in the form

$$L(u) \equiv -D_j(a^{kj}D_k u) + iQu + cu, \tag{2.6.18}$$

where $Qu \equiv (b^k - a^{kj}_{x_j})D_k u$.

THEOREM 2.6.1 (ENERGY ESTIMATE). *For any $s \geqslant 0$ and any function u in the class $C_0^\infty(K)$ the inequality*

$$\sum_{j=1}^{m}\{\|L^{0(j)}u\|_s^2 + \|L^0_{(j)}u\|_{s-1}^2\} + \|Qu\|_{s-\frac{1}{2}}^2$$

$$\leqslant C(K,s)\{\|Lu\|_0^2 + \|u\|_{2s}^2\} \tag{2.6.19}$$

holds, where the constant $C(K, s)$ depends on K and s.

PROOF. It follows from (2.6.6) and (2.6.9) that for $s \geqslant 0$

$$\sum_{j=1}^{m} \|L^{0(j)}v\|_0^2 \leqslant C_1\{|(Lv, v)| + \|v\|_0^2\} \leqslant 2C_1\{\|Lv\|_{-s}^2 + \|v\|_s^2\}. \tag{2.6.20}$$

We set $v = E_s u$, where $u \in C_0^\infty(K)$. From (2.6.20) we obtain

$$\sum_{j=1}^{m} \|L^{0(j)} E_s u\|_0^2 \leqslant C_2 \{\|LE_s u\|_{-s}^2 + \|u\|_{2s}^2\}. \tag{2.6.21}$$

On the basis of Theorem 2.2.6 we have

$$\|L^{0(j)} u\|_s^2 \leqslant 4\|L^{0(j)} E_s u\|_0^2 + 4\|[E_s, L^{0(j)}]u\|_0^2 + C_3 \|u\|_0^2 \tag{2.6.22}$$

for any function $u \in C_0^\infty(K)$.

Applying Lemma 2.6.3 and Theorem 2.2.1 on the order of a pseudodifferential operator, we obtain from (2.6.22) and (2.6.21) that

$$\sum_{j=1}^{m} \|L^{0(j)} u\|_s^2 \leqslant C_4 \{\|Lu\|_0^2 + \|u\|_{2s}^2\}. \tag{2.6.23}$$

From (2.6.13) and (2.1.19) it follows that

$$\sum_{j=1}^{m} \|L_{(j)}^0 v\|_0^2 \leqslant C_5 \left\{\sum_{l=1}^{m} \|LD_l v\|_{-s}^2 + \|v\|_{s+1}^2\right\},$$

where $\operatorname{supp} v \subset K_1$. In this inequality we set $v = E_{s-1} u$, where $u \in C_0^\infty(K)$ and $s \geqslant 0$. Applying Lemma 2.6.3, we obtain

$$\sum_{j=1}^{m} \|L_{(j)}^0 E_{s-1} u\|_0^2 \leqslant C_5 \left\{\sum_{l=1}^{m} (\|LD_l E_{s-1} u\|_{-s}^2 + \|E_{s-1} u\|_{s+1}^2\right\} \leqslant C_6 \{\|Lu\|_0^2 + \|u\|_{2s}^2\}. \tag{2.6.24}$$

From Theorems 2.2.6 and 2.2.3 we have

$$\begin{aligned} \|L_{(j)}^0 u\|_{s-1}^2 &\leqslant 2\|E_{s-1} L_{(j)}^0 u\|_0^2 + C_7 \|u\|_0^2 \\ &\leqslant 4\|L_{(j)}^0 E_{s-1} u\|_0^2 + 4\|[E_{s-1}, L_{(j)}^0]\, u\|_0^2 \\ &+ C_8 \|u\|_0^2 \leqslant 4\|L_{(j)}^0 E_{s-1} u\|_0^2 + C_9 \|u\|_s^2. \end{aligned}$$

Taking (2.6.24) into consideration, we deduce from the latter inequality that

$$\|L_{(j)}^0 u\|_{s-1}^2 \leqslant C_{10} \{\|Lu\|_0^2 + \|u\|_{2s}^2\}. \tag{2.6.25}$$

To complete the proof of Theorem 2.6.1 it remains to estimate $\|Qu\|_{s-1/2}^2$. For this purpose we express the operator Lu in the form (2.6.18) and examine the expression $(LE_s u, E_{-1/2}^* E_{-1/2} QE_s u)$, where E_s^* is the operator adjoint to E_s. After simple transformations we obtain

$$\begin{aligned} (LE_s u, E_{-\frac{1}{2}}^* E_{-\frac{1}{2}} QE_s u) &= i\|E_{-\frac{1}{2}} QE_s u\|_0^2 \\ &- (a^{kj} D_k E_s u, D_j E_{-\frac{1}{2}}^* E_{-\frac{1}{2}} QE_s u) + (cE_s u, E_{-\frac{1}{2}}^* E_{-\frac{1}{2}} QE_s u). \end{aligned} \tag{2.6.26}$$

We estimate the individual terms in (2.6.26). We have

$$I_1 \equiv |(LE_s u,\ E^*_{-\frac{1}{2}} E_{-\frac{1}{2}} QE_s u)|$$

$$\leqslant \frac{1}{2} (\|LE_s u\|^2_{-s} + \|E^*_{-\frac{1}{2}} E_{-\frac{1}{2}} QE_s u\|^2_s). \tag{2.6.27}$$

According to Lemma 2.6.3 and Theorem 2.2.1 we obtain

$$I_1 \leqslant C_{11} \{\|Lu\|^2_0 + \|u\|^2_{2s}\}.$$

Furthermore it is clear that

$$I_2 \equiv |(cE_s u,\ E^*_{-\frac{1}{2}} E_{-\frac{1}{2}} QE_s u)| \leqslant C_{12} \|u\|^2_s.$$

Since $a^{kj}(x)\xi_k\xi_j \geqslant 0$, we have

$$|(a^{kj}D_k u, D_j v)| \leqslant \frac{1}{2} \{(a^{kj}D_k u,\ D_j u) + (a^{kj}D_k v,\ D_j v)\} \tag{2.6.28}$$

for $u, v \in C_0^\infty(R^m)$. Hence

$$I_3 \equiv |(a^{kj}D_k E_s u,\ D_j E^*_{-\frac{1}{2}} E_{-\frac{1}{2}} QE_s u)|$$

$$\leqslant \frac{1}{2} \{(a^{kj}D_k E_s u,\ D_j E_s u) + (a^{kj}D_k E^*_{-\frac{1}{2}} E_{-\frac{1}{2}} QE_s u,\ D_j E^*_{-\frac{1}{2}} E_{-\frac{1}{2}} QE_s u)\}. \tag{2.6.29}$$

We estimate the integrals on the right of (2.6.29) by use of (2.6.9). We have

$$I_3 \leqslant |(LE_s u, E_s u)| + \left|\left(LE^*_{-\frac{1}{2}} E_{-\frac{1}{2}} QE_s u,\ E^*_{-\frac{1}{2}} E_{-\frac{1}{2}} QE_s u\right)\right|$$

$$+ C_{13} \left(\|E_s u\|^2_0 + \|E^*_{-\frac{1}{2}} E_{-\frac{1}{2}} QE_s u\|^2_0\right).$$

On the basis of this, Lemma 2.6.3, and Theorem 2.2.1, we obtain

$$I_3 \leqslant \|LE_s u\|^2_{-s} + \|E_s u\|^2_s + \|LE^*_{-\frac{1}{2}} E_{-\frac{1}{2}} QE_s u\|^2_{-s}$$

$$+ \|E^*_{-\frac{1}{2}} E_{-\frac{1}{2}} QE_s u\|^2_s + C_{14} \|u\|^2_s \leqslant C_{15} \{\|Lu\|^2_0 + \|u\|^2_{2s}\}.$$

From (2.6.26) follows

$$\|E_{-\frac{1}{2}} QE_s u\|^2_0 \leqslant I_1 + I_2 + I_3 \leqslant C_{16} \{\|Lu\|^2_0 + \|u\|^2_{2s}\}. \tag{2.6.30}$$

According to Theorem 2.2.6,

$$\|Qu\|^2_{s-\frac{1}{2}} \leqslant 4 \|E_{-\frac{1}{2}} E_s Qu\|^2_0 + C_{17} \|u\|^2_0$$

$$\leqslant C_{18} \{\|E_{-\frac{1}{2}} QE_s u\|^2_0 + \|E_{-\frac{1}{2}} [E_s,\ Q]\, u\|^2_0 + \|u\|^2_0\} \leqslant$$

$$\leqslant C_{19}\{\|E_{-\frac{1}{2}}QE_s u\|_0^2 + \|u\|_s^2\}. \tag{2.6.31}$$

Inequalities (2.6.31) and (2.6.30) yield

$$\|Qu\|^2_{s-\frac{1}{2}} \leqslant C_{20}\{\|Lu\|_0^2 + \|u\|_{2s}^2\}, \tag{2.6.32}$$

which is the desired result.

We consider the system of operators $\{Q_0, \cdots, Q_{2m}\}$ where $Q_0 \equiv Q$, $Q_j = L^{0(j)}$ for $j = 1, \cdots, m$ and $Q_j = E_{-1}L^0_{(j-m)}$ for $j = m+1, \cdots, 2m$. For any multi-index $I = (\alpha_1, \cdots, \alpha_k)$, where $\alpha_l = 0, 1, \cdots, 2m$ for $l = 1, \cdots, k$, we set $|I| = \Sigma_1^k \lambda_l$, the number λ_l being 1 if $\alpha_l = 1, \cdots, 2m$, and 2 if $\alpha_l = 0$. To each multi-index I we associate an operator

$$Q_I = \operatorname{ad} Q_{\alpha_1} \ldots \operatorname{ad} Q_{\alpha_{k-1}} Q_{\alpha_k},$$

where, as before $\operatorname{ad} AB = AB - BA = [A, B]$.

LEMMA 2.6.4. *For every compact set K, every $k \geqslant 1$, and every $s \geqslant 0$, there exists a constant $C(K, s, k)$ such that the inequality*

$$\sum_{|I|=k} \|Q_I u\|^2_{2^{1-k}+s-1} \leqslant C(K, s, k)\{\|Lu\|_0^2 + \|u\|^2_{2^k s}\} \tag{2.6.33}$$

holds for any function u in $C_0^\infty(K)$.

PROOF. We first verify that (2.6.33) is fulfilled for $k \leqslant 2$. For $k = 2$, inequality (2.6.33) is fulfilled by virtue of estimate (2.6.32) if $Q_I = Q_0$. If $Q_I = [Q_j, Q_l]$, where $j, l \neq 0$, then according to Theorem 2.2.8 we have

$$\|[Q_j, Q_l]u\|^2_{s-\frac{1}{2}} \leqslant C_1\left\{\sum_{j=1}^{2m} \|Q_j u\|_s^2 + \|u\|_s^2\right\}. \tag{2.6.34}$$

We apply Theorem 2.6.1 to estimate the right side of (2.6.34), obtaining

$$\|[Q_j, Q_l]u\|^2_{s-\frac{1}{2}} \leqslant C_2\{\|Lu\|_0^2 + \|u\|_{2s}^2\},$$

which means that (2.6.33) is valid for $k = 2$ in this case also.

We shall prove (2.6.33) in the general case by induction. We assume that it is satisfied for $k \leqslant k_0$ and show that it will then hold for $k = k_0 + 1$ as well. We first consider the case when $Q_I = [Q_{I_1}, Q_j]$, where $|I_1| = k_0$, $j \neq 0$. Then, according to Theorem 2.2.8 with $s_1 = -1 + 2^{1-k_0}$, we obtain

$$\sum_{j=1}^{2m} \sum_{|I_1|=k_0} \|[Q_j, Q_{I_1}]u\|^2_{s-1+2^{-k_0}} \leqslant$$

$$\leqslant C_3 \left\{ \sum_{j=1}^{2m} \|Q_j u\|_s^2 + \|Q_{I_1} u\|^2_{s-1+2^{1-k_0}} + \|u\|_s^2 \right\}. \tag{2.6.35}$$

According to Theorem 2.6.1 and the induction assumption the right side of (2.6.35) may be estimated by

$$C_4 \{\|Lu\|_0^2 + \|u\|^2_{2k_0 s}\},$$

which proves (2.6.33) in the case considered.

Now suppose $Q_I = [Q_0, Q_{I_0}]$, where $|I_0| = k_0 - 1 \geqslant 1$. The operator $Q_I = [Q_0, Q_j]$ corresponding to $k_0 = 1$ is estimated by (2.6.35). We write the operator $L(u)$ in the form (2.6.18). Then

$$L^*(u) = -D_j(a^{kj} D_k u) - iQu + c^* u,$$

where $c^* \in C_0^\infty(R^m)$ and Q_I may be represented in the form

$$Q_I w = [Q_0, Q_{I_0}]\, w = \frac{1}{i}(-D_j(a^{kj} D_k) - L^* + c^*)\, Q_{I_0} w$$
$$- \frac{1}{i} Q_{I_0}(L + D_j(a^{kj} D_k) - c)\, w. \tag{2.6.36}$$

In the equation (2.6.36) we set $w = E_s u$, multiply it by $\bar{v}$, where $v = E_\rho^* E_\rho [Q_0, Q_{I_0}] E_s u$ and $\rho = 2^{-k_0} - 1$, and integrate over R^m.

We thus obtain

$$\| E_{2^{-k_0}-1} [Q_0, Q_{I_0}]\, E_s u \|_0^2$$
$$= | -(L^* Q_{I_0} E_s u,\ v) + (-Q_{I_0} L E_s u,\ v) - (a^{kj} D_k Q_{I_0} E_s u,\ D_j v)$$
$$- (Q_{I_0} D_j (a^{kj} D_k) E_s u,\ v) + (c^* Q_{I_0} E_s u,\ v) + (Q_{I_0} c E_s u,\ v) |. \tag{2.6.37}$$

We estimate the integrals on the right of (2.6.37). Applying the Schwarz inequality (2.1.19), Lemma 2.6.3, and Theorem 2.2.3, we obtain

$$V_1 \equiv |(L E_s u,\ Q_{I_0}^* v)| \leqslant \| L E_s u \|_{-s}^2 + \|Q_{I_0}^* v\|_s^2$$
$$\leqslant C_5 \{\| Lu \|_0^2 + \| u \|_{2s}^2 + \| Q_{I_0} u \|^2_{2s+2^{1-k_0}-1}\}.$$

In exactly the same way we estimate

$$V_2 \equiv |(Q_{I_0} E_s u,\ Lv)| \leqslant \| Lv \|^2_{1-s-2^{1-k_0}} + \| Q_{I_0} E_s u \|^2_{s+2^{1-k_0}-1}$$
$$\leqslant C_6 \{\| Lu \|_0^2 + \| u \|_{2s}^2 + \| Q_{I_0} u \|^2_{2s+2^{1-k_0}-1}\}.$$

It is easy to see that

$$V_3 \equiv |(c^* Q_{I_0} E_s u,\ v)| + |(Q_{I_0} c E_s u,\ v)|$$
$$\leqslant C_7 \{\| c^* Q_{I_0} E_s u \|^2_{2^{1-k_0}-1} + \| v \|^2_{1-2^{1-k_0}} + \| Q_{I_0} c E_s u \|^2_{2^{1-k_0}-1}\}.$$

Applying Theorems 2.2.1 and 2.2.3, we obtain

$$V_3 \leqslant C_8\{\|u\|_s^2 + \|Q_{I_0}u\|^2_{s+2^{1-k_0-1}}\}.$$

To estimate the integral

$$V_4 \equiv |(a^{kj}D_kE_su,\ D_jQ^*_{I_0}v)|$$

we employ inequality (2.6.28). We have

$$V_4 \leqslant \{(a^{kj}D_kE_su,\ D_jE_su) + (a^{kj}D_kQ^*_{I_0}v,\ D_jQ^*_{I_0}v)\}.$$

Furthermore, according to (2.6.9) we have

$$V_4 \leqslant C_9\{|(LE_su,\ E_su)| + \|u\|_s^2 + |(LQ^*_{I_0}v,\ Q^*_{I_0}v)| + \|Q^*_{I_0}v\|_0^2\}.$$

Using (2.1.19) and the estimate proved in Lemma 2.6.3, we obtain

$$V_4 \leqslant C_{10}\{\|LE_su\|_{-s}^2 + \|u\|_{2s}^2 + \|LQ^*_{I_0}v\|^2_{-s-2^{1-k_0}} + \|Q^*_{I_0}v\|^2_{s+2^{1-k_0}}\}$$

$$\leqslant C_{11}\{\|Lu\|_0^2 + \|u\|_{2s}^2 + \|Q_{I_0}u\|^2_{2s+2^{2-k_0-1}}\}.$$

Here we have used the fact that $2^{2-k_0} - 1 \leqslant 0$. The integral on the right side of (2.6.37) of the form

$$V_5 \equiv |(a^{kj}D_kQ_{I_0}E_su,\ D_jv)|$$

may be transformed in the following fashion:

$$\begin{aligned} V_5 = |&(a^{kj}D_kQ_{I_0}E_su,\ D_j(E^*_\rho E_\rho - E^*_{2^{1-k_0-1}}E_{-1})[Q_0,Q_{I_0}]E_su) \\ &+ (a^{kj}D_kQ_{I_0}E_su,\ [D_j, E^*_{2^{1-k_0-1}}]E_{-1}[Q_0,Q_{I_0}]E_su) \\ &+ (a^{kj}D_kE_{2^{1-k_0-1}}Q_{I_0}E_su,\ D_jE_{-1}[Q_0,Q_{I_0}]E_su) \\ &+ ([E_{2^{1-k_0-1}}, \tfrac{1}{2}L_{0(j)}]Q_{I_0}E_su,\ D_jE_{-1}[Q_0,Q_{I_0}]E_su)|. \end{aligned} \tag{2.6.38}$$

The first three integrals on the right of (2.6.38) may be estimated in the same way as V_4 was estimated. We have

$$\begin{aligned} w_1 &\equiv |(a^{kj}D_kQ_{I_0}E_su,\ D_j(E^*_\rho E_\rho - E^*_{2^{1-k_0-1}}E_{-1})[Q_0,Q_{I_0}]E_su)| \\ &\leqslant |(a^{kj}D_k\{(E^*_\rho E_\rho - E^*_{2^{1-k_0-1}}E_{-1})[Q_0,Q_{I_0}]\}^*Q_{I_0}E_su,\ D_jE_su)| \\ &\quad + |([\{(E^*_\rho E_\rho - E^*_{2^{1-k_0-1}}E_{-1})[Q_0,Q_{I_0}]\}^*, \tilde{L}^0]Q_{I_0}E_su,\ E_su)|, \end{aligned}$$

where $\tilde{L}^0 \equiv D_j(a^{kj}D_k)$. Further, using (2.6.28), (2.6.9) and Lemma 2.6.3, and recalling that the order of the operator $E^*_\rho E_\rho - E^*_{2^{1-k_0-1}}E_{-1}$ does not surpass $-3 + 2^{1-k_0}$, we obtain the following inequality:

$$\begin{aligned} w_1 \leqslant C_{12}\{&\|L\{(E^*_\rho E_\rho - E^*_{2^{1-k_0-1}}E_{-1})[Q_0,Q_{I_0}]\}^*Q_{I_0}E_su\|^2_{1-s-2^{1-k_0}} \\ &+ \|\{(E^*_\rho E_\rho - E^*_{2^{1-k_0-1}}E_{-1})[Q_0,Q_{I_0}]\}^*Q_{I_0}E_su\|^2_{s+2^{1-k_0-1}} + \end{aligned}$$

$$+ \| u \|_{2s}^2 + \| LE_s u \|_{-s}^2 + \| Q_{I_0} u \|_{s+2^{1-k_0}-1}^2 \}$$
$$\leqslant C_{13} \{ \| Lu \|_0^2 + \| u \|_{2s}^2 + \| Q_{I_0} u \|_{s+2^{1-k_0}-1}^2 \}.$$

We estimate the integral

$$w_2 \equiv | (a^{kj} D_k Q_{I_0} E_s u, [D_j, E^*_{2^{1-k_0}-1}] E_{-1} [Q_0, Q_{I_0}] E_s u) |$$
$$\leqslant | (a^{kj} D_k Q_{I_0} E_s u, E_{-1} [Q_0, Q_{I_0}] E_s [D_j, E^*_{2^{1-k_0}-1}] u)$$
$$+ (a^{kj} D_k Q_{I_0} E_s u, [[D_j, E^*_{2^{1-k_0}-1}], E_{-1} [Q_0, Q_{I_0}] E_s] u) |.$$

We note that $u \in C_0^\infty(K)$, and that the function $\phi(x)$ entering into the symbol of the operator $E^*_{2^{1-k_0}-1}$ is equal to 1 on K; therefore according to Theorems 2.2.4 and 2.2.3 we have

$$\| [D_j, E^*_{2^{1-k_0}-1}] u \|_N^2 \leqslant C_N \| u \|_0^2,$$

where N is an arbitrary nonnegative number. Applying (2.1.19), we thus obtain

$$w_2 \leqslant C_{14} \{ \| u \|_0^2 + \| a^{kj} D_j Q_{I_0} E_s u \|_{-2+2^{1-k_0}}^2$$
$$+ \| [[D_j, E^*_{2^{1-k_0}-1}], E_{-1} [Q_0, Q_{I_0}] E_s] u \|_{2-2^{1-k_0}}^2 \}$$
$$\leqslant C_{15} \{ \| u \|_s^2 + \| Q_{I_0} u \|_{s-1+2^{1-k_0}}^2 \}.$$

According to inequalities (2.6.28) and (2.6.9), we have

$$w_3 \equiv | (a^{kj} D_k E_{2^{1-k_0}-1} Q_{I_0} E_s u, D_j E_{-1} [Q_0, Q_{I_0}] E_s u)$$
$$\leqslant C_{16} \{ | (L (E_{2^{1-k_0}-1} Q_{I_0} E_s u), E_{2^{1-k_0}-1} Q_{I_0} E_s u) |$$
$$+ \| E_{2^{1-k_0}-1} Q_{I_0} E_s u \|_0^2 + \| E_{-1} [Q_0, Q_{I_0}] E_s u \|_0^2$$
$$+ | (L (E_{-1} [Q_0, Q_{I_0}] E_s u), E_{-1} [Q_0, Q_{I_0}] E_s u) | \}.$$

Applying (2.1.19) and Lemma 2.6.3, we obtain

$$w_3 \leqslant C_{17} \{ \| Lu \|_0^2 + \| Q_{I_0} u \|_{2s+2^{2-k_0}-1}^2$$
$$+ \| Q_{I_0} u \|_{s+2^{1-k_0}-1}^2 + \| u \|_{2s}^2 \}.$$

We now estimate the last integral in (2.6.38):

$$w_4 \equiv | ([E_\nu, L^{0(j)}] Q_{I_0} E_s u, D_j E_{-1} [Q_0, Q_{I_0}] E_s u) |,$$

where $\nu = 2^{1-k_0} - 1$.

On the basis of Theorem 2.2.3,

$$w_4 \leqslant | ((L^{0(j)}_{(t)} E^{(t)}_\nu - L^{0(j)(t)} E_{\nu(t)}) Q_{I_0} E_s u, D_j E_{-1} [Q_0, Q_{I_0}] E_s u) |$$
$$+ \sum_{j=1}^m \left| (T Q_{I_0} E_s u, D_j E_{-1} [Q_0, Q_{I_0}] E_s u) \right| = w_5 + w_6, \qquad (2.6.39)$$

where T is an operator of order at most $2^{1-k_0} - 2$. As mentioned before, we use

$A^{(t)}$ and $A_{(t)}$ to denote operators with symbols $\partial A(x,\xi)/\partial\xi_t$ and $D_tA(x,\xi)$ respectively, where $A(x,\xi)$ is the symbol of the operator A. According to the Schwarz inequality (2.1.19),

$$w_6 \leqslant \sum_{j=1}^{m} \{\| TQ_{I_0}E_su\|_1^2 + \| D_jE_{-1}[Q_0, Q_{I_0}]E_su\|_{-1}^2\}$$
$$\leqslant C_{18}\left\{\| Q_{I_0}u\|^2_{s+2^{1-k_0}-1} + \| u\|_s^2\right\}.$$

We transform the integral w_5 in this manner:

$$w_5 = |\left(\left(L^{0(j)}_{(t)}D_jE^{(t)}_{\nu} - L^{0(j)(t)}D_jE_{\nu(t)}\right)Q_{I_0}E_su,\ E_{-1}[Q_0, Q_{I_0}]E_su\right)$$
$$+ \left(\left\{\left[D_j, L^{0(j)}_{(t)}\right]E^{(t)}_{\nu}Q_{I_0}E_su - \left[D_j, L^{0(j)(t)}\right]E_{\nu(t)}Q_{I_0}E_su\right\},\ E_{-1}[Q_0, Q_{I_0}]E_su\right)|.$$

It is clear that $L^{0(j)}_{(t)}D_j = 2L^{0}_{(t)}$, $L^{0(j)(t)}D_j = L^{0(t)}$. Hence, applying Theorems 2.2.1 and 2.2.3, we obtain

$$w_5 \leqslant |\left(\left(2L^0_{(t)}E^{(t)}_{\nu} - L^{0(t)}E_{\nu(t)}\right)Q_{I_0}E_su,\ E_{-1}[Q_0, Q_{I_0}]E_su\right)|$$
$$+C_{19}\left\{\| u\|_s^2 + \| Q_{I_0}u\|^2_{s+2^{1-k_0}-1}\right\}$$
$$\leqslant C_{20}\{\| L^0_{(t)}E^{(t)}_{\nu}Q_{I_0}E_su\|_0^2 + \| u\|_s^2$$
$$+ \| L^{0(t)}E_{\nu(t)}Q_{I_0}E_su\|_0^2 + \|Q_{I_0}u\|^2_{s+2^{1-k_0}-1}\}. \tag{2.6.40}$$

We further use inequalities (2.6.10) and (2.6.13) obtained in the proof of Lemma 2.6.2. Thus

$$\| L^0_{(t)}v\|_0^2 \leqslant C_{21}\{|(LD_kv, D_kv)| + \| v\|_1^2\}$$
$$\leqslant C_{22}\left\{\sum_{k=1}^{m} \| LD_kv\|^2_{-s-2^{1-k_0}} + \| v\|^2_{1+s+2^{1-k_0}} + \| v\|_1^2\right\}, \tag{2.6.41}$$
$$\| L^{0(t)}v\|_0^2 \leqslant C_{23}\{|(Lv, v)| + \| v\|_0^2\}$$
$$\leqslant C_{24}\left\{\| Lv\|^2_{-s-2^{1-k_0}} + \| v\|^2_{s+2^{1-k_0}}\right\}. \tag{2.6.42}$$

From Lemma 2.6.3 and inequalities (2.6.40)–(2.6.42) we obtain

$$w_5 \leqslant C_{25}\{\|Lu\|_0^2 + \|u\|_{2s}^2 + \|Q_{I_0}u\|^2_{2s-1+2^{2-k_0}} + \|Q_{I_0}u\|^2_{s+2^{1-k_0}-1}\}.$$

Thus, bearing in mind the estimates obtained for w_j and V_j, the following estimate follows from (2.6.37):

$$\|E_{2^{-k_0}-1}[Q_0, Q_{I_0}]E_su\|_0^2 \leqslant C_{26}\{\|Lu\|_0^2 + \|Q_{I_0}u\|^2_{2s+2^{2-k_0}-1} + \|u\|_{2s}^2\}. \tag{2.6.43}$$

Furthermore, according to Theorem 2.2.6,

$$\|[Q_0, Q_{I_0}]\,u\|^2_{s+2^{-k_0}-1} \leqslant C_{27}\{\|u\|_0^2 + \|E_{2^{-k_0}-1}E_s\,[Q_0, Q_{I_0}]\,u\|_0^2\}$$
$$\leqslant C_{28}\{\|u\|_s^2 + \|E_{2^{-k_0}-1}\,[Q_0, Q_{I_0}]\,E_s u\|_0^2\}. \qquad (2.6.44)$$

From (2.6.44) and (2.6.43) we obtain

$$\|[Q_0, Q_{I_0}]\,u\|^2_{s+2^{-k_0}-1} \leqslant C_{29}\{\|Lu\|_0^2 + \|Q_{I_0}u\|^2_{2s+2^{2-k_0}-1} + \|u\|^2_{2s}\}. \qquad (2.6.45)$$

The induction hypothesis now says

$$\|Q_{I_0}u\|^2_{2s+2^{1-(k_0-1)}-1} \leqslant C_{30}\{\|Lu\|_0^2 + \|u\|^2_{2^{k_0-1}\cdot 2s}\}.$$

From this inequality and (2.6.45) we conclude that

$$\|[Q_0, Q_{I_0}]\,u\|^2_{s+2^{-k_0}-1} \leqslant C_{31}\{\|Lu\|_0^2 + \|u\|^2_{2^{k_0}s}\}.$$

This means that (2.6.33) holds in the case considered. The theorem is proved.

We consider the operators Q_I generated by the system of operators $Q_0, \cdots, Q_{2m}$. According to Theorem 2.2.3, for each multi-index $I = (\alpha_1, \cdots, \alpha_k)$ the operator Q_I may be written in the form

$$Q_I = Q_I^0 + T_I, \qquad (2.6.46)$$

where the operator T_I has order at most zero, and Q_I^0 is the pseudodifferential operator, corresponding to $\sigma = 1$, with symbol $q_I^0(x, \xi)$.

DEFINITION. The system of operators $\{Q_0, \cdots, Q_{2m}\}$ on a compact set K is called a *system of rank* m if there exists a number $R(K)$ such that

$$1 + \sum_{|I| \leqslant R(K)} |q_I^0(x, \xi)|^2 \geqslant C_0(1 + |\xi|^2), \quad C_0 = \text{const} > 0, \qquad (2.6.47)$$

for all $x \in K$ and all $\xi \in R^m$.

THEOREM 2.6.2. *Let $\{Q_0, \cdots, Q_{2m}\}$ be a system of operators of rank m on K. Then there exist constants $C(K) > 0$ and $\epsilon(K)$ such that for any distribution $u \in D'(\Omega)$ satisfying the condition $Lu \in \mathcal{H}_s^{\mathrm{loc}}(\Omega)$ (i. e. $\psi Lu \in \mathcal{H}_s$ for any function $\psi \in C_0^\infty(\Omega)$) the estimate*

$$\|\varphi u\|^2_{s+\varepsilon(K)} \leqslant C(K, s)\{\|\varphi_1 Lu\|_s^2 + \|\varphi_1 u\|_\gamma^2\} \qquad (2.6.48)$$

holds, where the functions $\phi, \phi_1 \in C_0^\infty(K)$, $\phi_1 \equiv 1$ on $\operatorname{supp}\phi$ and $\gamma = \text{const} < s + \epsilon(K)$. If the rank of the system $\{Q_0, \cdots, Q_{2m}\}$ is equal to m on every compact set, then the operator L is hypoelliptic in Ω.

PROOF. We show that the conditions of Theorem 2.4.2 are fulfilled for the operator $L(u)$ in (2.6.1). Since by assumption inequality (2.6.47) is fulfilled, Theorem 2.2.9 implies the estimate

$$\|u\|_{s+1}^2 \leqslant C_1\Big\{\sum_{|I|\leqslant R(K)} \|Q_I^0 u\|_s^2 + \|u\|_s^2\Big\}$$

for any function $u \in C_0^\infty(K)$. It follows from this and from representation (2.6.46) for Q_I that

$$\|u\|_{s+1}^2 \leqslant C_2\Big\{\sum_{|I|\leqslant R(K)} (\|Q_I u\|_s^2 + \|T_I u\|_s^2) + \|u\|_s^2\Big\}$$
$$\leqslant C_3\Big\{\sum_{|I|\leqslant R(K)} \|Q_I u\|_s^2 + \|u\|_s^2\Big\}.$$

Applying Lemma 2.6.4 as in the proof of Theorem 2.5.2, we obtain the estimate

$$\|u\|_{s+1}^2 \leqslant C_4\{\|Lu\|_0^2 + \|u\|_s^2 + \|u\|_{2^{R(K)}(s+1-2^{1-R(K)})}^2\}. \tag{2.6.49}$$

for $s + 1 - 2^{1-R(K)} \geqslant 0$. We now choose s so that

$$s+1 > 2^{R(K)}(s+1-2^{1-R(K)}) \geqslant 0.$$

It follows that $2(2^{R(K)}-1)^{-1} > s+1 \geqslant 2^{1-R(K)}$. Such an s exists, because $2^{1-R(K)} < 2(2^{R(K)}-1)^{-1}$. We denote $s+1$ by $\epsilon(K)$. It is clear that $\epsilon(K) > 0$. From (2.6.49) and Theorem 2.1.12 it follows that

$$\|u\|_\varepsilon^2 \leqslant C_5\{\|Lu\|_0^2 + \|u\|_{-N}^2\} \tag{2.6.50}$$

for all $u \in C_0^\infty(K)$.

We have thus obtained inequality (2.4.18) of condition I of Theorem 2.4.2. According to Theorem 2.6.1 (the energy estimate),

$$\sum_{j=1}^m (\|L^{0(j)}u\|_s^2 + \|L_{(j)}^0 u\|_{s-1}^2) \leqslant C_6\{\|Lu\|_0^2 + \|u\|_{2s}^2\} \tag{2.6.51}$$

for all $u \in C_0^\infty(K)$ and $s \geqslant 0$.

It is easy to see that

$$L^{0(j)} = L^{(j)} - T_0, \tag{2.6.52}$$

where T_0 is an operator of order at most zero, and

$$L_{(j)} = L_{(j)}^0 + T_1, \tag{2.6.53}$$

where T_1 is an operator of order at most 1. From (2.6.51) it follows that

$$\sum_{j=1}^m (\|L^{(j)}u\|_s^2 + \|L_{(j)}u\|_{s-1}^2) \leqslant C_7\{\|Lu\|_0^2 + \|u\|_{2s}^2\}. \tag{2.6.54}$$

In (2.6.54) we set $2s = \epsilon(K)$. Then

$$\sum_{j=1}^m (\|L^{(j)}u\|_{\varepsilon/2}^2 + \|L_{(j)}u\|_{\varepsilon/2-1}^2) + \|u\|_\varepsilon^2 \leqslant$$

$$\leqslant C_8\{\|Lu\|_0^2 + \|u\|_{-N}^2\},\ u \in C_0^\infty(K). \tag{2.6.55}$$

In (2.6.55) we substitute $E_s u$ in place of u. Applying Theorems 2.2.6 and 2.2.3, we obtain from (2.6.55) that

$$\sum_{j=1}^m \left(\|L^{(j)}u\|^2_{s+\frac{\varepsilon}{2}} + \|L_{(j)}u\|^2_{s-1+\frac{\varepsilon}{2}}\right) + \|u\|^2_{s+\varepsilon}$$

$$\leqslant C_9\left\{\sum_{j=1}^m \left(\|E_sL^{(j)}u\|^2_{\varepsilon/2} + \|E_sL_{(j)}u\|^2_{\frac{\varepsilon}{2}-1}\right) + \|E_su\|^2_\varepsilon + \|u\|^2_{-N}\right\}$$

$$\leqslant C_{10}\{\|LE_su\|_0^2 + \|u\|^2_{-N_1} + \|u\|^2_{s+\varepsilon/2}\} \leqslant$$

$$\leqslant C_{11}\left\{\|u\|^2_{s+\frac{\varepsilon}{2}} + \sum_{j=1}^m \left(\|L^{(j)}u\|_s^2 + \|L_{(j)}u\|^2_{s-1}\right) + \|Lu\|_s^2\right\}.$$

Bearing in mind inequality (2.1.30) from Theorem 2.1.12, we finally obtain, for any $s \in R^1$,

$$\sum_{j=1}^m \left(\|L^{(j)}u\|^2_{s+\frac{\varepsilon}{2}} + \|L_{(j)}u\|^2_{s-1+\varepsilon/2}\right) \leqslant C_{12}\{\|Lu\|_s^2 + \|u\|^2_{-N}\}. \tag{2.6.56}$$

It is clear that inequalities (2.4.19) and (2.4.20) of Theorem 2.4.2 follow from (2.6.56). The theorem is proved.

The following theorem, analogous to Theorem 2.5.3, is also valid for equations of the type (2.6.1).

THEOREM 2.6.3. *Suppose the operator* L *of type* (2.6.1) *satisfies the following conditions in a domain* Ω:

1) *On any compact set* $K \subset \Omega \backslash M$ *the system of operators* $\{Q_0, \cdots, Q_{2m}\}$ *has rank* m; *here* M *is some bounded set lying in a finite number of* $(m-1)$*-dimensional smooth manifolds* $\mathfrak{M}$, *and is such that the closure of the set* M *lies in* Ω.

2) *At each point* $x \in M$ *the inequality* $d^{kj}n_kn_j + |a^{kj}\Phi_{x_kx_j} + b^k\Phi_{x_k}| > 0$ *is satisfied, where* n *is the vector normal to the surface* $\mathfrak{M}$ *at the point* x, *and* $\Phi(x_1, \cdots, x_m) = 0$, *with* $\operatorname{grad}\Phi \neq 0$, *is the equation for* $\mathfrak{M}$ *in a neighborhood of the point* x.

Then every distribution u *in* $D'(\Omega)$ *such that* $Lu \in \mathcal{H}_s^{\text{loc}}(\Omega)$ *satisfies the estimate*

$$\|\varphi u\|_s^2 \leqslant C_\gamma\{\|\varphi_1 Lu\|_s^2 + \|\varphi_1 u\|_\gamma^2\}, \tag{2.6.57}$$

where the functions $\phi, \phi_1 \in C_0^\infty(\Omega)$, $\phi_1 \equiv 1$ *on* $\operatorname{supp}\phi$, *and moreover either* $\phi \equiv 1$ *on* M *or* $\operatorname{supp}\phi \cap M = \varnothing$. *Here* $\gamma = \text{const} < s$.

A differential operator L *satisfying conditions* 1) *and* 2) *is globally hypoelliptic in* Ω.

PROOF. The proof of Theorem 2.6.3 is entirely analogous to that of Theorem 2.5.3. In order to prove that conditions I, II and IIIb of Theorem 2.4.2 are fulfilled, we proceed exactly as in the proof of Theorem 2.5.3 to obtain the estimate

$$\|u\|_0^2 \leqslant \frac{C_1}{\mu^2}\|Lu\|_0^2;\quad u \in C_0^\infty(G_\beta), \tag{2.6.58}$$

where C_1 does not depend on μ and the constant μ may be chosen as large as desired.

From representations (2.6.52) and (2.6.53) it follows that

$$\sum_{j=1}^m (\|L^{(j)}u\|_0^2 + \|L_{(j)}u\|_{-1}^2) \leqslant C_2\{\|L^{0(j)}u\|_0^2 + \|\overset{0}{L}_{(j)}u\|_{-1}^2 + \|u\|_0^2\}. \tag{2.6.59}$$

From (2.6.6) and (2.6.9) we obtain

$$\sum_{j=1}^m \|L^{0(j)}u\|_0^2 \leqslant C_3\{\|Lu\|_0\|u\|_0 + \|u\|_0^2\}.$$

From inequality (2.6.14) it follows that

$$\|\overset{0}{L}_{(j)}u\|_{-1}^2 \leqslant 2\|E_{-1}\overset{0}{L}_{(j)}u\|_0^2 + C_4\|u\|_0^2 \leqslant 2\|\overset{0}{L}_{(j)}E_{-1}u\|_0^2 + C_5\|u\|_0^2$$
$$\leqslant C_6\left\{\sum_{t=1}^m \|LD_tE_{-1}u\|_0\|D_tE_{-1}u\|_0 + \|u\|_0^2\right\}.$$

From the last inequality and applying Lemma 2.6.3, we deduce

$$\|\overset{0}{L}_{(j)}u\|_{-1}^2 \leqslant C_7\{\|u\|_0^2 + \|u\|_0\|Lu\|_0\}.$$

Further bearing in mind (2.6.58), we obtain that

$$\|L^{(j)}u\|_0^2 + \|L_{(j)}u\|_{-1}^2 \leqslant C_8\{\|u\|_0^2 + \|u\|_0\|Lu\|_0\} \leqslant \frac{C_9}{\mu}\|Lu\|_0^2$$

where $u \in C_0^\infty(G_\beta)$.

From this and from (2.6.58) we deduce exactly as in the proof of Theorem 2.5.3 that under the conditions of Theorem 2.6.3, the assertions I, II and IIIb of Theorem 2.4.2 are satisfied.

The following is a consequence of Theorem 2.6.3.

THEOREM 2.6.4. *Suppose the system of operators* $\{Q_0, \cdots, Q_{2m}\}$ *has rank* m *for every compact set* K *lying in* $\Omega\backslash x^0$, *and suppose the inequality* $\sum_{j=1}^m (a^{jj} + |b^j|) > 0$ *holds at the point* x^0. *Then the operator* L *is hypoelliptic in*

Ω and inequality (2.6.57) *holds for every compact set $K \subset \Omega$ and every distribution $u \in D'(\Omega)$ such that $\psi Lu \in \mathcal{H}_s(\Omega)$ for any $\psi \in C_0^\infty(\Omega)$.*

The proof of Theorem 2.6.4 is analogous to the proof of Theorem 2.5.4.

We note that the condition $\Sigma_1^m (a^{jj} + |b^j|) \neq 0$ at the point x^0 is essential for the validity of Theorem 2.6.4. In fact, the equation

$$|x|^2 \Delta u - (\nu m + \nu(\nu - 2))\, u = 0, \tag{2.6.60}$$

where $|x|^2 \equiv \Sigma_1^m x_j^2$, is satisfied by the function $u = |x|^\nu$, $\nu > 0$. If the number ν is not an integer, then clearly equation (2.6.60) is not hypoelliptic in a domain Ω containing the origin.

It is easy to see, on the basis of Theorem 2.6.4, that the equation

$$a(x)\Delta u + u_{x_1} = 0,$$

where $a(x) \in C^\infty(R^m)$, $a(x) > 0$ for $|x| \neq 0$ and $a^{(\alpha)}(0) = 0$ for all α ($|\alpha| \geqslant 0$), is hypoelliptic in a domain Ω containing the origin.

The equation

$$D_1^2 u + x_1^{2k} D_2^2 u + i x_2^l D_3 u = 0; \quad k, l \geqslant 0,$$

is hypoelliptic by virtue of Theorems 2.5.2 and 2.6.2.

The equation

$$D_1^2 u + (a_1(x_1) + a_2(x_2)) D_2^2 u = 0,$$

where $a_1(x_1)$ and $a_2(x_2)$ belong to $C^\infty(R^1)$, $a_1(x_1) > 0$ for $x_1 \neq 0$; $a_1(0) = 0$; $a_2(x_2) > 0$ for $|x_2| > 1$, and $a_2(x_2) = 0$ for $|x_2| \leqslant 1$, is (on the basis of Theorem 2.6.2) globally hypoelliptic in the domain $|x_1|^2 + |x_2|^2 \leqslant 2$.

REMARK. A class of equations of the form (2.6.1) may be constructed with left side not representable in the form (2.5.1). The construction given below is based on a theorem of Hilbert. In [48] (see also [47]), Hilbert constructed a polynomial $P(x, y)$ of two variables of sixth degree which may not be represented in the form of a finite sum of squares of polynomials.

We let

$$A(x, y; z) = z^6 P\left(\frac{x}{z}, \frac{y}{z}\right) + P_1(x, y, z), \tag{2.6.61}$$

where P_1 is an infinitely differentiable function which vanishes at $x = y = z = 0$ together with all its derivatives up to the sixth order inclusive.

LEMMA 2.6.5. *The infinitely differentiable function $A(x, y, z)$ given by* (2.6.61) *is not representable in the form of a finite sum of squares of infinitely differentiable functions in any neighborhood of the origin.*

PROOF. We assume the contrary. Suppose the equation

$$A(x, y, z) = \sum_{j=1}^{N} A_j^2(x, y, z) \tag{2.6.62}$$

held in a neighborhood Ω of the origin, where A_j are functions infinitely differentiable in Ω. We represent A_j in the form of a partial Taylor series expansion

$$A_j = \sum_{|\alpha| \leq 3} \frac{1}{\alpha!} \frac{\partial^{|\alpha|} A_j(0, 0, 0)}{\partial x^{\alpha_1} \partial y^{\alpha_2} \partial z^{\alpha_3}} x^{\alpha_1} y^{\alpha_2} z^{\alpha_3} + R_{j,3}, \tag{2.6.63}$$

where the functions $R_{j,3}$ vanish at the origin together with their derivatives up to the third order inclusive.

From (2.6.62) it easily follows that at the origin the derivatives $\partial^{|\alpha|} A_j / \partial \bar{x}^\alpha$ for $|\alpha| \leq 2$, $\bar{x} = (x, y, z)$ are equal to zero, and, moreover,

$$z^6 P\left(\frac{x}{z}, \frac{y}{z}\right) = \sum_{j=1}^{N} \left[\sum_{|\alpha|=3} \frac{\partial^{|\alpha|} A_j(0, 0, 0)}{\partial x^{\alpha_1} \partial y^{\alpha_2} \partial z^{\alpha_3}} x^{\alpha_1} y^{\alpha_2} z^{\alpha_3} \right]^2. \tag{2.6.64}$$

Setting $z = 1$ in (2.6.64), we obtain a representation of the Hilbert polynomial $P(x, y)$ in the form of a finite sum of squares of polynomials, which is impossible. The lemma is proved.

Lemma 2.6.5 was communicated to us by L. Hörmander and V. P. Palamodov. From this lemma it follows that an operator of the form

$$L(u) \equiv A \Delta u + Qu \tag{2.6.65}$$

is not representable in the form (2.5.1) in any neighborhood of the origin. Here Q is any differential operator of first order with infinitely differentiable coefficients. In fact, if

$$-(A\Delta u + Qu) \equiv \sum_{j=1}^{r} X_j^2 u + iX_0 u + cu, \tag{2.6.66}$$

then it follows from equating coefficients of second derivatives on the left and right sides of (2.6.66) that for any $k = 1, \cdots, m$

$$A(x, y, z) = \sum_{j=1}^{r} (a_j^k)^2,$$

which is impossible by virtue of Lemma 2.6.5. It is easy to produce examples of operators of the form (2.6.65) which satisfy the conditions of Theorem 2.6.2 in a neighborhood of the origin, and so are hypoelliptic.

Let $A(x, y, z) = z^6 P(x/z, y/z)$ and let Q be an operator of first order with constant coefficients such that for some k we have $(Q)^k A \neq 0$ at the origin. Then the operator (2.6.65) satisfies the conditions of Theorem 2.6.2 in a sufficiently small neighborhood of the origin.

In fact, in this case

$$L^{0(j)} = 2A(x, y, z)D_j$$

and, as is easily verified, the commutator Q_I^j for $|I| = 2k + 1$ of the form

$$Q_I^j = \mathrm{ad}Q \ldots \mathrm{ad}QL^{0(j)} = 2(Q)^k A \cdot D_j$$

has the symbol $Q_I^j(x, t, z, \xi) = 2(Q)^k A \cdot \xi_j$. Consequently

$$\sum_{j=1}^{m} |Q_I^j(x, y, z, \xi)|^2 + 1 \geqslant C_0(1 + |\xi|^2) \tag{2.6.67}$$

for all points (x, y, z) in a sufficiently small neighborhood Ω of the origin, with $C_0 = \mathrm{const} > 0$. Relation (2.6.67) implies that the assumptions of Theorem 2.6.2 are fulfilled in Ω.

§7. On the solution of the first boundary value problem in nonsmooth domains. The method of M. V. Keldyš

For hypoelliptic equations satisfying the conditions of Theorem 2.5.2, 2.5.3, 2.6.2, or 2.6.3, a solution of the first boundary value problem in nonsmooth domains may be constructed by the method of M. V. Keldyš, which is based on the application of barrier functions. By this method Keldyš [63] first investigated the boundary value problem (1.1.4), (1.1.5) in the case when equation (1.1.4) is elliptic at interior points of the domain.

LEMMA 2.7.1. *Let Ω_n $(n = 1, \cdots)$ be a sequence of domains such that $\bar{\Omega}_n \subset \Omega_{n+1} \subset \Omega$, such that the boundary S_n of the domain Ω_n belongs to the class $A^{(k)}$ for some sufficiently large k, and such that any closed set contained in Ω belongs to all the domains Ω_n, beginning with some n. Let g be a function continuous in $\bar{\Omega}$, and f bounded in Ω. Assume the following conditions are satisfied.*

1) *The solutions $u_{\epsilon,n}$ of the equation*

$$L_\varepsilon(u) \equiv \varepsilon\Delta u + L(u) = f \text{ in } \Omega_n,\ \varepsilon > 0, \tag{2.7.1}$$

with the condition

$$u_{\epsilon,n}|_{S_n} = g, \tag{2.7.2}$$

where

$$L(u) = a^{kj}u_{x_k x_j} + b^k u_{x_k} + cu, c<0, a^{kj}(x)\xi_k\xi_j \geqslant 0$$

in $\bar{\Omega}$, form a compact set in the space $C^{(s)}(\bar{\omega})$, where $s \geqslant 2$ and $\bar{\omega}$ is any closed domain contained in Ω.

2) *For each point P_1 of some set Σ' belonging to the boundary Σ of Ω*

there exists a barrier; i. e. a function $V(x)$ *which satisfies the following conditions*: a) $V(x)$ *is defined and continuous in the closure of the domain* $\omega_1 = \Omega \cap O(P_1)$, *where* $O(P_1)$ *is some neighborhood of the point* P_1, *and* $V(x)$ *belongs to the class* $C^{(2)}(\omega_1)$; b) $V(P_1) = 0$, *and* $V > 0$ *at the remaining points of* $\bar{\omega}_1$; c) $L(V) < C_1 = \text{const} < 0$ *in* ω_1.

3) *In* $\Omega \cup \Sigma'$ *there exists a continuous function* W *in class* $C^{(2)}(\Omega)$ *such that* $W \geqslant 0$ *in* $\Omega \cup \Sigma'$, $L(W) < 0$ *in* Ω, *and* $W(x) \to \infty$ *as* $x \to \Sigma \backslash \Sigma'$. *That is to say,* $W(x) > N$ *if* x *lies in a* δ*-neighborhood of the set* $\Sigma \backslash \Sigma'$ *and* δ *is sufficiently small; here* N *is arbitrary.*

Then there exists a unique bounded solution of the equation $L(u) = f$ *in* Ω, *continuous in* $\Omega \cup \Sigma'$, *coinciding with the function* g *on* Σ', *and belonging to the class* $C^{(s)}(\Omega)$.

PROOF. By virtue of the maximum principle, the solutions $u_{\epsilon,n}$ of problem (2.7.1), (2.7.2) are bounded uniformly in ϵ and n. From the collection of functions $u_{\epsilon,n}$ we choose a sequence $u_{\epsilon,n}$ converging as $\epsilon \to 0$ and $n \to \infty$ in the norm of the space $C^{(s)}(\Omega_1)$. From this sequence we choose a subsequence converging in the norm of the space $C^{(s)}(\Omega_2)$, etc. The diagonal sequence u_{ϵ_k,n_k} converges as $n_k \to \infty$ and $\epsilon_k \to 0$, uniformly in $\bar{\Omega}_l$ for each $l \geqslant 1$. The same is true of the derivatives of the sequence up to order s. Clearly the limit function $u(x)$ belongs to class $C^{(s)}(\Omega)$, is bounded in Ω, and satisfies the equation $L(u) = f$.

We shall show that $u(x)$ approaches a limit on Σ' equal to the function g. Let P_1 be an arbitrary point on Σ' and h_γ the ball with center at P_1 and radius γ; we suppose that γ is so small that all points x of this ball, belonging to $\bar{\Omega}$, satisfy $|g(x) - g(P_1)| < \delta$, where $\delta > 0$ is some given number, and $h_\gamma \subset O(P_1)$.

In the region $\Omega_n \cap h_\gamma$ we consider the functions

$$w_{\pm}(x) = \pm g(P_1) + \delta \mp u_{\varepsilon,n} + CV(x), \quad C = \text{const} > 0.$$

It is easy to see that in $\Omega_n \cap h_\gamma$,

$$L_\varepsilon(w_{\pm}) = \pm cg(P_1) + c\delta \mp f + C(\varepsilon \Delta V + L(V)).$$

Since $L(V) < C_1 < 0$ in ω_1 and $\omega_1 \supset \Omega_n \cap h_\gamma$ for small γ, we may choose C so large that $L_\epsilon(w_\pm) < 0$ in $\Omega_n \cap h_\gamma$, when ϵ is sufficiently small. In this choice the constant C does not depend on ϵ or n, although ϵ depends on n. The functions $w_\pm(x)$ are nonnegative on the boundary of $\Omega_n \cap h_\gamma$. In fact, at points of S_n we have

$$w_{\pm}(x) = \pm g(P_1) + \delta \mp g(x) + CV(x) > 0$$

due to the restriction on δ and γ; and at points of the boundary lying on the sphere of radius γ we have $w_\pm > 0$ for C sufficiently large, since at these points $V > C_2 > 0$ and the functions $u_{\epsilon,n}$ are bounded uniformly in ϵ and n. From the conditions $L_\epsilon(w_\pm) < 0$, $c < 0$ in the domain $\Omega_n \cap h_\gamma$ and $w_\pm > 0$ on its boundary, it follows by the maximum principle that $w_\pm > 0$ in $\Omega_n \cap h_\gamma$. This implies that

$$|u_{\epsilon,n}(x) - g(P_1)| \leqslant \delta + CV(x) \tag{2.7.3}$$

for all x in $\Omega_n \cap h_\gamma$. Since C does not depend on ϵ or n, we may choose a sequence ϵ_k, n_k such that $u_{\epsilon_k,n_k} \to u(x)$ as $\epsilon_k \to 0$ and $n_k \to \infty$, and may pass to the limit in (2.7.3) as $\epsilon_k \to 0$ and $n_k \to \infty$. We thus obtain that

$$|u(x) - g(P_1)| \leqslant \delta + CV(x) \tag{2.7.4}$$

for all $x \in \Omega \cap h_\gamma$. The constant δ is arbitrary and C depends on δ; hence (2.7.4) implies that $u(x) \to g(P_1)$ as $x \to P_1$, since $V(x) \to 0$.

We show now that a bounded function $u(x)$ continuous in $\Omega \cup \Sigma'$, equal to g on Σ' and satisfying the equation

$$L(u) = f \text{ in } \Omega, \quad u \in C^{(2)}(\Omega),$$

is unique.

If there existed two functions with the properties indicated above, their difference $v(x)$ would equal zero on Σ' and would satisfy $L(v) = 0$ in Ω, with $|v| \leqslant K_0$, $K_0 = \text{const}$. In Ω we consider the functions $\delta W \pm v$, $\delta = \text{const} > 0$. These functions are nonnegative on the boundary of the domain $\Omega \backslash G_\kappa$, where G_κ is the κ-neighborhood of the set $\Sigma \backslash \Sigma'$, if κ is sufficiently small. This follows from the facts that $W \geqslant 0$ in Ω, $v = 0$ on Σ', and $W(x) \to \infty$ for $x \to \Sigma \backslash \Sigma'$. Moreover we have

$$L(\delta W \pm v) < 0 \text{ in } \Omega \backslash G_\kappa;$$

hence according to the maximum principle $\delta W \pm v \geqslant 0$ in $\Omega \backslash G_\kappa$, and therefore $|v| \leqslant \delta W$ at the points of $\Omega \backslash G_\kappa$. Let x be an arbitrary point in Ω. Clearly $x \in \Omega \backslash G_\kappa$ for all sufficiently small κ. Since δ is arbitrary, the inequality $|v| \leqslant \delta W$ at the point x really implies that $v(x) = 0$, i.e. $v \equiv 0$ in Ω. The lemma is proved.

We now consider the first boundary value problem for the equation $Lu = -Pu = f$, where P is an operator of the type (2.5.1).

THEOREM 2.7.1. *Assume the system of operators $X_0, \cdots, X_r$ has rank m at each point x of the domain Ω, and that $-\gamma \equiv c < \text{const} < 0$ in Ω. Assume the function f is bounded and that $f \in \mathcal{H}_s^{\text{loc}}(\Omega)$, where $2(s-k) > m$ and $k \geqslant 2$. Assume the part Σ' of the boundary Σ has the following property: each point $P_1 \in \Sigma'$ has a neighborhood in which the domain Ω is contained in some domain $\tilde{\Omega}$ of class $A^{(2)}$ for which the point P_1 lies inside a set of type $\Sigma_3 \cup \Sigma_2$ and for* *which* $\beta < 0$ *at points of* Σ_2, *where* β *is the function defined by* (1.5.6) *for the operator* $L = -P$. *Assume either that* $\overline{\Sigma \setminus \Sigma'} \cap \overline{\Sigma}' = \varnothing$, *the set* $\Sigma \setminus \Sigma'$ *being the boundary of a domain of class* $A^{(2)}$ *and* $\Sigma \setminus \Sigma' \subset \Sigma_0 \cup \Sigma_1$, *or that for a suitable choice of the coordinates* $y_1, \cdots, y_m$ *the set* $\Sigma \setminus \Sigma'$ *lies in the plane* $y_m = 0$, *with* $\Sigma \setminus \Sigma' = \Sigma_0 \cup \Sigma_1$ *and* $y_m > 0$ *for all points in* $\Omega \cup \Sigma'$. *Then there exists a unique bounded solution* $u(x)$ *of the equation*

$$Lu = f \quad in \quad \Omega \tag{2.7.5}$$

such that $u \in \mathcal{H}_s^{\mathrm{loc}}(\Omega)$, $u \in C^{(k)}(\Omega)$, *and* $u(x)$ *is continuous on* Σ', *assuming the values of the given continuous function* g *on* Σ'.

PROOF. Under the conditions of the theorem, we shall verify that the domain Ω and equation (2.7.5) satisfy all the hypotheses of Lemma 2.7.1. First we show that if the portion of the boundary Σ near the boundary point P_1 is the boundary of a domain of class $A^{(2)}$ and belongs to $\Sigma_3 \cup \Sigma_2$ with $\beta < 0$ at points of Σ_2, then there exists a barrier $V(x)$ for the point P_1. For this purpose we pass to local coordinates $y_1, \cdots, y_m$ in a neighborhood of the point P_1 so that the boundary Σ lies in the plane $y_m = 0$ and $y_m > 0$ in Ω. In the new coordinates, equation (2.7.5) assumes the form

$$Lu = \alpha^{kj} u_{y_k y_j} + \beta^k u_{y_k} + cu = f.$$

By virtue of our assumptions, in a sufficiently small neighborhood of the point P_1, either $\alpha^{mm} > 0$ or $\beta = \beta^m < 0$; furthermore, $c < 0$. Let $\psi(y_1, \cdots, y_{m-1})$ be a smooth function defined on Σ in the 2δ-neighborhood of P_1 for sufficiently small δ having the properties that $\psi = 1$ at the point P_1, $\psi = 0$ outside the δ-neighborhood of P_1, and $0 < \psi < 1$ at the remaining points. In the domain $Q_{\delta,\tau}\{0 < y_m < \tau\psi\}$ we consider the function $V = e^{\kappa\tau} - e^{\kappa(\tau\psi - y_m)}$ with $\kappa, \tau = \mathrm{const} > 0$. Clearly $V(P_1) = 0$, $V(x) > 0$ for $x \neq P_1$, and $V \in C^{(2)}(\overline{Q}_{\delta,\tau})$. Since with respect to the coordinates $x_1, \cdots, x_m$ the boundary Σ of the domain Ω in a neighborhood of P_1 is the boundary of a domain of class $A^{(2)}$, it follows that α^{kj} and β^k are bounded. We need to verify that $L(V) < 0$. We have

$$L(V) = -e^{\kappa(\tau\psi - y_m)} \Bigg[\alpha^{mm}\kappa^2 + \sum_{k,j=1}^{m-1} \left(\alpha^{kj}\kappa^2\tau^2\psi_{y_k}\psi_{y_j} + \alpha^{kj}\kappa\tau\psi_{y_k y_j} \right) - 2\sum_{k=1}^{m-1} \alpha^{km}\kappa^2\tau\psi_{y_k} + \sum_{k=1}^{m-1} \beta^k\kappa\tau\psi_{y_k} - \beta^m\kappa + c \Bigg] + ce^{\kappa\tau} < 0,$$

if τ is chosen sufficiently small and κ sufficiently large.

By the assumptions of Theorem 2.7.1, in some neighborhood of each point P_1 on Σ' the domain Ω is contained in some domain $\widetilde{\Omega}$ with respect to which we may construct a barrier $V(x)$ at the point P_1. It is clear that this same function $V(x)$ will serve as barrier at the point P_1 for the domain Ω, and therefore condition 2) of Lemma 2.7.1 is satisfied for the domain Ω.

We define W to be a function equal to a positive constant K outside the δ-neighborhood of $\Sigma\setminus\Sigma'$, and given by the following equation inside that neighborhood (here $\Sigma\setminus\Sigma'$ lies in the plane $y_m = 0$):

$$W = -\ln y_m \cdot h(y_m) + K,$$

where $h(y_m) = 1$ for $0 \leqslant y_m \leqslant \delta/2$, $h(y_m) = 0$ for $y_m \geqslant \delta$ and $h(y_m)$ is a smooth function such that $0 \leqslant h(y_m) \leqslant 1$. Then

$$L(W) \equiv \alpha^{kj} W_{y_k y_j} + \beta^k W_{y_k} + cW = \alpha^{mm} y_m^{-2} \cdot h(y_m)$$

$$- 2\alpha^{mm} y_m^{-1} h'(y_m) - \alpha^{mm} \ln y_m \cdot h''(y_m) - \beta^m y_m^{-1} h(y_m)$$

$$- \beta^m \ln y_m h'(y_m) + Kc - \ln y_m \cdot h(y_m) \cdot c.$$

If $\alpha^{mm} = O(y_m^2)$ in a neighborhood of $\Sigma\setminus\Sigma'$, then for sufficiently large K we obtain that $L(W) < 0$, since $\beta^m \geqslant 0$ and $c < 0$. If $\alpha^{mm} = O(y_m)$, then by assumption $\alpha^{mm}_{y_m} - \beta^m \leqslant 0$, and therefore $L(W) < 0$ for sufficiently large K.

If $\Sigma\setminus\Sigma'$ forms one component of the boundary so that $\overline{\Sigma\setminus\Sigma'} \cap \Sigma' = \varnothing$, then the function W is defined by the equation

$$W(x) = -\ln r \cdot h(r) + K,$$

where r is the distance from the point x to $\Sigma\setminus\Sigma'$, and $h(r)$ is the function defined above. It is easy to see that for sufficiently large K and small δ the function W satisfies condition 3) of Lemma 2.7.1.

We now prove that condition 1) of Lemma 2.7.1 is satisfied in the case when $L(u) \equiv -P(u)$. By assumption, at each point x of the domain Ω the system of operators $X_0, \cdots, X_r$ corresponding to the operator P has rank m. Since for (2.7.1) the corresponding system of operators $\epsilon\partial/\partial x_j$ $(j = 1, \cdots, m), X_0, \cdots, X_r$ contains the operator $X_0, \cdots, X_r$, by Theorem 2.5.2 we have

$$\|\varphi u\|^2_{s+\delta} \leqslant C(\omega, s)\{\|\varphi_1 L_\epsilon u\|^2_s + \|\varphi_1 u\|^2_0\}, \tag{2.7.6}$$

where $\delta = \text{const} > 0$, the functions $\phi, \phi_1 \in C_0^\infty(\omega)$, $\bar{\omega} \subset \Omega$, $\phi_1 \equiv 1$ in a neighborhood of the support of ϕ, the function $u \in D'(\Omega)$ and $L_\epsilon(u) \in \mathcal{H}_s^{\text{loc}}$, and the constant C does not depend on ϵ. Substituting into (2.7.6) the functions $u_{\epsilon,n}$ for $n \geqslant n_0$, which functions are solutions of the problem (2.7.1), (2.7.2), we obtain the following for sufficiently large n_0:

$$\|\varphi u_{\varepsilon,n}\|_{s+\delta}^2 \leqslant C(\omega, s)\{\|\varphi_1 f\|_s^2 + \|\varphi_1 u_{\varepsilon,n}\|_0^2\}, \tag{2.7.7}$$

where $\delta = \text{const} > 0$.

Since $f \in \mathcal{H}_s^{\text{loc}}$ and $u_{\epsilon,n}$ are bounded in Ω uniformly wtth respect to ϵ and n by virtue of the maximum principle, it follows from (2.7.7) that

$$\|\varphi u_{\varepsilon,n}\|_{s+\delta} \leqslant C_1,$$

where C_1 does not depend on ϵ or n. From Sobolev's theorem (see Theorem 2.1.11) it follows that the family of functions $\phi u_{\epsilon,n}$ is compact in the space $\mathcal{H}_s$. Theorem 2.1.8 implies that if $2(s-k) > m$, then the family $\phi u_{\epsilon,n}$ is compact in the space $C^{(k)}(\omega)$.

Since ϕ is an arbitrary function in $C_0^\infty(\Omega)$, we know that condition 1 of Lemma 2.7.1 is satisfied by $u_{\epsilon,n}$. The theorem is thereby proved.

REMARK 1. Theorem 2.7.1 is also true in the case when the operator L satisfies the hypotheses of Theorem 2.6.1. In this case in order to obtain an estimate of the type (2.7.7) we need to use the inequality

$$\varepsilon^2\|\varphi u\|_{s+2}^2 \leqslant C_1\{\|\varphi_1 L_\varepsilon u\|_s^2 + \|\varphi_1 u\|_s^2\},\ u \in \mathcal{H}_{s+2}^{\text{loc}},\ L_\varepsilon(u) \in \mathcal{H}_s^{\text{loc}} \tag{2.7.8}$$

for $s \geqslant 0$ and the estimate

$$\|\varphi u\|_{s+\delta}^2 \leqslant C_2\{\|\varphi_1 L u\|_s^2 + \|\varphi_1 u\|_s^2\},\ \delta = \text{const} > 0, \tag{2.7.9}$$

proved in Theorem 2.6.2, where $\phi, \phi_1 \in C_0^\infty(\Omega)$ and $\phi_1 \equiv 1$ on supp ϕ. Thus

$$\|\varphi u\|_{s+\delta}^2 \leqslant C_3\{\|\varphi_1 L_\varepsilon u\|_s^2 + \varepsilon^2\|\varphi_1 \Delta u\|_s^2 + \|\varphi_1 u\|_s^2\}, \tag{2.7.10}$$

since $L(u) = L_\epsilon(u) - \epsilon\Delta u$. Estimates (2.7.10) and (2.7.8) easily yield an inequality of the form (2.7.6). Inequality (2.7.8) may be obtained by a procedure analogous to that used in obtaining (2.6.23).

The methods of Chapter I, §8, may be used in studying the smoothness of weak solutions of the problem (1.1.4), (1.1.5) in the closed region $\bar{\Omega}$ for hypoelliptic equations satisfying the conditions of Theorems 2.5.3 or 2.6.3. Clearly, a consequence of these theorems is the local smoothness of generalized solutions of the first boundary value problem in any subdomain of the domain Ω for which the assumptions of Theorem 2.5.3 or Theorem 2.6.3 are fulfilled.

§8. On hypoellipticity of second order differential operators with analytic coefficients*

We consider the case when the coefficients of the operator (2.6.1) are real analytic functions. Necessary and sufficient conditions for hypoellipticity of the operators

---

*Editor's note. This section, added in translation at the authors' request, consists of the last eight or nine pages of their recent survey article [98], with only slight changes.

(2.6.1) and (2.5.1) may then be formulated in a simple manner. In our treatment the following lemma on analytic vector fields (see [145], [146]) will play a significant role.

Let $Y_1(x), \cdots, Y_l(x)$ be analytic vector fields given in a domain Ω of the space $\mathbf{R}^m$, i.e. $Y_j(x) = (a_j^1(x), \cdots, a_j^m(x))$, where the a_j^s are analytic functions in Ω, $s = 1, \cdots, m$; $j = 1, \cdots, l$. By Y_j we denote the differential operator

$$Y_j = \sum_{s=1}^{m} a_j^s(x) \frac{\partial}{\partial x_s}, \qquad x \in \Omega,$$

corresponding to the vector field $Y_j(x)$. For each multi-index $I = (\alpha_1, \cdots, \alpha_t)$ we define the differential operator $Y_I = \text{ad } Y_{\alpha_1} \cdots \text{ad } Y_{\alpha_{t-1}} Y_{\alpha_t}$, where $\alpha_s = 1, \cdots, l$ for $s = 1, \cdots, t$. By $\mathfrak{L}(Y_1, \cdots, Y_l)$ we denote the set of all possible linear combinations of the operators Y_I with coefficients analytic in Ω.

The rank of the system of operators $\{Y_1, \cdots, Y_l\}$ at the point x^0 is defined as the maximal number of linearly independent operators in the set $\mathfrak{L}(Y_1, \cdots, Y_l)$, considered at the point x^0.

LEMMA 2.8.1. *Let $Y_1(x), \cdots, Y_l(x)$ be analytic vector fields given in a domain Ω of the space $\mathbf{R}^m$, and suppose that at a point $x^0 \in \Omega$ the rank of the system of operators $\{Y_1, \cdots, Y_l\}$ is equal to k, where $1 \leqslant k < m$. Then in a neighborhood of the point x^0 there exists an analytic manifold V of dimension k such that $x^0 \in V$ and at each point of this manifold all vectors corresponding to operators from $\mathfrak{L}(Y_1, \cdots, Y_l)$ are tangent to V at the same point.*

PROOF. We shall suppose that x^0 is the origin. We first show that in a neighborhood U_0 of x^0 there exists a nondegenerate coordinate transformation

$$y_s = F_s(x_1, \ldots, x_m) \qquad (s = 1, \ldots, m), \tag{2.8.1}$$

given by analytic functions F_s, and also k differential operators $Y_{I_1}, \cdots, Y_{I_k}$, belonging to $\mathfrak{L}(Y_1, \cdots, Y_l)$, such that in the new coordinates y these operators may be expressed in the form

$$Y_{I_j} = e_j + \sum_{s=1}^{j-1} y_s Y_j^s \qquad (j = 1, \cdots, k), \tag{2.8.2}$$

where $e_j = \partial/\partial y_j$ and Y_j^s are certain operators corresponding to an analytic vector field in a neighborhood of the point x^0. Since the rank of the system $\{Y_1, \cdots, Y_l\}$ at the point x^0 is equal to $k \geqslant 1$, we know that there exists a vector field $Y_{I_1}(x)$ such that $Y_{I_1}(0) \neq 0$ and $Y_{I_1} \in \mathfrak{L}(Y_1, \cdots, Y_l)$. It is easy to show that in this case in a neighborhood U_1 of the point x^0 there exists an analytic coordinate transformation of the type (2.8.1) such that in this neighborhood $Y_{I_1} \equiv \partial/\partial y_1$. Therefore, if $k = 1$, then the relation (2.8.2) is obtained. If $k > 1$, then clearly there exists a vector field $\widetilde{Y}_{I_2}(y)$, such that $\widetilde{Y}_{I_2} \in \mathfrak{L}(Y_1, \cdots, Y_l)$ and furthermore $\widetilde{Y}_{I_2} = q_2^1(y)e_1 + Y_{I_2}$, $Y_{I_2}(0) \neq 0$, where $Y_{I_2}(y)$ is orthogonal to $e_1(y)$ at all points of some neighborhood U_2 of the point x^0. It is clear that $Y_{I_2} \in \mathfrak{L}(Y_1, \cdots, Y_l)$ and in some neighborhood U_3 of x^0,

$$Y_{I_2}(y) = Y_{I_2}(0, y_2, \ldots, y_m) + y_1 Y_2^1(y),$$

where $Y_2^1(y)$ is an analytic vector field. In the space $y_2, \cdots, y_m$ we perform a coordinate transformation $y_j' = y_j'(y_2, \cdots, y_m)$, $j = 2, \cdots, m$, for which the operator corresponding to the vector field $Y_{I_2}(0, y_2, \cdots, y_m)$ is transformed into $e_2 = \partial/\partial y_2'$. Thus in the new coordinates $y_1' = y_1$, $y_j' = y_j'(y_2, \cdots, y_m)$ $(j = 2, \cdots, m)$, which we still denote by y, we have

$$Y_{I_1} = e_1, \quad Y_{I_2} \equiv e_2 + y_1 Y_2^1.$$

For $k > 2$ the relation (2.8.2) shall be proved by induction. We assume that we have chosen the coordinate system $y_1, \cdots, y_m$ and s operators from $\mathfrak{L}(Y_1, \cdots, Y_l)$ so that

$$Y_{I_j} = e_j + \sum_{\rho=1}^{j-1} y_\rho Y_j^\rho \qquad (j = 1, \ldots, s), \tag{2.8.3}$$

where $e_j = \partial/\partial y_j$ and $Y_j^\rho(y)$ are analytic vector fields. We now show that if $s < k$, then the same is valid for $s + 1$ as well. According to the assumptions of the lemma, in a neighborhood of x^0 there exists a vector field $\widetilde{Y}_{I_{s+1}}(y)$ such that $\widetilde{Y}_{I_{s+1}} \in \mathfrak{L}(Y_1, \cdots, Y_l)$ and

$$\widetilde{Y}_{I_{s+1}} = \sum_{j=1}^{s} q_{s+1}^j(y)\, e_j + Y'_{s+1}, \tag{2.8.4}$$

where $Y'_{s+1}(y)$ is orthogonal to $e_j(y)$ for $j \leqslant s$, q_{s+1}^j are analytic functions, and $Y'_{s+1}(0) \neq 0$. By Taylor's formula we have

$$Y'_{s+1}(y) = Y'_{s+1}(0, \ldots, 0, y_{s+1}, \ldots, y_m) + \sum_{j=1}^{s} y_j Y'_{j,s+1}(y),$$

where $Y'_{j,s+1}$ are certain analytic vector fields. In the plane of $y_{s+1}, \cdots, y_m$ we effect a coordinate transformation under which the operator corresponding to the vector field $Y'_{s+1}(0, \cdots, 0, y_{s+1}, \cdots, y_m)$ becomes $e_{s+1} = \partial/\partial y'_{s+1}$. Inequality (2.8.4) then implies that in the new coordinates, which we still denote by y,

$$\tilde{Y}_{I_{s+1}} = \sum_{j=1}^{s} q_{s+1}^{j}(y)\, e_j + e_{s+1} + \sum_{j=1}^{s} y_j Y'_{j,\, s+1}. \tag{2.8.5}$$

Using equations (2.8.3) (valid by the induction hypothesis) to represent e_j for $j \leqslant s$, we obtain from (2.8.5) that

$$Y_{I_{s+1}} = e_{s+1} + \sum_{j=1}^{s} y_j Y_{s+1}^{j}.$$

where $Y_{I_{s+1}} \in \mathfrak{L}(Y_1, \cdots, Y_l)$ and $Y_{s+1}^{j}(y)$ are certain vector fields analytic in a neighborhood of x^0. This is the required result.

The hyperplane $E^k = \{y_{k+1} = 0, \cdots, y_m = 0\}$ is a manifold V in the y coordinates for which relation (2.8.2) is valid. For this we show that if Y_I is any operator in $\mathfrak{L}(Y_1, \cdots, Y_l)$ with respect to the coordinates y, then the scalar product of $e_j(y)$ with $Y_I(y)$ is equal to zero at points of E^k belonging to some neighborhood of x^0, when $j \geqslant k+1$. Since $f(y) = (e_j(y), Y_I(y))$ is an analytic function of y, it is sufficient to prove that all derivatives of this function with respect to $y_1, \cdots, y_k$ are equal to zero at the origin.

We shall show that in some neighborhood U_0 of the point x^0,

$$\frac{\partial f}{\partial y_s} = (e_j(y),\ Y_I^s(y)) + \sum_{\rho=1}^{s-1} y_\rho A_s^\rho, \qquad j \geqslant k+1, \quad s \leqslant k, \tag{2.8.6}$$

where $Y_I^s \in \mathfrak{L}(Y_1, \cdots, Y_l)$, and A_s^ρ are certain analytic functions. We have

$$\frac{\partial f}{\partial y_1} = (e_j(y),\ (\mathrm{ad}\, e_1 Y_I)(y)) = (e_j(y),\ Y_I^1(y)),$$

since $\mathrm{ad}\, e_1 Y_I \in \mathfrak{L}(Y_1, \cdots, Y_l)$. We proceed by induction on s. We assume that (2.8.6) is valid for $s \leqslant s_0 < k$ and show that it is valid also for $s = s_0 + 1$. Using the relation (2.8.2) for $j = s_0 + 1$, we obtain

$$\begin{aligned}
\frac{\partial f}{\partial y_{s_0+1}} &= (e_j(y),\ (\mathrm{ad}\, e_{s_0+1} Y_I)(y)) \\
&= \Big(e_j(y),\ \Big(\mathrm{ad}\Big(Y_{I_{s_0+1}} - \sum_{\rho=1}^{s_0} y_\rho Y_{s_0+1}^{\rho}\Big) Y_I\Big)(y)\Big) \\
&= (e_j(y),\ (\mathrm{ad}\, Y_{I_{s_0+1}} Y_I)(y)) - \sum_{\rho=1}^{s_0} (e_j(y),\ (\mathrm{ad}\, y_\rho Y_{s_0+1}^{\rho} Y_I)) \\
&= (e_j(y),\ Y_I'(y)) - \sum_{\rho=1}^{s_0} a_\rho (e_j(y),\ Y_{s_0+1}^{\rho}(y)) - \sum_{\rho=1}^{s_0} y_\rho A_\rho,
\end{aligned} \tag{2.8.7}$$

where a_ρ and A_ρ are certain analytic functions, and $Y'_I \in \mathfrak{L}(Y_1, \cdots, Y_l)$. From (2.8.2) for $j = s_0 + 1$ it follows that

$$Y^\rho_{s_0+1} = \operatorname{ad} e_\rho Y_{I_{s_0+1}} + \sum_{\rho=1}^{s_0} y_\rho Y_{\rho, s_0+1}, \tag{2.8.8}$$

where Y_{ρ,s_0+1} is an operator corresponding to an analytic vector field. Hence from (2.8.7) and (2.8.8) it follows that

$$\frac{\partial f}{\partial y_{s_0+1}} = (e_j(y), Y'_I(y)) + \sum_{\rho=1}^{s_0} a_\rho (e_j(y), (\operatorname{ad} e_\rho Y_{I_{s_0+1}})(y)) + \sum_{\rho=1}^{s_0} y_\rho A'_\rho,$$

where A'_ρ are analytic functions. Since for $\rho \leqslant s_0$ we have

$$(e_j(y), (\operatorname{ad} e_\rho Y_{I_{s_0+1}})(y)) = \frac{\partial}{\partial y_\rho}(e_j(y), Y_{I_{s_0+1}}(y)), \tag{2.8.9}$$

and $Y_{I_{s_0+1}} \in \mathfrak{L}(Y_1, \cdots, Y_l)$, it follows from the induction assumption that the left side of (2.8.9) may be written in the form (2.8.6). Hence (2.8.6) is also valid for $s = s_0 + 1$.

We now show that for each multi-index $\alpha = (\alpha_1, \cdots, \alpha_k)$ an equation of the form

$$\frac{\partial^{|\alpha|} f(y)}{\partial y_1^{\alpha_1} \ldots \partial y_k^{\alpha_k}} = (e_j(y), Y_I^\alpha(y)) + \sum_{\rho=1}^{k-1} y_\rho A^\rho_\alpha \tag{2.8.10}$$

holds, where $Y_I^\alpha \in \mathfrak{L}(Y_1, \cdots, Y_l)$ and A^ρ_α are analytic functions. For $|\alpha| = 1$ this equation has already been proved. We assume that (2.8.10) has been proved for all α of the form $(\alpha_1, \cdots, \alpha_s, 0, \cdots, 0)$, $s < k$, and prove it for all α of the form $(\alpha_1, \cdots, \alpha_s, \alpha_{s+1}, 0, \cdots, 0)$.

If $\alpha_{s+1} = 1$, then by the induction hypothesis and (2.8.6) it follows that

$$\frac{\partial}{\partial y_{s+1}}\left(\frac{\partial^{|\alpha'|} f}{\partial y_1^{\alpha_1} \ldots \partial y_s^{\alpha_s}}\right) = \frac{\partial}{\partial y_{s+1}}(e_j(y), Y_I^{\alpha'}(y)) + \sum_{\rho=1}^{s-1} y_\rho A^\rho_{\alpha', 1}$$

$$= (e_j(y), Y_I^\alpha(y)) + \sum_{\rho=1}^{s} y_\rho A^\rho_\alpha,$$

where $\alpha' = (\alpha_1, \cdots, \alpha_s, 0, \cdots, 0)$, $Y_I^{\alpha'} \in \mathfrak{L}(Y_1, \cdots, Y_l)$ and $A^\rho_{\alpha',1}$ are analytic functions. We further assume that for $p \leqslant p_0$

$$\frac{\partial^p}{\partial y^p_{s+1}} \frac{\partial^{|\alpha'|} f}{\partial y_1^{\alpha_1} \ldots \partial y_s^{\alpha_s}} = (e_j(y), Y^{\alpha'}_{I, p}(y)) + \sum_{\rho=1}^{s} y_\rho A^\rho_{\alpha', p}, \tag{2.8.11}$$

and show that this equation is also valid for $p = p_0 + 1$. We have

$$\frac{\partial^{p_0+1}}{\partial y_{s+1}^{p_0+1}}\frac{\partial^{|\alpha'|}f}{\partial y_1^{\alpha_1}\dots\partial y_s^{\alpha_s}}=\frac{\partial}{\partial y_{s+1}}(e_j(y),\,Y_{I,\,p_0}^{\alpha'}(y))+\sum_{\rho=1}^{s}y_\rho\tilde{A}_{\alpha',\,p_0}^{\rho}, \tag{2.8.12}$$

where $Y_{I,p_0}^{\alpha'}\in\mathfrak{L}(Y_1,\cdots,Y_l)$ and $\tilde{A}_{\alpha',p_0}^{\rho}$ are analytic functions. From (2.8.12) and (2.8.6) it follows that (2.8.11) is valid for $p=p_0+1$.

We note that at the origin $(e_j(y),Y_I(y))=0$ if $Y_I\in\mathfrak{L}(Y_1,\cdots,Y_l)$, and $j\geqslant k+1$, since at the origin $e_y(y)$ for $j=1,\cdots,k$ belong to vector fields corresponding to operators from $\mathfrak{L}(Y_1,\cdots,Y_l)$ and by assumption they form a basis for $\mathfrak{L}(Y_1,\cdots,Y_l)$ at the origin. From this remark and from formula (2.8.10) it follows that all derivatives of the function $f(y)$ with respect to $y_1,\cdots,y_k$ are equal to zero at $y=0$, which is the desired result.

We consider the vector fields $Q(x)$ and $L^{0(j)}(x)$ with associated operators Q and $L^{0(j)}$, where

$$L^{0(j)}(x)=(a^{j1}(x),\,\dots,\,a^{jm}(x)),\qquad Q(x)=(b^1(x)-a_{x_k}^{1k}(x),\,\dots,\,b^m(x)-a_{x_k}^{mk}(x))$$

and $a^{kj}(x)$ and $b^k(x)$ are coefficients of the operator (2.6.1).

LEMMA 2.8.2. *Suppose the rank of the system of operators* $\{Q,L^{0(1)},\cdots,L^{0(m)}\}$ *is equal to* k *at a point* $x^0\in\Omega$, *where* $1\leqslant k<m$, *and that the coefficients of the operator* (2.6.1) *are analytic in the domain* Ω. *Then there exists a change of independent variables of the form* (2.8.1) *with analytic functions* F_s *such that the operator* (2.6.1) *may be represented in the form*

$$L(u)\equiv\alpha^{kj}(y)\,u_{y_ky_j}+\beta^k(y)\,u_{y_k}+cu=L_1(u)+L_2(u)+cu \tag{2.8.13}$$

in a neighborhood of the point x^0, *where the operator* L_1 *contains differentiations only with respect to the variables* $y_1,\cdots,y_k$, *and the coefficients of the operator* L_2 *are equal to zero on the hyperplane* $E^k\{y_{k+1}=0,\cdots,y_m=0\}$.

PROOF. According to Lemma 2.8.1 there exists an analytic k-dimensional manifold V in a neighborhood of the point x^0 such that $x^0\in V$ and at each point of V all vectors corresponding to operators in $\mathfrak{L}(Q,L^{0(1)},\cdots,L^{0(m)})$ are tangent to V at that point. In a neighborhood of x^0 we take a coordinate transformation of the form (2.8.1) such that in the new coordinates $y_1,\cdots,y_m$ the manifold V coincides with the k-dimensional hyperplane $E^k=\{y_{k+1}=0,\cdots,y_m=0\}$. Since the operators Q and $L^{0(j)}$ $(j=1,\cdots,m)$ belong to $\mathfrak{L}(Q,L^{0(1)},\cdots,L^{0(m)})$, it follows from Lemma 2.8.1 that at points of V we have

$$(Q(x),\,\mathrm{grad}\,F_s)=0,\ (L^{0(j)}(x),\,\mathrm{grad}\,F_s)=0\ \text{ for }\ s\geqslant k+1,\quad j=1,\,\dots,\,m. \tag{2.8.14}$$

This yields the fact that $\alpha^{sj} = a^{l\rho} F_{sx_l} F_{jx_\rho} = 0$ on E^k if either $s \geqslant k+1$ or $j \geqslant k+1$. It is easy to see that

$$\beta^s = L(F_s) = (a^{l\rho} F_{sx_l})_{x_\rho} + (Q(x), \operatorname{grad} F_s) + cF_s$$
$$= \frac{\partial}{\partial y_\nu}(a^{l\rho} F_{sx_l} F_{\nu x_\rho}) - (L^{0(\rho)}, \operatorname{grad} F_s) \frac{\partial}{\partial y_\nu}(F_{\nu x_\rho}) + (Q(x), \operatorname{grad} F_s) + cF_s. \tag{2.8.15}$$

The last three terms in (2.8.15) are equal to zero for $s \geqslant k+1$ at points of V, since $F_s = 0$ on V for $s \geqslant k+1$ and (2.8.14) holds. Clearly $a^{l\rho} F_{sx_l} F_{\nu x_\rho} = 0$ on V for $s \geqslant k+1$. Therefore $(\partial/\partial y_\nu)(a^{l\rho} F_{sx_l} F_{\nu x_\rho}) = 0$ if $\nu \leqslant k$, $s \geqslant k+1$. For $\nu \geqslant k+1$ we have

$$|a^{l\rho} F_{sx_l} F_{\nu x_\rho}| \leqslant (a^{l\rho} F_{sx_l} F_{sx_\rho})^{1/2} (a^{l\rho} F_{\nu x_l} F_{\nu x_\rho})^{1/2}. \tag{2.8.16}$$

Since $a^{kj}\xi_k \xi_j \geqslant 0$ for all ξ, it follows easily from estimate (2.8.16) that $\operatorname{grad}(a^{l\rho} F_{sx_l} F_{\nu x_\rho}) = 0$ on V for $\nu, s \geqslant k+1$, which means that $\beta^s = 0$ on E^k for $s \geqslant k+1$. The lemma is proved.

THEOREM 2.8.1. *If the coefficients of the operators* (2.6.1) *are analytic in* Ω *and the system of operators* $\{Q, L^{0(1)}, \cdots, L^{0(m)}\}$ *has rank* k *at some point* x^0 *in* Ω, *where* $1 \leqslant k < m$, *then the operator* L *is not hypoelliptic in* Ω.

PROOF. In a neighborhood U_0 of x^0 we choose a coordinate transformation such as was indicated in Lemma 2.8.2. Then the operator (2.6.1) assumes the form (2.8.13). We consider the operator L^*. Clearly $L^*u = L_1^*u + L_2^*u + cu$, where L_1^* contains differentiations only with respect to the variables $y_1, \cdots, y_k$, and

$$L_2^*u = M_1u + M_2u + c_2^*u, \text{ where } M_1 = \sum_{\rho=k+1}^{m} \sum_{l=1}^{k} \alpha^{l\rho}_{y_\rho} \frac{\partial}{\partial y_l},$$

so that M_1 contains differentiations only with respect to $y_1, \cdots, y_k$, and the coefficients of the operator M_2 are equal to zero on E^k. This follows from the fact that $\alpha^{l\rho} = 0$ on E^k if either $l \geqslant k+1$ or $\rho \geqslant k+1$, and $\operatorname{grad} \alpha^{l\rho} = 0$ if $l, \rho \geqslant k+1$.

On the hyperplane E^k we consider the operator

$$L_V^*u = L_1^*u + M_1u + (c_2^* + c)u.$$

It is clear that the operator L_V adjoint to the operator L_V^* has the form

$$L_Vu = L_1u + M_1^*u + (c_2^* + c)u,$$

where M_1^* is a first order differential operator. We assume that the equation $L_Vu = 0$ has a nontrivial solution $v(y_1, \cdots, y_k)$ in the plane E^k in a neighborhood G_0 of some point $y^0 \in U_0$. Then $(v, L_V^*\phi)_0 = 0$ for each infinitely differentiable function

$\phi(y_1, \cdots, y_k)$ with support in G_0. Here $(u, v)_0 = \int_{E^k} uv\, dy_1 \cdots dy_k$. We consider, in a neighborhood Ω_0 of the point y^0, the distribution $u(y)$ defined by the equation

$$(u, \varphi) = (v, \varphi(y_1, \ldots, y_k, 0, \ldots, 0))_0,$$

where $\phi \in C_0^\infty(\Omega_0)$. Clearly $u = 0$ off the hyperplane E^k, and for $\phi \in C_0^\infty(\Omega_0)$ we have

$$(u, L^*\varphi) = (v, (L_1^*\varphi + M_1\varphi + (c_2^* + c)\varphi)(y_1, \ldots, y_k, 0, \ldots, 0))_0$$
$$= (v, L_V^*\varphi(y_1, \ldots, y_k, 0, \ldots, 0))_0 = 0.$$

This means that the distribution $u(x)$ is a solution of the equation $Lu = 0$ in a neighborhood Ω_0 of the point y^0 and $u(x)$ is not an infinitely differentiable function in Ω_0.

We now show that if $k \geqslant 1$, then a nontrivial solution of the equation $L_V v = 0$ exists on the plane E^k in a neighborhood G_0 of some point $y^0 \in U_0$. If one of the coefficients $\alpha^{l\rho} \not\equiv 0$ in $E^k \cap U_0$, then such a solution exists in a neighborhood of some point $y^0 \in E^k \cap U_0$ by virtue of the Cauchy-Kowalewski Theorem. However, if all coefficients $\alpha^{l\rho} \equiv 0$ on E^k in a neighborhood U_0 of x^0, then it is easy to see that $\operatorname{grad} \alpha^{l\rho} \equiv 0$ on E^k and hence that the coefficients of the operator M_1 are equal to zero on $E^k \cap U_0$. Since

$$\sum_{l,\rho=1}^{m} |\alpha^{l\rho}| + \sum_{\rho=1}^{m} |\beta^\rho| \not\equiv 0$$

on E^k, we have in this case that one of the coefficients $\beta^s \not\equiv 0$ for $s \leqslant k$ on $E^k \cap U_0$. Therefore in a neighborhood of some point $y^0 \in E^k \cap U_0$ there exists a nontrivial solution of $L_V v = 0$ by virtue of the Cauchy-Kowalewski Theorem applied to a first order equation. The theorem is proved.

From Theorems 2.6.2 and 2.8.1 we have the following result.

THEOREM 2.8.2. *Suppose all the coefficients* a^{lj}, b^j, c *of the differential operator* (2.6.1) *are real analytic functions in* Ω, *and suppose that*

$$\sum_{k,j=1}^{m} |a^{kj}| + \sum_{j=1}^{m} |b^j| \neq 0 \tag{2.8.17}$$

everywhere in Ω. *Then for hypoellipticity of the operator* (2.6.1) *it is necessary and sufficient that the rank of the system* $\{Q, L^{0(1)}, \cdots, L^{0(m)}\}$ *be equal to* m *at each point of the domain* Ω.

Examples show that the conclusion of this theorem is not necessarily true if condition (2.8.17) is not satisfied. The equation

$$|x|^2 \Delta u - (\nu m + \nu(\nu - 2))\, u = 0 \tag{2.8.18}$$

is satisfied by the function $u = |x|^\nu$, $\nu > 0$, and if ν is not an integer, equation (2.8.18) is clearly not hypoelliptic in a neighborhood of the origin. On the other hand, it is shown in a paper of L. Hörmander [52] that if γ is an integer and $\gamma > 1$, then the equation

$$|x|^{2\gamma} \Delta u - u = 0 \tag{2.8.19}$$

is hypoelliptic in every domain $\Omega \in \mathbf{R}^m$.

In the case when the coefficients of the operator (2.5.1) are analytic functions in Ω, M. Derridj [25] proved the following theorem, which together with Theorem 2.5.2 provides a necessary and sufficient condition for hypoellipticity of the operator (2.5.1) under the assumption that at each point of Ω not all of the coefficients of the operators $X_0, \cdots, X_r$ are equal to zero.

THEOREM 2.8.3. *Suppose all coefficients* a_j^l *and* c $(j = 0, 1, \cdots, r;\ l = 1, \cdots, m)$ *of the differential operator* (2.5.1) *are analytic functions in* Ω, *and suppose, at some point* $x^0 \in \Omega$, *the rank of the system of operators* $\{X_0, \cdots, X_r\}$ *is equal to* k, *where* $0 < k < m$. *Then the operator* (2.5.1) *is not hypoelliptic in* Ω.

Theorems 2.5.2 and 2.8.3 yield the following result:

THEOREM 2.8.4. *Suppose all coefficients* a_j^l *and* c $(j = 0, 1, \cdots r;\ l = 1, \cdots, m)$ *of the differential operator* (2.5.1) *are analytic functions in* Ω, *and that*

$$\sum_{j=0}^{r} \sum_{l=1}^{m} |a_j^l| \neq 0 \tag{2.8.20}$$

for all points in Ω. *Then a necessary and sufficient condition for hypoellipticity of the operator* (2.5.1) *in* Ω *is that the rank of the system* $\{X_0, \cdots, X_r\}$ *be equal to* m *at each point of the domain* Ω.

PROOF OF THEOREM 2.8.3. It is sufficient to show that if the operator (2.5.1) is hypoelliptic in Ω, then the rank of the system of operators $\{X_0, \cdots, X_r\}$ is equal to m at all points of Ω. We assume the contrary. Suppose the system $\{X_0, \cdots, X_r\}$ has rank k at a point $x^0 \in \Omega$ with $1 \leqslant k < m$. We write the operator (2.5.1) in the form (2.6.1). It is easy to see that if (2.8.20) holds for the operator (2.5.1), then condition (2.8.17) holds for the corresponding operator (2.6.1), and, moreover,

$$Q = X_0 - \sum_{s=1}^{r} a^j_{sx_j} X_s, \qquad L^{0(j)} = 2 \sum_{s=1}^{r} a_s^j X_s.$$

Thus for the operator (2.6.1) so obtained the rank of the system $\{Q, L^{0(1)}, \cdots, L^{0(m)}\}$ at the point x^0 is also less than m, which by Theorem 2.8.2 implies that the derived operator (2.6.1) is not hypoelliptic. This would contradict the condition that (2.5.1) is hypoelliptic.

CHAPTER III
ADDITIONAL TOPICS

§1. Qualitative properties of solutions of second order equations with nonnegative characteristic form

A maximum principle for second order equations with nonnegative characteristic form was proved in §§1 and 5 of Chapter I.

For the Laplace equation, a well-known property called the strong maximum principle holds: a function u which is harmonic in a domain Ω and which assumes its greatest value at an interior point of Ω is a constant. For the heat equation $u_{xx} - u_t = 0$ considered in a rectangle Q with sides parallel to the coordinate axes, the strong maximum principle takes the following form: a solution $u(t, x)$ assuming its greatest value at a point (t_0, x_0) interior to Q is constant in Q for $t \leqslant t_0$. For second order elliptic equations, the strong maximum principle was proved by Hopf [49] (see also [81, 95]), and for second order parabolic equations by Nirenberg [85].

The strong maximum principle for general second order equations with nonnegative characteristic form was studied by Pucci [108, 109] and by A. D. Aleksandrov [1–5]. For equations of the form (2.5.1), it was studied by Bony [15–17].

We shall assume that the coefficients of the equation

$$L(u) \equiv a^{kj} u_{x_k x_j} + b^k u_{x_k} + cu = 0, \quad a^{kj}\xi_k\xi_j \geqslant 0 \tag{3.1.1}$$

are bounded in Ω and that $u(x) \in C^{(2)}(\Omega)$. At each point x we consider the eigenvectors corresponding to positive eigenvalues of the matrix $\|a^{kj}\|$, and let $E(x)$ be the linear space spanned by these vectors. We shall call this space the plane of ellipticity at the point x. A curve l is called a line of ellipticity for the equation (3.1.1) if in a neighborhood of each of its points there exists a vector field $(Y_1(x), \cdots, Y_m(x))$ such that $Y_j \in C^{(1)}$, the vector $(Y_1(x), \cdots, Y_m(x))$ lies in the plane $E(x)$, $a^{kj}(x) Y_k(x) Y_j(x) \geqslant \text{const} > 0$ at each point x of this neighborhood, and the curve l is a trajectory of the system of equations $dx_j/dt = Y_j(x)$, $j = 1, \cdots, m$.

The set $\mathfrak{M}$ is called a set of elliptic connectivity of the equation (3.1.1) if any two points of $\mathfrak{M}$ may be joined by a curve consisting of a finite number of arcs of lines of ellipticity, and if there exists no set properly containing $\mathfrak{M}$ with this same property. If the entire domain Ω is a set of elliptic connectivity, equation (3.1.1) is called elliptically connected in Ω. Clearly an elliptic equation is elliptically connected. It is easy to produce examples of equations which are elliptically connected but not elliptic in a domain Ω. For example, the equation

$$L(u) \equiv u_{x_1x_1} + \cos^2 x_1 u_{x_2x_2} + 2\sin x_1 \cos x_1 u_{x_2x_3} + \sin^2 x_1 u_{x_3x_3} = 0 \tag{3.1.2}$$

in the domain $\Omega\{|x_1| < \pi/2\}$ is elliptically connected, although it is easily verified not to be elliptic at any point of the domain. In order to show that (3.1.2) is elliptically connected in Ω, we shall find the plane of ellipticity $E(x)$. This plane is generated by the eigenvectors $(1, 0, 0)$ and $(0, 1, \tan x_1)$. If $A = \partial/\partial x_1$ and

$$B = \frac{\partial}{\partial x_2} + \tan x_1 \frac{\partial}{\partial x_3},$$

then

$$[A, B] \equiv AB - BA = \frac{1}{\cos^2 x_1} \frac{\partial}{\partial x_3}.$$

The vectors $(1, 0, 0)$, $(0, 1, \tan x_1)$ and $(0, 0, 1/\cos^2 x_1)$ are linearly independent at each point of Ω. Therefore the elliptic connectivity of (3.1.2) follows from the following theorem of P. K. Raševskiĭ [115].

THEOREM 3.1.1. *Assume that among the operators*

$$Z_j \equiv \sum_{s=1}^{m} \alpha^{sj}(x) \frac{\partial}{\partial x_s}, \quad j = 1, \ldots, n, \tag{3.1.3}$$

and their commutators at each point of Ω there exist m linearly independent operators (the coefficients α^{kj} are assumed sufficiently smooth). Then any two points of Ω may be joined by a line consisting of a finite number of arcs of trajectories of the system of differential equations

$$\frac{dx_s}{dt} = \alpha^{sj}(x), \quad s = 1, \ldots, m,$$

where j assumes the values $1, \cdots, n$.

The following basic theorem on "propagation of zeros" holds.

THEOREM 3.1.2. *If $u \geqslant 0$ and $L(u) \leqslant 0$ in the domain Ω and if $u = 0$ at a point $x^0 \in \Omega$, then $u = 0$ on the set of elliptic connectivity containing the point x^0.*

The following theorem on the strong maximum principle for equation (3.1.1) follows immediately from Theorem 3.1.2.

THEOREM 3.1.3. *Suppose* $L(u) \geqslant 0$ *in the domain* Ω *and suppose the coefficient* c *and* $M = \sup_\Omega u$ *are related by the inequality* $Mc \leqslant 0$. *If* $u(x^0) = M$ *and* $x^0 \in \Omega$, *then either* $u \equiv 0$ *or* $u \equiv M$ *and* $c \equiv 0$ *on the set of elliptic connectivity containing the point* x^0.

In particular, if the equation $L(u) = 0$ is elliptically connected in Ω and $c \equiv 0$, then $u \equiv M$ in Ω. We shall not give the detailed proof of Theorems 3.1.2 and 3.1.3 contained in the papers by A. D. Aleksandrov [1–5], but shall merely indicate the basic steps of that proof. (Later we shall give a detailed proof of the theorem on the strong maximum principle for equations of type (2.5.1) due to Bony [15] and an analogous proof for equation (3.1.1).)

Theorem 3.1.2 is proved with the aid of the following auxiliary results.

LEMMA 3.1.1. *Suppose* Ω_1 *is a domain contained in the halfspace* $x_1 > 0$ *whose boundary contains a domain* G *in the plane* $x_1 = 0$. *Let* G_1 *be a closed subset of* G, *and* x^0 *a point of the set* G_1. *Let* $u(x)$ *be a nonnegative function in class* $C^{(2)}(\overline{\Omega}_1)$, *positive in* $\overline{\Omega}_1 \backslash G_1$, *and such that*

$$u(x^0) = 0, \quad u_{x_k}(x^0) = 0, \quad k = 1, \ldots, m.$$

Then for every $N > 0$ *there are points* $x = (x_1, \cdots, x_m)$ *in* Ω_1, *arbitrarily close to the plane* $x_1 = 0$ *satisfying*

1) $d^2u(x) \geqslant N u_{x_1}(x)\, dx_1^2$,

2) $u_{x_1}(x) > \dfrac{1}{x_1} u(x)$,

3) $u_{x_2}(x) = \ldots = u_{x_m}(x) = 0$.

This lemma easily yields the following.

LEMMA 3.1.2. *Let the domain* Ω_1 *and the function* $u(x)$ *satisfy the hypotheses of Lemma* 3.1.1. *Furthermore, suppose that the coefficient* a^{11} *in* (3.1.1) *is different from zero at the point* x^0. *Then there exist points of* Ω, *in arbitrarily small neighborhoods of* x^0, *at which* $L(u) > 0$.

The proof of Theorem 3.1.2 proceeds by the principle of contradiction. Supposing that there is a point x in Ω such that $u(x) = 0$ and $u \neq 0$ at some point on a line of ellipticity passing through x, one may find an ellipsoid contained in Ω, inside of which $u > 0$, with $u = 0$ at some point x^0 on its boundary. Furthermore $a^{kj}(x^0)n_k n_j > 0$, where $\overline{n} = (n_1, \cdots, n_m)$ is the unit normal to the ellipsoid at the point x^0. We now choose new independent variables $y_1, \cdots, y_m$ such that the direction of the y_1 axis coincides with the direction of $\bar{n}$, and such that the boundary of the ellipsoid in a neighborhood of the point x^0 lies in the plane $y_1 = 0$. Then according to Lemma 3.1.2 there are points in a neighborhood of x^0 for which $L(u) > 0$, which contradicts the hypothesis of Theorem 3.1.2.

For a wide class of equations (3.1.1) with $b^k \equiv 0$ and $c \equiv 0$, A. D. Aleksandrov also proved that if $\mathfrak{M}$ is the set of elliptic connectivity containing the point x^0, then in a sufficiently small neighborhood Ω_1 of x^0 there is a function $u(x)$ in class $C^{(2)}(\Omega)$ such that $u = 0$ on $\mathfrak{M}$, $u > 0$ on $\Omega_1 \setminus \mathfrak{M}$, and $L(u) = 0$ in Ω_1.

Under certain special assumptions, a theorem analogous to the theorem on the strong maximum principle for parabolic equations is valid also for (3.1.1). We shall assume that in a neighborhood of each point P in Ω, local coordinates $y_1, \cdots, y_m$ may be introduced in such a manner that equation (3.1.1) assumes the form

$$L(u) \equiv \sum_{k,\, j=1}^{n} \alpha^{kj} u_{y_k y_j} + \sum_{k=1}^{m} \beta^k u_{y_k} + cu = 0, \quad \alpha^{kj}\xi_k\xi_j \geqslant 0, \tag{3.1.4}$$

where $n < m$ and the plane given by the equations $y_{n+1} = \text{const}, \cdots, y_m = \text{const}$ is a set of elliptic connectivity for (3.1.4).

We assume that in this neighborhood the components of the vector $\bar{\beta} = (0, \cdots, 0, \beta_{n+1}, \cdots, \beta_m)$ satisfy a Lipschitz condition and $\bar{\beta} \neq 0$.

Let $y = (y_1, \cdots, y_m)$. The curve l is called a line of parabolicity for (3.1.4) if l is a trajectory of the system of differential equations

$$\frac{dy}{dt} = \bar{\beta}(y). \tag{3.1.5}$$

The set $\mathfrak{M}$ is called the set of parabolic connectivity for the point x^0 for equation (3.1.1), if the point x^0 may be joined to any point of $\mathfrak{M}$ by a curve consisting of a finite number of arcs of lines of ellipticity of equation (3.1.1) and arcs of lines of parabolicity, followed in the direction of the vector $\bar{\beta}$; and, moreover, if there exists no set containing $\mathfrak{M}$ and possessing these same properties.

THEOREM 3.1.4. *If equation* (3.1.1) *may be transformed into the form* (3.1.4) *in a neighborhood of each point of* Ω, *if* $u \geqslant 0$ *and* $L(u) \leqslant 0$ *in* Ω, *and if* $u = 0$ *at a point* x^0 *belonging to* Ω, *then* $u = 0$ *on the set of parabolic connectivity of the point* x^0. *If* $c \equiv 0$ *and* $\bar{\beta} \neq 0$ *in* (3.1.4), *then in a sufficiently small neighborhood* Ω_1 *of any point* x^1 *in* Ω *there exists a function* $u(y)$ *of class* $C^{(2)}(\Omega_1)$ *such that* $L(u) = 0$ *in* Ω_1, $u = 0$ *on the set of parabolic connectivity of the point* x^1, *and* $u > 0$ *outside this set.*

THEOREM 3.1.5. *Suppose equation* (3.1.1) *may be transformed into the form* (3.1.4) *in a neighborhood of each point of* Ω. *Suppose* $L(u) \geqslant 0$ *in* Ω *and that the coefficient* c *and* $M = \sup_\Omega u$ *are related by the inequality* $Mc \leqslant 0$. *If* $u(x^0) = M$ *and* $x^0 \in \Omega$, *then either* $u \equiv 0$ *or* $u \equiv M$ *and* $c \equiv 0$ *on the set of parabolic connectivity of the point* x^0

The proof of Theorem 3.1.4 is carried through on the basis of Lemma 3.1.1 and results analogous to Lemma 3.1.2.

In the papers of A. D. Aleksandrov, Theorems 3.1.2–3.1.5 are proved under more general hypotheses regarding the coefficients of equation (3.1.1) and weaker hypotheses on the smoothness of $u(x)$. In particular, the theorem analogous to Theorem 3.1.2 is proved for functions $u(x)$ in the Sobolev space $W_m^2(\Omega)$. He also studies the set of zeros of a function $u(x)$ in Ω with the properties that $u \geqslant 0$, $L(u) \leqslant 0$ in Ω, and $u(x^0) = 0$ and $u_{x_j}(x^0) = 0$, $j = 1, \cdots, m$, where x^0 is some point on the boundary of Ω.

Following the papers of Bony [15–17], we shall now prove a strong maximum principle for equations of the form

$$P(u) = \sum_{j=1}^{r} X_j^2 u + X_0 u + cu = 0; \; c \leqslant 0, \tag{3.1.6}$$

where X_j $(j = 0, 1, \cdots, r)$ are first order differential operators

$$X_j \equiv \sum_{k=1}^{m} a_j^k(x) \frac{\partial}{\partial x_k}. \tag{3.1.7}$$

For simplicity we assume the coefficients $a_j^k(x)$ to be infinitely differentiable (or sufficiently smooth) functions in Ω. By $\overline{X}_j(x)$ we denote the vector field

$$\overline{X}_j(x) = (a_j^1(x), \ldots, a_j^m(x))$$

in the domain Ω. We shall write the operator $P(u)$ also in the form

$$P(u) \equiv a^{kj} u_{x_k x_j} + b^k u_{x_k} + cu,$$

where $c(x)$ is bounded in Ω and $c \leqslant 0$.

We shall use the same method to prove a strong maximum principle for general second order equations (3.1.1).

We shall denote by F the set of points in Ω where $u(x) = M$ and $M = \sup_\Omega u$. We shall assume that $M \geqslant 0$ and $u \subset C^{(2)}(\Omega)$. The proof of the strong maximum principle is based on the following auxiliary results.

LEMMA 3.1.3. *Let the point* x^0 *belong to* F *and suppose there exists a ball*

Q bounded by a sphere S with center at x^1 and radius ρ such that $x^0 \in S$, $Q \cup S \subset \Omega$ and $(Q \cup S) \cap F = x^0$. Assume $P(u) \geqslant 0$ in Ω. Then for any $j = 1, \cdots, r$ the vector $\bar{X}_j$ is orthogonal to the vector $x^1 - x^0 = (x_1^1 - x_1^0, \cdots, x_m^1 - x_m^0)$.

PROOF. We set $a(x, \xi) \equiv a^{kj}(x)\xi_k\xi_j$ and consider $a(x^0, x^1 - x^0)$. Since

$$a(x, \xi) = \sum_{j=1}^{r} (\bar{X}_j(x), \xi)^2 \equiv \sum_{j=1}^{r} \left(\sum_{k=1}^{m} a_j^k(x)\xi_k \right)^2,$$

it follows from the equation $a(x^0, x^1 - x^0) = 0$ that $(\bar{X}_j(x^0), x^1 - x^0) = 0$. Hence to prove Lemma 3.1.3 it suffices to show that $a(x^0, x^1 - x^0) = 0$.

We assume the contrary. Suppose $a(x^0, x^1 - x^0) > 0$. We consider the function

$$v(x) = e^{-q|x-x^1|^2} - e^{-q\rho^2},$$

where $q = \text{const} > 0$. Clearly at the point x^0

$$P(v) = e^{-q\rho^2}\left[4q^2 a(x^0, x^1 - x^0) - 2q \sum_{k=1}^{m} (a^{kk}(x^0) + b^k(x^0)(x_k^0 - x_k^1))\right].$$

If q is sufficiently large and $a(x^0, x^1 - x^0) > 0$, then $P(v) > 0$ at the point x^0. Consequently there exists a ball Q_1 with center at the point x^0 such that $P(v) > 0$ in Q_1 and $Q_1 \subset \Omega$. Let S_1 be the boundary of Q_1. We set $w(x) = u(x) + \epsilon v(x)$, where $\epsilon = \text{const} > 0$. Since $P(u) \geqslant 0$ in Ω by assumption, it follows that $P(w) > 0$ in Q_1. At the points of S_1 belonging to F, the function $v(x) \leqslant \text{const} < 0$, since the distance from the point x^1 to the set $S_1 \cap F$ is clearly larger than ρ. Hence there is a neighborhood σ_1 of the set $S_1 \cap F$ in which the inequality $w(x) < M$ holds. On the set $S_1 \backslash \sigma_1$ we have $u(x) < M - \delta$, where $\delta = \text{const} > 0$. So if ϵ is sufficiently small, we have $w < M$ on $S_1 \backslash \sigma_1$.

Hence for sufficiently small $\epsilon > 0$ we have $w(x) < M$ at all points of S_1. It is clear that $w(x^0) = M$. From this it follows that the function $w(x)$ takes on its greatest nonnegative value at the interior point Q_1, where $P(w) > 0$. Since $c \leqslant 0$ in Ω, according to Lemma 1.1.2 the function $w(x)$ may not assume its greatest non-negative value at an interior point of the domain Q_1. This contradiction proves Lemma 3.1.3.

LEMMA 3.1.4. *Let the trajectory $x(t)$ of the vector field $\bar{X}_j(x)$ $(j = 1, \cdots, r)$ be such that $x(t_0) \in F$, and let $P(u) \geqslant 0$ in Ω. Then if $\delta(t)$ is the distance from the point $x(t)$ to F and if $\delta(t) > 0$ for $t_0 < t \leqslant t_1$, it follows that for every $t \in [t_0, t_1]$*

$$\liminf_{h \to 0} \frac{\delta(t+h) - \delta(t)}{|h|} \geqslant -K\delta(t), \tag{3.1.8}$$

where $K = \text{const} > 0$.

PROOF. Let $h_n \to 0$ as $n \to \infty$ and $x^n = x(t+h_n)$. By y^n we denote the projection of the point x^n on F, i.e. $|x^n - y^n|$ is equal to the infimum of the distances from x^n to the points of F. We consider x^n sufficiently near to $x(t) = x$. We choose a subsequence y^n such that $y^n \to y$ as $n \to \infty$. It is clear that y is the projection of the point x on F. It is easily seen that

$$\frac{1}{|h_n|}(\delta(t+h_n) - \delta(t)) = \frac{1}{|h_n|}(|y^n - x^n| - |y - x|) \geqslant -|\overline{X}_j(x)|\,|\cos\gamma| - K_1 h_n \tag{3.1.9}$$

holds, where γ is the angle between $\overline{X}_j(x)$ and the vector $(y-x)$, with $K_1 = \text{const} > 0$. According to Lemma 3.1.3 the vector $\overline{X}_j(y)$ is orthogonal to the vector $y - x$. Hence

$$\cos\gamma = \sin\gamma_1,$$

where γ_1 is the angle between $\overline{X}_j(x)$ and $\overline{X}_j(y)$. Clearly

$$|\overline{X}_j(x)\sin\gamma_1| \leqslant |\overline{X}_j(x) - \overline{X}_j(y)| \leqslant K|x - y| = K\delta(t).$$

This estimate and (3.1.9) yield relation (3.1.8).

LEMMA 3.1.5. *If the function $f(t)$ is continuous on the segment $[t_1, t_2]$ and the relation*

$$\liminf_{h\to 0} \frac{f(t+h)-f(t)}{|h|} \geqslant -K,\ K = \text{const} > 0, \tag{3.1.10}$$

holds for each $t \in [t_1, t_2]$, then the function $f(t)$ satisfies a Lipschitz condition with constant K on the segment $[t_1, t_2]$.

PROOF. We assume that $f(t)$ does not satisfy a Lipschitz condition with constant K. Then there exist points s_1 and s_2 and some positive number ϵ such that

$$\left|\frac{f(s_1)-f(s_2)}{s_1-s_2}\right| \geqslant K + \varepsilon.$$

Since the function $f(t)$ is continuous for $t \in [t_1, t_2]$, there exists a value $\rho_0 > 0$ such that for $|s - t'| \leqslant \rho_0$ and $|s_2 - t''| \leqslant \rho_0$ the relation

$$\left|\frac{f(t')-f(t'')}{t'-t''}\right| \geqslant K + \frac{\varepsilon}{2} \tag{3.1.11}$$

holds. From (3.1.10) it follows that with each point $\tau \in [t_1, t_2]$ there is a neighborhood $|t - \tau| \leqslant \rho(\tau)$ such that

$$\left|\frac{f(t)-f(\tau)}{t-\tau}\right| \leqslant K + \frac{\varepsilon}{4}$$

holds for all t in this neighborhood. We may suppose that $\rho(\tau) \leqslant \rho_0/2$. We now choose a finite covering of the interval $[t_1, t_2]$ by neighborhoods of this type and suppose that $\tau^1 < \tau^2 < \cdots < \tau^n$ are the centers of these neighborhoods. Clearly

$$\left|\frac{f(\tau^k)-f(\tau^s)}{\tau^k-\tau^s}\right| \leqslant K + \frac{\varepsilon}{4}$$

for all $k, s = 1, \cdots, n$; $k \neq s$. Hence for some t' and t'',

$$\left|\frac{f(t')-f(t'')}{t'-t''}\right| \leqslant K + \frac{\varepsilon}{4},$$

which contradicts inequality (3.1.11). The lemma is proved.

The following result follows from Lemmas 3.1.4 and 3.1.5.

THEOREM 3.1.6. *Assume that* $P(u) \geqslant 0$, *and that the trajectory* $x(t)$ *of the vector field* $\bar{X}_j(x)$ $(j = 1, \cdots, r)$ *contains a point* $x(t_0)$ *of the set* F. *Then the entire trajectory* $x(t)$ *belongs to* F.

PROOF. We assume the contrary. Then there exists an interval $[t_1, t_2]$ such that $x(t_1) \in F$, and $x(t)$ does not belong to F for $t_1 < t \leqslant t_2$. It follows from Lemmas 3.1.4 and 3.1.5 that

$$|\delta(t+h) - \delta(t)| \leqslant K\,|h| \max_{[t',t'']} \delta(t)$$

for any t and $t+h$ belonging to the interval $[t', t'']$, where $t_1 \leqslant t' \leqslant t'' \leqslant t_2$. If $|h| \leqslant 1/2K$, then for $t \in [t_1, t_1 + h]$ we have

$$\delta(t) \leqslant \frac{1}{2} \max_{[t_1,t_1+h]} \delta(t)$$

which means that $\delta(t) = 0$ for $|t - t_1| \leqslant h$, contradicting the choice of the point t_1. The theorem is proved.

A consequence of Theorem 3.1.6 is a strong maximum principle for equation (3.1.1).

THEOREM 3.1.7. *Assume that* $u(x) \in C^{(2)}(\Omega)$, *that* $P(u) \geqslant 0$ *in* Ω, *and that the system of operators* $\{X_1, \cdots, X_r\}$ *has rank* m *at each point of* Ω *(see the definition of rank of a system of operators in* §5 *of Chapter* II). *If* $u(x)$ *assumes its greatest nonnegative value* M *at the point* x^0 *in* Ω, *then* $u \equiv M$ *in* Ω.

PROOF. The condition that the system $\{X_1, \cdots, X_r\}$ has rank m, together with Theorem 3.1.1, yields that the point x^0 may be joined to any point of the domain Ω by a line consisting of a finite number of arcs of trajectories of the vector fields $\bar{X}_j(x)$, where $j = 1, \cdots, r$. Hence by virtue of Theorem 3.1.6 the function $u(x)$ must be constant in Ω. The theorem is proved.

Under definite conditions, a strong maximum principle analogous to Theorem 3.1.5 is also valid for equations of the form (3.1.6).

We first establish some auxiliary results. We shall say that the operator $P(u)$ satisfies condition A in the domain Ω if in a neighborhood $\Omega(x^0)$ of each point x^0 in Ω there exists a local coordinate system $y_1, \cdots, y_m$ with origin at x^0 such that in $\Omega(x^0)$ the operator $P(u)$ assumes the form

$$P(u) = \sum_{k,j=1}^{n} \alpha^{kj} u_{y_k y_j} + \sum_{k=1}^{m} \beta^k u_{y_k} + cu; \quad n < m, \tag{3.1.12}$$

and has the property that if the point $y^0 = (y_1^0, \cdots, y_m^0)$ belongs to F, then all points of $\Omega(x^0)$ for which

$$y_{n+1} = y_{n+1}^0, \ldots, y_m = y_m^0$$

also belong to F.

LEMMA 3.1.6. *Let the point x^0 belong to F, and assume that there exists a ball Q bounded by a sphere S with center at a point x^1 and radius ρ, such that $x^0 \subset S$, $Q \cup S \subset \Omega$ and $(Q \cup S) \cap F = x^0$. Assume that the operator $P(u)$ satisfies condition A and that $P(u) \geqslant 0$ in Ω. Then*

$$\left(\overline{X}_0(x^0), x^1 - x^0\right) \leqslant 0. \tag{3.1.13}$$

PROOF. We may assume that the local coordinates $y_1, \cdots, y_m$ have been chosen in the neighborhood $\Omega(x^0)$ of the point x^0 in accordance with condition A so that the point x^1 has coordinates $(0, \cdots, 0, 1)$ and the y_m axis has the direction of the vector $x^1 - x^0$. We consider the cylinder

$$y_{n+1}^2 + \ldots + y_{m-1}^2 + (y_m - 1)^2 \leqslant 1.$$

Clearly there are no points of F in the part of $\Omega(x^0)$ lying inside this cylinder. Let

$$w(y) = \varepsilon \left(y_1^2 + \ldots + y_n^2\right) + y_{n+1}^2 + \ldots + (y_m - 1)^2,$$

where $\epsilon = \text{const} > 0$, and let

$$v(y) = e^{1 - w(y)} - 1.$$

It is easy to see that at the point x^0,

$$P(v) = -2\varepsilon \sum_{k=1}^{n} \alpha^{kk} + 2\beta^m.$$

Since by assumption the operator assumes the form (3.1.12) in a neighborhood of x^0, we have $\beta^m(x^0) = \lambda(\overline{X}_0(x^0), x^1 - x^0)$, where $\lambda = \text{const} > 0$. We assume that (3.1.13) is not satisfied. Then $\beta^m(y) > 0$ in some neighborhood Ω_1 of the point

x^0, and in this neighborhood $P(v) > 0$, if ϵ is sufficiently small. We consider the function

$$V(y) = u(y) + \alpha v(y), \quad \alpha = \text{const} > 0.$$

Clearly $P(V) > 0$ in Ω_1. Let S_1 be the boundary of Ω_1. Since the distance between $S_1 \cap F$ and the set of points where $v(y) \geqslant 0$ is positive, for sufficiently small α we have $V(y) < M$ on S_1. It is easily seen that $V(x^0) = M$, so that $V(y)$ must take on its greatest nonnegative value inside Ω_1. This contradicts the fact that $P(V) > 0$ in Ω_1 and $c \leqslant 0$.

LEMMA 3.1.7. *Let the trajectory $x(t)$ of the vector field $\overline{X}_0(x)$ be such that $x(t_0) \in F$. Assume that the operator P satisfies condition A and that $P(u) \geqslant 0$ in Ω. If $\delta(t) > 0$ for $t_0 < t \leqslant t_1$, where $\delta(t)$ is the distance from the point $x(t)$ to F, then for any $h > 0$ and $t \in [t_0, t_1]$ such that $t - h \in [t_0, t_1]$ we have*

$$\liminf_{h \to 0} \frac{\delta(t-h) - \delta(t)}{h} \geqslant -K\delta(t), \tag{3.1.14}$$

where $K = \text{const} > 0$.

PROOF. As in the proof of Lemma 3.1.4, we consider the sequence $x^n = x(t - h_n)$, where $h_n > 0$, $h_n \to 0$ as $n \to \infty$, and t and $t - h_n$ belong to the interval $[t_0, t_1]$. Let y^n be a projection of x^n on the set F, and let $y^n \to y$ as $n \to \infty$, where y is the projection of the point $x = x(t)$ onto F. Then

$$\begin{aligned} \delta(t - h_n) - \delta(t) &= |x^n - y^n| - |x - y| \\ &\geqslant -h_n\left(\overline{X}_0(x), \frac{x-y}{|x-y|}\right) - K_1 h_n^2 \\ &= -h_n\left(\overline{X}_0(y), \frac{x-y}{|x-y|}\right) + h_n\left((\overline{X}_0(y) - \overline{X}_0(x)), \frac{x-y}{|x-y|}\right) - K_1 h_n^2. \end{aligned}$$

According to Lemma 3.1.6,

$$\left(\overline{X}_0(y), \frac{x-y}{|x-y|}\right) \leqslant 0.$$

Therefore

$$\delta(t - h_n) - \delta(t) \geqslant -h_n K\,|x - y| - K_1 h_n^2;$$

relation (3.1.4) follows from this last inequality.

The following strong maximum principle follows from Lemmas 3.1.6 and 3.1.7.

THEOREM 3.1.8. *Let $P(u) \geqslant 0$ in Ω and let the operator P satisfy condition A. If $x(t)$ is a trajectory of the vector field $\overline{X}_0(x)$ such that $x(t_0) \in F$, then $x(t) \in F$ for all $t \geqslant t_0$.*

PROOF. We assume the contrary. Then there exists a point x^0 such that $x^0 \in F$, whereas the points of the trajectory $x(t)$ of the vector field $\bar{X}_0(x)$ defined by the condition $x(t_1) = x^0$ do not belong to F for $t_1 < t \leqslant t_2$. According to Lemma 3.1.7, relation (3.1.14) holds for the distance $\delta(t)$ of the points $x(t)$ to the set F.

We show that it follows from (3.1.14) that $\delta(t) = 0$ for sufficiently small $t - t_1$. Exactly as Lemma 3.1.5 was proved, we obtain that for any values t and $t - h$ in the interval $[t', t'']$, where $t_1 \leqslant t' \leqslant t'' \leqslant t_2$, the relation

$$\delta(t) - \delta(t - h) \leqslant Kh \max_{[t', t'']} \delta(t) \tag{3.1.15}$$

holds. Here $K = \text{const} > 0$ and $h > 0$. If $t - h = t_1$ and $h \leqslant 1/2K$, then for $t \in [t_1, t_1 + h]$ it follows from (3.1.15) that

$$\delta(t) \leqslant \frac{1}{2} \max_{[t_1, t_1+h]} \delta(t). \tag{3.1.16}$$

Since $\delta(t) \geqslant 0$, it follows from (3.1.16) that $\delta(t) = 0$ for $t_1 \leqslant t \leqslant t_1 + h$. The contradiction so obtained proves Theorem 3.1.8.

THEOREM 3.1.9. *Assume that the rank of the system* $\{X_1, \cdots, X_r\}$ *is equal to* n *at each point of* Ω, *where* $n < m$, *and assume that* $P(u) \geqslant 0$ *in* Ω. *If* $u(x)$ *assumes its greatest nonnegative value* M *at a point* x^0 *belonging to* Ω, *then* $u(x) = M$ *at each point of* Ω *which may be joined with the point* x^0 *by a line consisting of a finite number of arcs of trajectories of the vector fields* $\bar{X}_j(x)$ $(j = 0, 1, \cdots, r)$, *with the stipulation that when such a line is followed away from the point* x^0, *any arcs of trajectories of the field* $\bar{X}_0(x)$ *are followed in the direction of the vector* $\bar{X}_0(x)$.

PROOF. Under the conditions of the theorem, it follows from Frobenius' theorem [25] and from Theorem 3.1.6 that the operator $P(u)$ satisfies condition A in Ω. The conclusion of Theorem 3.1.9 then follows immediately from Theorem 3.1.6 and 3.1.8.

In his papers [15–17], Bony also proved uniqueness theorems for the solution of the Cauchy problem for equations of the form (3.1.1) with analytic coefficients, and a Harnack theorem.

Let us consider now equation (3.1.1).* We write it in the form

$$L(u) \equiv (a^{kj}(x)u_{x_j})_{x_k} + Qu + c(x)u = 0, \quad a^{kj}(x)\xi_k\xi_j \geqslant 0,$$

where

---

* *Editor's note.* **The remainder of this section has been added in translation.**

$$Q = \sum_{k=1}^{m} (b^k(x) - a^{kj}_{x_j}(x)) \frac{\partial}{\partial x_k}, \quad c \leq 0.$$

Let $a(x, \xi) \equiv a^{kj}(x)\xi_k \xi_j$. Consider in a domain Ω the operators Q and

$$L^{(j)} \equiv \sum_{k=1}^{m} a^{kj}(x) \frac{\partial}{\partial x_k} \quad (j = 1, \cdots, m)$$

and the corresponding vector fields $\bar{Q}(x)$ and $\bar{L}^{(j)}(x) = (a^{1j}(x), \cdots, a^{mj}(x)), j = 1, \cdots, m$.

Lemmas 3.1.3–3.1.5 are valid if the operator $P(u)$ is replaced by the operator $L(u)$ and the vector fields $\bar{X}_j(x)$ $(j = 1, \cdots, r)$ are replaced by $\bar{L}^{(j)}(x)$ $(j = 1, \cdots, m)$. We only note that for the proof of Lemma 3.1.3 we first prove that $a(x^0, x^1 - x^0) = 0$. From this relation and the condition $a(x, \xi) \geq 0$ in Ω, it follows that $(\bar{L}^{(j)}(x_0), (x^1 - x^0)) = 0$.

Lemmas 3.1.6 and 3.1.7 are valid if the operator $P(u)$ is replaced by the operator $L(u)$, the vector field $\bar{X}_0(x)$ is replaced by $\bar{Q}(x)$, and the vector fields $\bar{X}_j(x)$ $(j = 1, \cdots, r)$ by $\bar{L}^{(j)}(x)$ $(j = 1, \cdots, m)$. We note that for the proof of Lemma 3.1.6 we have to show that $\beta^m(x^0) = \lambda(\bar{Q}(x^0), x^1 - x^0)$, $\lambda = \text{const} > 0$. It follows easily from the assumption that $L(u)$ satisfies the condition A.

Therefore we have proved the following theorems.

THEOREM 3.1.10. *Assume that $L(u) \geq 0$, and that the trajectory $x(t)$ of the vector field $\bar{L}^{(j)}(x)$ $(j = 1, \cdots, r)$ contains a point $x(t_0)$ of the set F. Then the entire trajectory $x(t)$ belongs to F.*

THEOREM 3.1.11. *Assume that $L(u) \geq 0$ in Ω, and that the system of operators $\{L^{(1)}, \cdots, L^{(m)}\}$ has rank m at each point of Ω. If $u(x)$ assumes its greatest nonnegative value M at the point x^0 in Ω, then $u \equiv M$ in Ω.*

THEOREM 3.1.12. *Let $L(u) \geq 0$ in Ω and let the operator L satisfy condition A. If $x(t)$ is a trajectory of the vector field $\bar{Q}(x)$ such that $x(t_0) \in F$, then $x(t) \in F$ for all $t \geq t_0$.*

THEOREM 3.1.13. *Assume that the rank of the system $\{L^{(1)}, \cdots, L^{(m)}\}$ is equal to n at each point of Ω, where $n < m$; and assume that $L(u) \geq 0$ in Ω. If $u(x)$ assumes its greatest nonnegative value M at a point x^0 belonging to Ω, then $u(x) = M$ at each point x of Ω which may be joined with the point x^0 by a line consisting of a finite number of arcs of trajectories of the vector fields $\bar{L}^{(j)}(x)$ $(j = 1, 2, \cdots, m)$ and $\bar{Q}(x)$ with the stipulation that when such a line is followed* *away from the point x^0, any arcs of trajectories of the field $\bar{Q}(x)$ are followed in the direction of the vector $\bar{Q}(x)$.*

Theorems 3.1.11 and 3.1.13 are analogous to Theorems 3.1.3 and 3.1.5 of A. D. Aleksandrov.

§2. The Cauchy problem for degenerating second order hyperbolic equations

We consider the class of second order equations whose characteristic form has one negative eigenvalue at each point of the domain considered, the remaining eigenvalues being positive or zero. It is natural to call such equations weakly hyperbolic. They are often also called degenerating hyperbolic equations. The Cauchy problem for such equations may be studied by methods similar to those we have applied in Chapters I and II in our study of second order equations with nonnegative characteristic form. The construction of the solution of Cauchy's problem is carried out by the method of hyperbolic regularization. Lemma 1.7.1 plays an important role in establishing a priori estimates for the solution.

We shall consider the Cauchy problem for the equation

$$u_{tt} - (a^{kj}(t, x) u_{x_k})_{x_j} + b^k(t, x) u_{x_k} + b^0(t, x) u_t + c(t, x) u = f(t, x) \tag{3.2.1}$$

in the domain $G_T\{0 \leqslant t \leqslant T,\ x \in R^m\}$ with initial conditions

$$u|_{t=0} = \varphi(x), \quad u_t|_{t=0} = \psi(x), \tag{3.2.2}$$

where $x \in R^m$ and $a^{kj}(t, x)\xi_k\xi_j \geqslant 0$ in G_T for all real $\xi = (\xi_1, \cdots, \xi_m)$.

In this section we obtain sufficient conditions for the problem (3.2.1), (3.2.2) to be correctly posed; in §3 we will obtain some necessary conditions as well. In the particular case when $x \in R^1$, $a^{kj}\xi_k\xi_j = 0$ for $t = 0$ and $a^{kj}\xi_k\xi_j > 0$ for $t \neq 0, t > 0$, the problem (3.2.1), (3.2.2) has been studied in many papers (see, for example, [10, 12, 106] and others; a detailed bibliography is given in [123]). Interest in this problem has arisen particularly in connection with the investigation of Tricomi's problem for equations of mixed type [131]. Necessary conditions for the Cauchy problem to be correctly posed are considered in papers [74, 72, 73, 82, 127, 62, 116, 149], and others.

We introduce the notation

$$G_\tau = \{0 \leqslant t \leqslant \tau,\ x \in R^m\}, \quad [\Phi, \Psi]_{G_\tau} = \int_{G_\tau} \Phi\Psi\, dt\, dx,$$

$$(\Phi, \Psi)_{t=\tau} = \int_{R^m} \Phi(\tau, x)\Psi(\tau, x)\, dx,$$

$$\|v\|_{\tau;s}=\left\{\sum_{|l|+\rho\leqslant s}\left(D^l\frac{\partial^\rho v}{\partial t^\rho},\ D^l\frac{\partial^\rho v}{\partial t^\rho}\right)_{t=\tau}\right\}^{1/2},$$

where

$$D^l=\frac{\partial^{|l|}}{\partial x_1^{l_1}\ldots\partial x_m^{l_m}},\qquad |l|=l_1+\ldots+l_m.$$

(Note that this definition of D^l differs by a factor $(-i)^{|l|}$ from that used in Chapter II.) We denote

$$\|v\|_{G_\tau;s}=\left\{\int_0^\tau\|v\|^2_{\sigma;s}d\sigma\right\}^{1/2};$$

$$\|v\|_{\tau;q,\mu,s}=\left\{\sum_{\substack{\rho\leqslant q,\,|l|\leqslant\mu\\ q+\mu\leqslant s}}\left(D^l\frac{\partial^\rho v}{\partial t^\rho},\ D^l\frac{\partial^\rho v}{\partial t^\rho}\right)_{t=\tau}\right\}^{1/2},$$

$$\|v\|_{G_\tau;q,\mu,s}=\left\{\int_0^\tau\|v\|^2_{\sigma;q,\mu,s}d\sigma\right\}^{1/2}.$$

By $\mathcal{H}^s(G_T)$ we denote the class of functions obtained by closing the set of infinitely differentiable functions in G_T with compact support in x with respect to the norm $\|v\|_{G_T;s}$.

LEMMA 3.2.1. *Let $u_\epsilon(t,x)$ be a solution of the hyperbolic equation*

$$L_\varepsilon(u)=u_{tt}-\varepsilon\Delta u-(a^{kj}_\varepsilon u_{x_k})_{x_j}+b^k_\varepsilon u_{x_k}+b^0_\varepsilon u_t+c_\varepsilon u=f_\varepsilon$$
$$(\varepsilon=\text{const}>0)\tag{3.2.3}$$

in the domain $G^\epsilon=\{0\leqslant t\leqslant T-\epsilon,\ x\in R^m\}$ satisfying the conditions

$$u|_{t=0}=\varphi_\varepsilon(x),\quad u_t|_{t=0}=\psi_\varepsilon(x),\tag{3.2.4}$$

where a^{kj}_ϵ, b^k_ϵ, b^0_ϵ, c_ϵ, f_ϵ, ϕ_ϵ, ψ_ϵ are infinitely differentiable functions in G^ϵ, with f_ϵ, ϕ_ϵ, and ψ_ϵ having compact support in x. Suppose that the inequality

$$\alpha t\,(b^k_\varepsilon\xi_k)^2\leqslant Aa^{kj}_\varepsilon\xi_k\xi_j+a^{kj}_{\varepsilon t}\xi_k\xi_j\tag{3.2.5}$$

holds for all ξ and for t in the range $0\leqslant t\leqslant t_0$, t_0 being some small positive number less than $T-\epsilon$, where α is a constant with $\alpha>(2p+6)^{-1}$ (p is an integer with $p\geqslant -1$) and A is a constant. Furthermore, assume that (3.2.5) *holds for all ξ and for t in the range $t_0\leqslant t\leqslant T-\epsilon$ for some (possibly different) constants α and A. Then for $0\leqslant\tau\leqslant T-\epsilon$ we have the estimate*

$$\| u_\varepsilon \|^2_{\tau,s} \leqslant M_1 \{ \| \varphi_\varepsilon \|^2_{0;s+p+4} + \| \psi_\varepsilon \|^2_{0;s+p+3} + \max_{0 \leqslant \mu \leqslant t_0} \| f_\varepsilon \|^2_{\mu;p+1,s,p+1+s} + \| f_\varepsilon \|^2_{\tau;\, s-2} + \| f_\varepsilon \|^2_{G_\tau;0,s,s} + \| f_\varepsilon \|^2_{0;p,p+s+2,p+s+2} \}, \tag{3.2.6}$$

where the constant M_1 *depends on the maxima of the absolute values of the derivatives of the functions* a_ϵ^{kj}, $a_{\epsilon x_k}^{kj}$, b_ϵ^k, b_ϵ^0, $b_{\epsilon t}^0$ *and* c_ϵ *with respect to* t *and* x *up to order* $s-2$, *of their derivatives with respect to* x *up to order* s, *of their derivatives of the form* $D^l\partial^\rho/\partial t^\rho$ *at* $t=0$ *with orders satisfying* $\rho \leqslant p$, $|l| \leqslant p+2+s$, $\rho+|l| \leqslant p+2+s$, *and finally of their derivatives of the form* $D^l\partial^\rho/\partial t^\rho$, *where* $|l| \leqslant s$, $\rho \leqslant p+1$, *in the range* $0 \leqslant t \leqslant t_0$. *Here it is assumed that* $s \geqslant 2$.

PROOF. Let

$$v_p = \varphi_\varepsilon + t\psi_\varepsilon + \frac{t^2}{2!}\frac{\partial^2 u_\varepsilon}{\partial t^2}\bigg|_{t=0} + \dots + \frac{t^{p+2}}{(p+2)!}\frac{\partial^{p+2} u_\varepsilon}{\partial t^{p+2}}\bigg|_{t=0}, \tag{3.2.7}$$

where the derivatives of u_ϵ at $t=0$ are expressed by means of equation (3.2.3) and the equations obtained from it by differentiation with respect to t, account being taken of the initial conditions (3.2.4). It is clear that v_p depends on a_ϵ^{kj}, $a_{\epsilon x_k}^{kj}$, b_ϵ^k, b_ϵ^0, c_ϵ, f_ϵ and their derivatives with respect to t and x up to order p at $t=0$, and on ϕ_ϵ and ψ_ϵ with their derivatives up to orders $p+2$ and $p+1$ respectively.

For the function $u = u_\epsilon - v_p$ we obtain the equation

$$L_\varepsilon(u) = f - L_\varepsilon(v_p) \equiv F(t, x). \tag{3.2.8}$$

It is easy to see that F and its derivatives with respect to t up to order p inclusive vanish at $t=0$. Let $w = \int_t^\tau u(\sigma, x)\, d\sigma$. We multiply (3.2.8) by $we^{\theta t}$, where θ is a positive constant to be chosen below, and integrate over the domain G_τ. Let $\tau \leqslant t_0$. We transform the individual terms of the resulting equation

$$[L_\varepsilon(u), e^{\theta t}w]_{G_\tau} = [F, e^{\theta t}w]_{G_\tau} \tag{3.2.9}$$

by integration by parts. Clearly

$$[u_{tt}, e^{\theta t}w]_{G_\tau} = \frac{1}{2}(u, e^{\theta t}u)_{t=\tau} + \left[ue^{\theta t}, \theta^2 w - \frac{3}{2}\theta u\right]_{G_\tau},$$

$$[\varepsilon\Delta u, e^{\theta t}w]_{G_\tau} = -\frac{1}{2}\varepsilon\theta\,[w_{x_k}, e^{\theta t}w_{x_k}]_{G_\tau} - \frac{1}{2}\varepsilon\,(w_{x_k}, w_{x_k})_{t=0},$$

$$[(a_\varepsilon^{kj}u_{x_k})_{x_j}, e^{\theta t}w]_{G_\tau} = -\frac{1}{2}[(\theta a_\varepsilon^{kj} + a_{\varepsilon t}^{kj})\, w_{x_k}, e^{\theta t}w_{x_j}]_{G_\tau} - \frac{1}{2}(a_\varepsilon^{kj}w_{x_k}, w_{x_j})_{t=0}.$$

We further obtain

$$[b_\varepsilon^0 u_t, e^{\theta t} w]_{G_\tau} = [u, e^{\theta t} b_\varepsilon^0 u - (b_\varepsilon^0 e^{\theta t})_t w]_{G_\tau};$$

$$[b_\varepsilon^k u_{x_k}, e^{\theta t} w]_{G_\tau} = -[b_{\varepsilon x_k}^k u, e^{\theta t} w]_{G_\tau} - [b_\varepsilon^k w_{x_k}, e^{\theta t} u]_{G_\tau}.$$

This last equation yields the estimate

$$|[b_\varepsilon^k u_{x_k}, e^{\theta t} w]_{G_\tau}| \leqslant M_2 \tau^2 [u, t^{-1} e^{\theta t} u]_{G_\tau}$$

$$+ \frac{\alpha}{2} [t b_\varepsilon^k w_{x_k}, e^{\theta t} b_\varepsilon^k w_{x_k}]_{G_\tau} + \frac{1}{2\alpha} [u, t^{-1} e^{\theta t} u]_{G_\tau}, \tag{3.2.10}$$

where M_2 is a constant depending on $\sup_{G_T} |b_{\varepsilon x_k}^k|$. We now estimate $[F, e^{\theta t} w]_{G_T}$. Integrating by parts and using the fact that F and its derivatives with respect to t up to order p inclusive vanish at $t = 0$, we obtain

$$[F, e^{\theta t} w]_{G_\tau} = \left[\frac{\partial^{p+1} F}{\partial t^{p+1}}, W_{p+1}\right]_{G_\tau},$$

where

$$W_{\mu+1} = \int_t^\tau W_\mu(\sigma, x)\, d\sigma, \quad \mu = 0, 1, \ldots, p; \quad W_0 = e^{\theta t} w.$$

Since

$$|W_{p+1}|^2 \leqslant e^{2\theta T} \tau^{2p+3} \int_0^T u^2(\sigma, x)\, d\sigma,$$

it follows that

$$|[F, e^{\theta t} w]_{G_\tau}| \leqslant \frac{\tau^{2p+6}}{4\delta} e^{\theta T} |||F|||_{p+1, t_0}^2 + \delta [u, e^{\theta t} t^{-1} u]_{G_\tau}, \tag{3.2.11}$$

where the constant $\delta > 0$ is chosen so that

$$\alpha^{-1} + 2\delta < 2p + 6, \quad |||F|||_{p+1,\tau}^2 = \max_{0 \leqslant \sigma \leqslant \tau} \left(\frac{\partial^{p+1} F}{\partial t^{p+1}}, \frac{\partial^{p+1} F}{\partial t^{p+1}}\right)_{t=\sigma}.$$

Using the estimates (3.2.10), (3.2.11), and condition (3.2.5), and setting θ equal to A, we deduce from (3.2.9) that for $\tau \leqslant t_0$,

$$\tau y'(\tau) \leqslant (\alpha^{-1} + 2\delta) y(\tau) + M_3 \tau y(\tau) + M_4 \tau^{2p+6} |||F|||_{p+1, t_0}^2, \tag{3.2.12}$$

where $y(\tau) = [u, t^{-1} e^{\theta t} u]_{G_\tau}$, the constant M_3 depending on the maxima in G_{t_0} of the absolute values of the functions b_ϵ^0, $b_{\epsilon t}^0$, $b_{\epsilon x_k}^k$, c_ϵ and on T, and $M_4 = e^{\theta T} \delta^{-1}/2$. From (3.2.12) it follows that

$$y(\tau) \leqslant \tau^{(\alpha^{-1} + 2\delta)} e^{M_3 \tau} |||F|||_{p+1, t_0}^2 M_4 \int_0^\tau e^{-M_3 \tau} \tau^{2p+6-1-\alpha^{-1}-2\delta} d\tau$$

$$\leqslant M_5 \tau^{2p+6} |||F|||_{p+1, t_0}^2, \tag{3.2.13}$$

since by assumption $2p + 6 - 1 - \alpha^{-1} - 2\delta > -1$. Therefore

$$(u, e^{A\tau}u)_{t=\tau} \leqslant M_6 \tau^{2p+6} |||F|||^2_{p+1,\, t_0}, \quad M_6 = \text{const} > 0. \tag{3.2.14}$$

We now use induction to estimate the derivatives of u with respect to x up to order s inclusive. We suppose that we have obtained an estimate of the form

$$(D^l u, D^l u)_{t=\tau} \leqslant E_l \tau^{2p+6} \sum_{|\beta| \leqslant |l|} |||D^\beta F|||^2_{p+1,\, t_0}, \tag{3.2.15}$$

for all $|l| \leqslant q-1$; here E_l depends on the maxima of the absolute values of $a_\epsilon^{kj}, a_{\epsilon x_k}^{kj}, b_\epsilon^k, b_\epsilon^0, b_{\epsilon t}^0, c_\epsilon$ and of their derivatives with respect to x up to order $q-1$ inclusive. We show that this same estimate is also valid for $|l| \leqslant q \leqslant s$.

For this purpose we apply the operator D^l to the right and left sides of (3.2.8), multiply the resulting equation by $e^{\theta_1 t} D^l w$, integrate over G_τ, and sum over all l with $|l| = q$. We thus obtain

$$[D^l L_\varepsilon(u), e^{\theta_1 t} D^l w]_{G_\tau} = [D^l F, e^{\theta_1 t} D^l w]_{G_\tau}, \quad |l| = q. \tag{3.2.16}$$

We transform the integrals in (3.2.16) by integration by parts in the same manner as was done in the derivation of (3.2.14). We have

$$|[D^l F, e^{\theta_1 t} D^l w]_{G_\tau}| \leqslant \delta\, [D^l u, e^{\theta_1 t} t^{-1} D^l u]_{G_\tau} + M_7 \tau^{2p+6} |||D^l F|||^2_{p+1,\, t_0};$$

$$[D^l (a_\varepsilon^{kj} u_{x_k})_{x_j}, e^{\theta_1 t} D^l w]_{G_\tau} = -\frac{1}{2} (a_\varepsilon^{kj} D^l w_{x_k}, e^{\theta_1 t} D^l w_{x_j})_{t=\tau}$$

$$-\frac{1}{2} [(\theta_1 a_\varepsilon^{kj} + a_{\varepsilon t}^{kj}) D^l w_{x_k}, e^{\theta_1 t} D^l w_{x_j}]_{G_\tau}$$

$$- \sum_{1 \leqslant |\gamma| \leqslant q} C_\gamma [D^\gamma a_\varepsilon^{kj} D^{l-\gamma} u_{x_k}, e^{\theta_1 t} D^l w_{x_j}]_{G_\tau};$$

$$C_\gamma = \text{const}. \tag{3.2.17}$$

We estimate the last term of this equation. Integrals of the form

$$[a_{\varepsilon x_\rho}^{kj} D^{l-\gamma} u_{x_k}, e^{\theta_1 t} D^l w_{x_j}]_{G_\tau},$$

with $|\gamma| = 1$ may be estimated using Lemma 1.7.1, according to which the inequality

$$(a_{\varepsilon x_\rho}^{kj} v_{x_k x_j})^2 \leqslant M a_\varepsilon^{kj} v_{x_k x_\nu} v_{x_j x_\nu} \tag{3.2.18}$$

holds for all functions $v \in C_{(2)}(R^m)$, the constant M depending only on the second derivatives of the functions a_ϵ^{kj}. Integrating by parts and taking account of (3.2.18), we obtain for $|\gamma| = 1$ and $|l| = q$ that

$$|[a_{\varepsilon x_\rho}^{kj} D^{l-\gamma} u_{x_k}, e^{\theta_1 t} D^l w_{x_j}]_{G_\tau}|$$

$$= | [a_{\varepsilon x_\rho}^{kj} D^l u, e^{\theta_1 t} D^{l-\gamma} w_{x_k x_j}]_{G_\tau} + [a_{\varepsilon x_\rho x_k}^{kj} D^{l-\gamma} u, e^{\theta_1 t} D^l w_{x_j}]_{G_\tau}$$

$$+ [a_{\varepsilon x_\rho x_\rho}^{kj} D^{l-\gamma} u, e^{\theta_1 t} D^{l-\gamma} w_{x_k x_j}]_{G_\tau}| =$$

$$= | [a^{kj}_{\varepsilon x_p}(D^{l-\gamma}w)_{x_k x_j}, e^{\theta_1 t}D^l u]_{G_\tau} + A^q_1$$
$$\leqslant A^q_1 + M_8 [a^{kj}_\varepsilon D^{l-\gamma}w_{x_k x_\gamma}, e^{\theta_1 t} D^{l-\gamma}w_{x_j x_\gamma}]_{G_\tau},$$

where the symbol A^q_j stands for integrals admitting the estimate

$$| A^q_j | \leqslant N_j \sum_{|l| \leqslant q} \tau [D^l u, t^{-1}e^{\theta_1 t}D^l u]_{G_\tau}$$

with constants N_j depending only on the coefficients of (3.2.3). The constant N_1 depends on a^{kj} and their derivatives with respect to x up to third order inclusive. We use integration by parts to transform the integrals in the last term of (3.2.17) which correspond to $|\gamma| \geqslant 2$. We have

$$\sum_{|\gamma| \geqslant 2} C_\gamma [D^\gamma a^{kj}_\varepsilon D^{l-\gamma}u_{x_k}, e^{\theta_1 t}D^l w_{x_j}]_{G_\tau}$$
$$= - \sum_{|\gamma| \geqslant 2} C_\gamma](D^\gamma a^{kj}_\varepsilon D^{l-\gamma}u_{x_k})_{x_j}, e^{\theta_1 t}D^l w]_{G_\tau} = A^q_2.$$

The constant N_2 depends on the maxima of the absolute values of a^{kj}_ϵ, $a^{kj}_{\epsilon x_k}$ and their derivatives with respect to x up to order q. It is clear that

$$| [D^l(b^k_\varepsilon u_{x_k}), e^{\theta_1 t}D^l w]_{G_\tau} | = |]b^k_\varepsilon D^l u_{x_k}, e^{\theta_1 t}D^l w]_{G_\tau} + A^q_3 |$$
$$\leqslant | [b^k_\varepsilon D^l u, e^{\theta_1 t}D^l w_{x_k}]_{G_\tau} + A^q_4 |$$
$$\leqslant | A^q_4 | + \left[\frac{\alpha}{2} t b^k_\varepsilon D^l w_{x_k}, e^{\theta_1 t} b^k_\varepsilon D^l w_{x_k}\right]_{G_\tau} + \frac{1}{2\alpha}[D^l u, e^{\theta_1 t}t^{-1}D^l u]_{G_\tau}.$$

The constant N_4 depends on b^k_ϵ and on their derivatives with respect to x up to order q.

Integrating by parts, we obtain

$$[D^l(b^0_\varepsilon u_t), e^{\theta_1 t}D^l w]_{G_\tau} = [D^l(b^0_\varepsilon u), e^{\theta_1 t}(D^l u - \theta_1 D^l w)]_{G_\tau}$$
$$- [D^l(b^0_{\varepsilon t}u), e^{\theta_1 t}D^l w]_{G_\tau} = A^q_5.$$

By using the estimates obtained for the individual terms of (3.2.16) as well as condition (3.2.5), by choosing the constant $\theta_1 > 0$ sufficiently large, and by using the induction assumption (3.2.15), we deduce from (3.2.16) that

$$\tau y'_q \leqslant (\alpha^{-1} + 2\delta) y_q + K_q \tau y_q + \widetilde{K}_q \tau^{2p+6} \sum_{|\gamma| \leqslant q} |||D^\gamma F|||^2_{p+1,\, t_0}, \tag{3.2.19}$$

where $y_q = \Sigma_{|l|=q} [D^l u, e^{\theta_1 t} t^{-1} D^l u]_{G_\tau}$ and where the constants K_q and $\widetilde{K}_q$ depend on the derivatives with respect to x of a^{kj}_ϵ, $a^{kj}_{\epsilon x_k}$, b^k_ϵ, b^0_ϵ, $b^0_{\epsilon t}$ and c_ϵ up to order q inclusive, as well as on the derivatives of a^{kj}_ϵ and $a^{kj}_{\epsilon x_k}$ with respect to x

up to order two, when $q = 1$. It follows from (3.2.19) that for $|l| = q \leqslant s$

$$(D^l u,\ D^l u)_{t=\tau} \leqslant E_q \tau^{2p+6} \sum_{|\gamma| \leqslant q} |||D^\gamma F|||_{p+1,\ t_0}, \tag{3.2.20}$$

which is the required estimate.

In order to estimate the derivatives of the form $D^l u_t$ for $|l| \leqslant s - 1$, we consider the equation

$$[D^l L_\varepsilon(u),\ e^{-\theta_2 t} D^l u_t]_{G_\tau} = [D^l F,\ e^{-\theta_2 t} D^l u_t]_{G_\tau},\ \theta_2 = \text{const}, \tag{3.2.21}$$

and transform its individual terms by integration by parts. We obtain

$$[D^l u_{tt},\ e^{-\theta_2 t} D^l u_t]_{G_\tau} = \frac{1}{2}(D^l u_t,\ e^{-\theta_2 t} D^l u_t)_{t=\tau} + [\theta_2 e^{-\theta_2 t} D^l u_t, D^l u_t]_{G_\tau};$$

$$\varepsilon [D^l \Delta u,\ e^{-\theta_2 t} D^l u_t]_{G_\tau} = -\frac{\varepsilon}{2}(D^l u_{x_k},\ e^{-\theta_2 t} D^l u_{x_k})_{t=\tau} - \frac{\varepsilon}{2}[D^l u_{x_k},\ \theta_2 e^{-\theta_2 t} D^l u_{x_k}]_{G_\tau}.$$

It is easily seen that

$$[D^l (a_\varepsilon^{kj} u_{x_k})_{x_j},\ e^{-\theta_2 t} D^l u_t]_{G_\tau} = -\frac{1}{2}(a_\varepsilon^{kj} D^l u_{x_k},\ e^{-\theta_2 t} D^l u_{x_j})_{t=\tau} + \frac{1}{2}[(a_{\varepsilon t}^{kj} - \theta_2 a_\varepsilon^{kj}) D^l u_{x_k},\ e^{-\theta_2 t} D^l u_{x_j}]_{G_\tau} + \sum_{|\gamma| \geqslant 1} C_\gamma [(D^\gamma a_\varepsilon^{kj} D^{l-\gamma} u_{x_k})_{x_j},\ e^{-\theta_2 t} D^l u_t]_{G_\tau},\ C_\gamma = \text{const}.$$

Clearly the inequalities

$$|[D^l b_\varepsilon^k u_{x_k},\ D^l u_t e^{-\theta_2 t}]_{G_\tau}| \leqslant [D^l u_t,\ e^{-\theta_2 t} D^l u_t]_{G_\tau} + M_8 \sum_{|\gamma| \leqslant |l|+1} [D^\gamma u,\ e^{-\theta_2 t} D^\gamma u]_{G_\tau};$$

$$|[D^l b_\varepsilon^0 u_t,\ D^l u_t e^{-\theta_2 t}]_{G_\tau}| \leqslant M_9 \sum_{|\gamma| \leqslant |l|} [D^\gamma u_t,\ e^{-\theta_2 t} D^\gamma u_t]_{G_\tau}$$

hold with constants M_8 and M_9 depending on the derivatives with respect to x of b_ε^k and b_ε^0, respectively, up to order $|l|$. Considering the inequality (3.2.20) already proved, and choosing the constant θ_2 sufficiently large, we obtain from (3.2.21), using induction on $|l|$, that for $|l| \leqslant s - 1$

$$(D^l u_t,\ D^l u_t)_{t=\tau} \leqslant M_{10} \tau^{2p+6} \sum_{|\gamma| \leqslant s} |||D^\gamma F|||_{p+1,\ t_0} + M_{11} \sum_{|l| \leqslant s-1} [D^l F,\ D^l F]_{G_\tau}. \tag{3.2.22}$$

The constants M_{10} and M_{11} here depend on a_ϵ^{kj}, $a_{\epsilon x_k}^{kj}$, b_ϵ^k, c_ϵ, b_ϵ^0, $b_{\epsilon t}^0$ and on their derivatives with respect to x up to order s inclusive.

Applying the operator $D^l\partial^\rho/\partial t^\rho$ with $\rho \geqslant 0$ and $|l| + \rho \leqslant s - 2$ to the right and left sides of (3.2.8), we obtain equations by means of which the derivatives of the form $D^l\partial^{\rho+2}u/\partial t^{\rho+2}$ ($\rho \geqslant 0$, $|l| + \rho \leqslant s - 2$) may be expressed in terms of derivatives of u which have already been estimated above. We thus obtain (3.2.6) for $\tau \leqslant t_0$.

We obtain estimate (3.2.6) for $t_0 \leqslant \tau \leqslant T - \epsilon$ in an analogous manner. In order to estimate $(u, u)_{t=\tau}$ we consider (3.2.9) and transform its terms on the left side exactly as was done for $\tau \leqslant \tau_0$. However, instead of inequality (3.2.10) we consider the inequality

$$|[b_\varepsilon^k u_{x_k}, e^{\theta t}w]_{G_\tau}| \leqslant M_{12}[u, e^{\theta t}u]_{G_\tau}$$

$$+\frac{\alpha}{2}[tb_\varepsilon^k w_{x_k}, e^{\theta t}b_\varepsilon^k w_{x_k}]_{G_{t_0}} + \frac{1}{2\alpha}[u, t^{-1}e^{\theta t}u]_{G_{t_0}}$$

$$+\frac{\alpha_1}{2}[b_\varepsilon^k w_{x_k}, e^{\theta t}b_\varepsilon^k w_{x_k}]_{G_\tau\setminus G_{t_0}} + \frac{1}{2\alpha_1}[u, e^{\theta t}u]_{G_\tau\setminus G_{t_0}},$$

where α_1 is a constant such that $\alpha_1 \leqslant \alpha t$ for $t \geqslant t_0$, and we use estimate (3.2.13) for $y(t_0) = [u, t^{-1}e^{\theta t}u]_{G_{t_0}}$. We estimate the right side of (3.2.9) using the inequality

$$|[F, e^{\theta t}w]_{G_\tau}| \leqslant [F, e^{\theta t}F]_{G_\tau} + M_{13}[u, e^{\theta t}u]_{G_\tau},$$

where the constant M_{13} depends only on T. Setting $z = [u, e^{\theta t}u]_{G_\tau}$ and choosing the constant θ sufficiently large, for $\tau \geqslant t_0$ we obtain from (3.2.9) the inequality

$$z'(\tau) \leqslant M_{14}z(\tau) + M_{15}|||F|||^2_{p+1,t_0} + M_{16}[F, F]_{G_\tau}, \tag{3.2.23}$$

where M_{14}, M_{15} and M_{16} depend on the maxima of the absolute values of b_ϵ^0, $b_{\epsilon x_k}^k$ and $b_{\epsilon t}^0$, and on t_0. From (3.2.23) follows the required estimate.

In an analogous manner we estimate $(D^l u, D^l u)_{t=\tau}$, $|l| \leqslant s$, for $\tau > t_0$. Estimates for the derivatives $D^l u_t$ and for derivatives of the form $D^l\partial^\rho u/\partial t^\rho$ with $\rho \geqslant 2$, $|l| + \rho \leqslant s$ and $\tau \geqslant t_0$ are obtained exactly as was done for $\tau \leqslant t_0$.

From the estimate for u and the relation $u = u_\epsilon - v_p$ follows the required estimate for u_ϵ.

THEOREM 3.2.1. *Suppose the coefficients of the equation* (3.2.1) *satisfy the inequality*

$$\alpha t(b^k\xi_k)^2 \leqslant Aa^{kj}\xi_k\xi_j + a_t^{kj}\xi_k\xi_j \tag{3.2.24}$$

for all ξ, *and for* t *in the range* $0 \leqslant t \leqslant t_0$, t_0 *being some small positive number less than* T, *where* α *is a constant satisfying* $\alpha > (2p + 6)^{-1}$, p *being an integer with* $p \geqslant -1$, *and* A *is some constant. Suppose also that* (3.2.24) *holds for all* ξ, *and for* t *in the range* $t_0 \leqslant t \leqslant T$, *with possibly different constants* α *and* A. *Suppose* $a^{kj}, a^{kj}_{x_k}, b^k, b^0, c, b^0_t$ *and their derivatives up to order* s $(s \geqslant 2)$ *with respect to* x *and up to order* $s-2$ *in* x *and* t *are bounded in* G_T; *moreover suppose that for* $0 \leqslant t \leqslant t_0$ *the derivatives of the form* $D^l \partial^\rho / \partial t^\rho$ *of* $a^{kj}, a^{kj}_{x_k}, b^k, b^0$ *and* c, *where* $\rho \leqslant p+1$ *and* $|l| \leqslant s$, *are bounded, as well as the derivatives of these same functions of the form* $D^l \partial^\rho / \partial t^\rho$ *with* $\rho \leqslant p$, $|l| \leqslant p+s$ *and* $\rho + |l| \leqslant p+s$, *evaluated at* $t=0$. *Assume that the functions* $f(t, x)$, $\phi(x)$ *and* $\psi(x)$ *have compact support in* x. *Then there exists a unique solution* $u(t, x)$ *of the problem* (3.2.1), (3.2.2). *This solution satisfies the following estimate for almost all* τ *in the interval* $0 \leqslant \tau \leqslant T$:

$$\begin{aligned} \| u \|^2_{\tau, s} \leqslant M_1 \{ \| \varphi \|^2_{0; s+p+4} + \| \psi \|^2_{0; s+p+3} \\ + \max_{0 \leqslant \sigma \leqslant t_0} \| f \|^2_{\sigma; p+1, s, p+1+s} + \| f \|^2_{\tau; s-2} \\ + \| f \|^2_{G_\tau; 0, s, s} + \| f \|^2_{0; p, p+s+2, p+s+2} \}, \end{aligned} \tag{3.2.25}$$

provided f, ϕ *and* ψ *have norms entering into* (3.2.25) *which are finite*; *here* M_1 *is a constant depending on the coefficients of the equation* (3.2.1) *and on their derivatives indicated above. If* $2(s-2) \geqslant m+1$, *then there exists a unique classical solution of* (3.2.1), (3.2.2).

PROOF. We introduce the mollifying operator

$$P_\varepsilon [v] (t, x) = \int_{R^m} \omega_\varepsilon (x-y) v(t, y) \, dy,$$

where $\omega_\epsilon(x-y)$ is an infinitely differentiable function of x and y depending only on the distance $r(x, y)$ between the points x and y, and satisfying $\omega_\epsilon \geqslant 0$ with $\omega_\epsilon = 0$ for $r \geqslant \epsilon$. Let

$$\widetilde{P}_\varepsilon [v] (t, x) = \int_0^\infty \widetilde{\omega}_\varepsilon (\tau - t) P_\varepsilon [v] (\tau, x) \, d\tau; \quad \int_0^\infty \int_{R^m} \widetilde{\omega}_\varepsilon (\sigma) \omega_\varepsilon (x) \, dx \, d\sigma = 1,$$

where $\widetilde{\omega}_\epsilon$ is an infinitely differentiable function of σ such that $\widetilde{\omega}_\epsilon(\sigma) = 0$ for $\sigma \leqslant 0$ and for $\sigma \geqslant \epsilon$, with $\widetilde{\omega}_\epsilon \geqslant 0$ for all σ. We consider the equation (3.2.3) with conditions (3.2.4), where

$$\begin{aligned} \varphi_\varepsilon = P_\varepsilon [\varphi], \ \psi_\varepsilon = P_\varepsilon [\psi], \ a^{kj}_\varepsilon = \widetilde{P}_\varepsilon [a^{kj}], \ b^k_\varepsilon = \widetilde{P}_\varepsilon]b^k], \\ b^0_\varepsilon = \widetilde{P}_\varepsilon [b^0], \ c_\varepsilon = \widetilde{P}_\varepsilon [c], \ f_\varepsilon = \widetilde{P}_\varepsilon [f]. \end{aligned} \tag{3.2.26}$$

These functions are defined for $t \leqslant T - \epsilon$. We show that if condition (3.2.24) of Theorem 3.2.1 is fulfilled, then for sufficiently small ϵ and $0 \leqslant t \leqslant T - \epsilon$, condition (3.2.5) is satisfied for the mollified function. Applying the operator $\widetilde{P}_\epsilon$ to the right and left sides of (3.2.24), we obtain

$$\widetilde{P}_\varepsilon[\alpha t(b^k\xi_k)^2] \leqslant A\widetilde{P}_\varepsilon[a^{kj}\xi_k\xi_j] + \widetilde{P}_\varepsilon[a_t^{kj}\xi_k\xi_j].$$

It is clear that

$$\widetilde{P}_\varepsilon[\alpha t(b^k\xi_k)^2] \geqslant \alpha t\widetilde{P}_\varepsilon[(b^k\xi_k)^2] \geqslant \alpha t(\widetilde{P}_\varepsilon[b^k\xi_k])^2 \geqslant \alpha t(b_\varepsilon^k\xi_k)^2,$$

$$\widetilde{P}_\varepsilon[a_t^{kj}\xi_k\xi_j] = \frac{\partial}{\partial t}\widetilde{P}_\varepsilon[a^{kj}\xi_k\xi_j] = a_{\varepsilon t}^{kj}\xi_k\xi_j;\ \widetilde{P}_\varepsilon[a^{kj}\xi_k\xi_j] = a_\varepsilon^{kj}\xi_k\xi_j$$

and therefore the coefficients of (3.2.3) satisfy condition (3.2.5) for $t \leqslant T - \epsilon$.

We consider the Cauchy problem in the domain $G_{T-\epsilon}$ for the equation (3.2.3) with coefficients given by (3.2.26), and initial conditions (3.2.4). Since (3.2.3) is hyperbolic for $\epsilon > 0$, it is known that the problem (3.2.3), (3.2.4) has a solution $u_\epsilon(t, x)$ infinitely differentiable in $G_{T-\epsilon}$. Since condition (3.2.5) is satisfied, it follows that $u_\epsilon(t, x)$ satisfies (3.2.6). Therefore one may extract a sequence from the set $u_\epsilon(t, x)$ which converges in the norm of $\mathcal{H}^{s-1}(G_T)$ and converges weakly in $\mathcal{H}^s(G_T)$ as $\epsilon \to 0$. It is easy to see that the limit function $u(t, x)$ satisfies inequality (3.2.25).

REMARK 1. The requirement of smoothness on the coefficients of (3.2.1) in Theorem 3.2.1 may be weakened by applying Sobolev's imbedding theorems in the derivation of (3.2.20) and (3.2.22), as was done for hyperbolic equations in [125] and for equations of the type (3.2.1) in [27].

REMARK 2. In the case of equations of the form (3.2.1) with two independent variables

$$u_{tt} - \lambda^2(t)K^2(t, x)u_{xx} + a(t, x)u_x + b(t, x)u_t + c(t, x)u = f(t, x), \tag{3.2.27}$$

where $\lambda(0) = 0$, $\lambda'(t) \geqslant 0$, $\lambda(t) > 0$ for $t > 0$ and $K(t, x) \geqslant \text{const} > 0$, it follows from Theorem 3.2.1 that the Cauchy problem (3.2.27), (3.2.2) is correctly posed if

$$\alpha t a^2 \leqslant 2\lambda_t\lambda K^2 + \lambda^2(K^2)_t \tag{3.2.28}$$

and if the coefficients of the equation, the function f, and the initial functions ϕ and ψ are sufficiently smooth. If $\lambda = t^\beta$ and $K = 1$, then condition (3.2.28) takes the form

$$t^{1-\beta}|a| < (2\beta(2p+6))^{1/2}. \tag{3.2.29}$$

The problem (3.2.27), (3.2.2) was studied in [10, 12, 83, 106]. In [83] a method of integral equations was used to obtain correctness conditions for the problem (3.2.27), (3.2.2), which are close to (3.2.29). In [10, 42] it was shown in particular that when $\lambda(t) = t^\beta$ and $\beta > 1$, the problem (3.2.27), (3.2.2) may turn out to be incorrectly posed.

In [18], an explicit solution of the Cauchy problem for the equation

$$u_{tt} - t^2u_{xx} = au_x\ (t > 0,\ 0 \leqslant x \leqslant 1),$$

was obtained, where $a = \text{const}$, $a = 4n + 1$, $n \geqslant 0$ is an integer, the initial conditions being of the form

$$u|_{t=0} = \mu(x), \quad u_t|_{t=0} = 0.$$

This unique solution is

$$u(t, x) = \sum_{k=0}^{n} \frac{\sqrt{\pi}\, t^{2k}}{k!\,(n-k)!\,\Gamma\left(k+\frac{1}{2}\right)} \frac{\partial^k \mu\left(x+\frac{t^2}{2}\right)}{\partial x^k}.$$

This formula shows the dependence between the quantity a and the required degree of smoothness of the initial functions. Such a dependence is accounted for by inequality (3.2.29).

REMARK 3. From condition (3.2.24) of Theorem 3.2.1 it follows that if $a^{kj}(t_0, x_0)\xi_k\xi_j = 0$ for some vector ξ at a point (t_0, x_0), then for this same ξ the equation $a^{kj}(t, x_0)\xi_k\xi_j = 0$ holds for all $t \leqslant t_0$, since according to (3.2.24) we have that the function $Z(t) = a^{kj}(t, x_0)\xi_k\xi_j$ satisfies $Z_t + AZ \geqslant 0$ with $Z = 0$ at $t = t_0$.

In order to allow consideration of more general hyperbolic equations, degenerating inside the domain, we prove the following lemma.

LEMMA 3.2.2. *Let* $u_\epsilon(t, x)$ *be a solution of the hyperbolic equation* (3.2.3) *in the domain* $G_{T-\epsilon}$ *with conditions* (3.2.4), *where* a_ϵ^{kj}, b_ϵ^k, c_ϵ, b_ϵ^0, f_ϵ, ϕ_ϵ *and* ψ_ϵ *are infinitely differentiable in* $G_{T-\epsilon}$, *and* f_ϵ, ϕ_ϵ *and* ψ_ϵ *have compact support in* x. *Assume that for every* ξ *the inequality*

$$\alpha(T-t-\varepsilon)(b_\varepsilon^k\xi_k)^2 \leqslant Aa_\varepsilon^{kj}\xi_k\xi_j - a_{\varepsilon t}^{kj}\xi_k\xi_j + a_\varepsilon^{kj}\xi_k\xi_j(T-t-\varepsilon)^{-1}\alpha^{-1} \tag{3.2.30}$$

holds in the range $t_1 \leqslant t \leqslant T - \epsilon$, t_1 *being some number less than* T, *where* α *is a constant satisfying* $\alpha > (2p+1)^{-1}$ (p *a nonnegative integer*) *and* A *is a constant. Furthermore, assume that this same inequality holds with some (possibly different) constants* $\alpha > 0$ *and* $A > 0$ *in the range* $0 \leqslant t \leqslant t_1$. *Then the estimate*

$$\|u_\varepsilon\|^2_{G^\varepsilon;\,s} \leqslant C_1\{\|f_\varepsilon\|^2_{G^\varepsilon;\,s+p-2} + \|f_\varepsilon\|^2_{G^\varepsilon;\,0,\,s+p,\,s+p} + \|\varphi_\varepsilon\|_{0;\,s+p+1} + \|\psi_\varepsilon\|_{0;\,s+p}\} \tag{3.2.31}$$

holds with a constant C_1 *depending on the maxima of the absolute values of the coefficients* a_ϵ^{kj}, $a_{\epsilon x_k}^{kj}$, b_ϵ^k, b_ϵ^0, c_ϵ *and of their derivatives up to order* $s + p - 2$ *with respect to* x *and* t, *and of their derivatives up to order* $s + p$ *with respect to* x; *here* $G^\epsilon = G_{T-\epsilon}$.

PROOF. Multiply equation (3.2.3) by $u_t(T - t - \epsilon)^N e^{-\theta t}$ and integrate over the domain $G^\epsilon\{0 \leqslant t \leqslant T - \epsilon,\ x \in R^m\}$. Here $\theta = \text{const} > 0$ and $N = 2p + 1$.

We transform the terms of the resulting equation by integration by parts as follows:

$$[u_{tt}, u_t(T-t-\varepsilon)^N e^{-\theta t}]_{G^\varepsilon} = \frac{1}{2}[Nu_t, u_t(T-t-\varepsilon)^{N-1} e^{-\theta t}]_{G^\varepsilon}$$

$$+\frac{1}{2}[u_t, u_t(T-t-\varepsilon)^N \theta e^{-\theta t}]_{G^\varepsilon} - \frac{1}{2}(u_t, u_t(T-\varepsilon)^N)_{t=0};$$

$$[\varepsilon u_{x_k x_k}, u_t(T-t-\varepsilon)^N e^{-\theta t}]_{G^\varepsilon}$$

$$= -\frac{\varepsilon}{2}[Nu_{x_k}, u_{x_k}(T-t-\varepsilon)^{N-1} e^{-\theta t}]_{G^\varepsilon}$$

$$-\frac{\varepsilon}{2}[u_{x_k}, (T-t-\varepsilon)^N u_{x_k}\theta e^{-\theta t}]_{G^\varepsilon}$$

$$+\frac{1}{2}\varepsilon(u_{x_k}, u_{x_k}(T-\varepsilon)^N)_{t=0};$$

$$[(a_\varepsilon^{kj}u_{x_k})_{x_j}, u_t(T-t-\varepsilon)^N e^{-\theta t}]_{G^\varepsilon}$$

$$= \frac{1}{2}[a_{\varepsilon t}^{kj}u_{x_k}, u_{x_j}(T-t-\varepsilon)^N e^{-\theta t}]_{G^\varepsilon}$$

$$-\frac{1}{2}[a_\varepsilon^{kj}u_{x_k}, N(T-t-\varepsilon)^{N-1}u_{x_j}e^{-\theta t}]_{G^\varepsilon}$$

$$-\frac{1}{2}[a_\varepsilon^{kj}u_{x_k}, u_{x_j}(T-t-\varepsilon)^N \theta e^{-\theta t}]_{G^\varepsilon}$$

$$+\frac{1}{2}(a_\varepsilon^{kj}u_{x_k}, u_{x_j}(T-\varepsilon)^N)_{t=0}.$$

It is easy to see that the estimate

$$|[b_\varepsilon^k u_{x_k}, u_t(T-t-\varepsilon)^N e^{-\theta t}]_{G^\varepsilon}|$$

$$\leqslant \frac{1}{2(N-\delta)}[b_\varepsilon^k u_{x_k}, (T-t-\varepsilon)^{N+1}b_\varepsilon^k u_{x_k}e^{-\theta t}]_{G^\varepsilon\setminus G_{t_1}}$$

$$+\frac{N-\delta}{2}[u_t, u_t e^{-\theta t}(T-t-\varepsilon)^{N-1}]_{G^\varepsilon\setminus G_{t_1}}$$

$$+\frac{\alpha}{2}[b_\varepsilon^k u_{x_k}, (T-t-\varepsilon)^{N+1}b_\varepsilon^k u_{x_k}e^{-\theta t}]_{G_{t_1}}$$

$$+\frac{1}{2\alpha}[u_t, u_t e^{-\theta t}(T-t-\varepsilon)^{N-1}]_{G_{t_1}}$$

is valid if $\delta > 0$ is a small number such that $(2p+1)^{-1} < 1/(N-\delta) < \alpha$, α being the constant in inequality (3.2.30). Furthermore,

$$[b_\varepsilon^0 u_t, u_t(T-t-\varepsilon)^N e^{-\theta t}]_{G^\varepsilon} + [c_\varepsilon u, u_t(T-t-\varepsilon)^N e^{-\theta t}]_{G^\varepsilon}$$

$$\leqslant C_2[u_t, u_t(T-t-\varepsilon)^N e^{-\theta t}]_{G^\varepsilon} + C_3[u, u(T-t-\varepsilon)^N e^{-\theta t}]$$

$$[f_\varepsilon, u_t(T-t-\varepsilon)^N e^{-\theta t}]_{G^\varepsilon} \leqslant [f_\varepsilon, f_\varepsilon(T-t-\varepsilon)^N e^{-\theta t}]_{G^\varepsilon}$$

$$+[u_t, u_t(T-t-\varepsilon)^N e^{-\theta t}]_{G^\varepsilon}; \qquad C_2, C_3 = \text{const.}$$

It is seen that for any smooth function v and for $N \geqslant 3$ the relation

$$-(T-\varepsilon)^{N-2}v^2|_{t=0} = \int_0^{T-\varepsilon} 2(T-t-\varepsilon)^{N-2}v_t v\,dt - \int_0^{T-\varepsilon}(N-2)(T-t-\varepsilon)^{N-3}v^2\,dt$$

holds. From this we have that for $N \geqslant 3$

$$\left[\left(N-\frac{3}{2}\right)(T-t-\varepsilon)^{N-3}v, v\right]_{G^\varepsilon} \leqslant T^{N-2}(v, v)_{t=0} + 2\left[(T-t-\varepsilon)^{N-1}v_t, v_t\right]_{G^\varepsilon}. \tag{3.2.32}$$

In particular, for any $N \geqslant 0$

$$\left[(T-t-\varepsilon)^N v, v\right]_{G^\varepsilon} \leqslant C_4\left\{\left[(T-t-\varepsilon)^N v_t, v_t\right]_{G^\varepsilon} + (v, v)_{t=0}\right\}. \tag{3.2.33}$$

Considering the inequalities obtained above, condition (3.2.30), and the expressions obtained after integration by parts, and also choosing $\theta > 0$ sufficiently large, we obtain from the equation

$$\left[L_\varepsilon(u), e^{-\theta t}(T-t-\varepsilon)^N u_t\right]_{G^\varepsilon} = \left[f_\varepsilon, e^{-\theta t}(T-t-\varepsilon)^N u_t\right]_{G^\varepsilon}$$

that

$$\left[u, u(T-t-\varepsilon)^{N-1}e^{-\theta t}\right]_{G^\varepsilon} + \left[u_t, u_t(T-t-\varepsilon)^{N-1}e^{-\theta t}\right]_{G^\varepsilon} \leqslant C_5\left\{\left[f_\varepsilon, f_\varepsilon(T-t-\varepsilon)^N e^{-\theta t}\right]_{G^\varepsilon} + \left(\varphi_{\varepsilon x_k}, \varphi_{\varepsilon x_k}\right)_{t=0} + (\psi_\varepsilon, \psi_\varepsilon)_{t=0} + (\varphi_\varepsilon, \varphi_\varepsilon)_{t=0}\right\},$$

where the constant C_5 depends on a_ε^{kj}, b_ε, c_ε, T, δ and α.

We now assume that we have obtained estimates of the type

$$\left[D^l u, D^l u(T-t-\varepsilon)^{N-1}e^{-\theta t}\right]_{G^\varepsilon} + \left[D^l u_t, D^l u_t(T-t-\varepsilon)^{N-1}e^{-\theta t}\right]_{G^\varepsilon} \leqslant C_l\left\{\sum_{|\gamma|\leqslant|l|}\left[D^\gamma f_\varepsilon, D^\gamma f_\varepsilon(T-t-\varepsilon)^N e^{-\theta t}\right]_{G^\varepsilon} + \sum_{|\gamma|=|l|+1}\left(D^\gamma\varphi_\varepsilon, D^\gamma\varphi_\varepsilon\right)_{t=0} + \sum_{|\gamma|\leqslant|l|}\left(D^\gamma\psi_\varepsilon, D^\gamma\psi_\varepsilon\right)_{t=0}\right\} \tag{3.2.34}$$

for $|l| \leqslant q-1$, where the coefficients C_l depend on a_ε^{kj}, $a_{\varepsilon x_k}^{kj}$, b_ε^k, b_ε^0, c_ε, and on their derivatives with respect to x up to order $q-1$. We show that estimate (3.2.34) is also valid for $|l| = q \leqslant s+p$. For this purpose we examine the equation

$$[D^l L_\varepsilon(u), D^l u_t(T-t-\varepsilon)^N e^{-\theta_1 t}]_{G^\varepsilon} = [D^l f_\varepsilon, D^l u_t(T-t-\varepsilon)^N e^{-\theta_1 t}]_{G^\varepsilon} \tag{3.2.35}$$

and transform its terms by integration by parts. We have

$$[D^l u_{tt}, D^l u_t(T-t-\varepsilon)^N e^{-\theta_1 t}]_{G^\varepsilon} = \frac{N}{2}[D^l u_t, D^l u_t(T-t-\varepsilon)^{N-1}e^{-\theta_1 t}]_{G^\varepsilon} +$$

$$+\frac{1}{2}[D^l u_t, \theta_1 D^l u_t (T-t-\varepsilon)^N e^{-\theta_1 t}]_{G^\varepsilon} - \frac{1}{2}(D^l u_t, D^l u_t (T-\varepsilon)^N)_{t=0}.$$

In an analogous manner we obtain

$$\begin{aligned}
&[\varepsilon D^l u_{x_k x_k}, D^l u_t (T-t-\varepsilon)^N e^{-\theta_1 t}]_{G^\varepsilon} \\
&= -\frac{\varepsilon}{2}[N D^l u_{x_k}, D^l u_{x_k} (T-t-\varepsilon)^{N-1} e^{-\theta_1 t}]_{G^\varepsilon} \\
&\quad - \frac{1}{2}\varepsilon [D^l u_{x_k}, D^l u_{x_k} (T-t-\varepsilon)^N e^{-\theta_1 t}\theta_1]_{G^\varepsilon} \\
&\quad + \frac{\varepsilon}{2}(D^l u_{x_k}, D^l u_{x_k} (T-\varepsilon)^N)_{t=0}.
\end{aligned}$$

It is easy to see that

$$\begin{aligned}
&[D^l(a_\varepsilon^{kj} u_{x_k})_{x_j}, D^l u_t (T-t-\varepsilon)^N e^{-\theta_1 t}]_{G^\varepsilon} = \frac{1}{2}(a_\varepsilon^{kj} D^l u_{x_k}, D^l u_{x_j} (T-\varepsilon)^N)_{t=0} \\
&\quad - \frac{N}{2}\left[a_\varepsilon^{kj} D^l u_{x_k}, D^l u_{x_j} (T-t-\varepsilon)^{N-1} e^{-\theta_1 t}\right]_{G^\varepsilon} \\
&\quad + \frac{1}{2}[a_{\varepsilon t}^{kj} D^l u_{x_k}, D^l u_{x_j} (T-t-\varepsilon)^N e^{-\theta_1 t}]_{G^\varepsilon} \\
&\quad - \frac{1}{2}[a_\varepsilon^{kj} D^l u_{x_k}, D^l u_{x_j} \theta_1 (T-t-\varepsilon)^N e^{-\theta_1 t}]_{G^\varepsilon} \\
&\quad - \sum_{|\gamma|=1} [a_{\varepsilon x_\rho}^{kj} D^{l-\gamma} u_{x_k}, D^l u_{t x_j} (T-t-\varepsilon)^N e^{-\theta_1 t}]_{G^\varepsilon} \\
&\quad + \sum_{|\gamma|\geqslant 2} C_\gamma [(D^\gamma (a_\varepsilon^{kj}) D^{l-\gamma} u_{x_k})_{x_j}, \\
&\quad D^l u_t (T-t-\varepsilon)^N e^{-\theta_1 t}]_{G^\varepsilon}, \quad C_\gamma = \text{const.} \qquad (3.2.36)
\end{aligned}$$

We estimate the next to last term in (3.2.36) by use of the inequality (3.2.18). For $|\gamma| = 1$ we have

$$\begin{aligned}
&\left|[a_{\varepsilon x_\rho}^{kj} D^{l-\gamma} u_{x_k}, D^l u_{t x_j} (T-t-\varepsilon)^N e^{-\theta_1 t}]_{G^\varepsilon}\right| \\
&= \Big|[a_{\varepsilon x_\rho x_j}^{kj} D^{l-\gamma} u_{x_k}, D^l u_t (T-t-\varepsilon)^N e^{-\theta_1 t}]_{G^\varepsilon} \\
&\quad + [a_{\varepsilon x_\rho}^{kj} D^{l-\gamma} u_{x_k x_j}, D^l u_t (T-t-\varepsilon)^N e^{-\theta_1 t}]_{G^\varepsilon}\Big| \\
&\leqslant M\left[a_\varepsilon^{kj} D^{l-\gamma} u_{x_k x_\rho}, D^{l-\gamma} u_{x_\rho x_j} (T-t-\varepsilon)^N e^{-\theta_1 t}\right]_{G^\varepsilon} \\
&\quad + C_6 [D^l u_t, D^l u_t (T-t-\varepsilon)^N e^{-\theta_1 t}]_{G^\varepsilon} \\
&\quad + C_7 \sum_{|l|\leqslant q} [D^l u, D^l u (T-t-\varepsilon)^N e^{-\theta_1 t}]_{G^\varepsilon},
\end{aligned}$$

where M, C_6 and C_7 depend on the derivatives of a_ε^{kj} with respect to x up to second order inclusive. The last term in (3.2.36) clearly does not surpass

$$C_8 \sum_{|\gamma|\leqslant q}\left\{\left[D^\gamma u, D^\gamma u (T-t-\varepsilon)^N e^{-\theta_1 t}\right]_{G^\varepsilon} + [D^\gamma u_t, D^\gamma u_t (T-t-\varepsilon)^N e^{-\theta_1 t}]_{G^\varepsilon}\right\},$$

where C_8 depends on a_ϵ^{kj}, $a_{\epsilon x_k}^{kj}$, and on their derivatives with respect to x up to order q.

Furthermore,

$$
\begin{aligned}
&|[D^l b_\varepsilon^k u_{x_k}, D^l u_t (T-t-\varepsilon)^N e^{-\theta_1 t}]_{G^\varepsilon}| \\
&\leqslant |[b_\varepsilon^k D^l u_{x_k}, D^l u_t (T-t-\varepsilon)^N e^{-\theta_1 t}]_{G^\varepsilon}| \\
&+ \Big| \sum_{1 \leqslant |\gamma| \leqslant q} C_\gamma [D^\gamma b_\varepsilon^k D^{l-\gamma} u_{x_k}, D^l u_t (T-t-\varepsilon)^N e^{-\theta_1 t}]_{G^\varepsilon} \Big| \\
&\leqslant \frac{1}{2(N-\delta)} [b_\varepsilon^k D^l u_{x_k}, b_\varepsilon^k D^l u_{x_k} (T-t-\varepsilon)^{N+1} e^{-\theta_1 t}]_{G^\varepsilon \setminus G_{t_1}} \\
&+ \frac{1}{2}(N-\delta)[D^l u_t, D^l u_t (T-t-\varepsilon)^{N-1} e^{-\theta_1 t}]_{G^\varepsilon \setminus G_{t_1}} \\
&+ \frac{\alpha}{2} [b_\varepsilon^k D^l u_{x_k}, b_\varepsilon^k D^l u_{x_k} (T-t-\varepsilon)^{N+1} e^{-\theta_1 t}]_{G_{t_1}} \\
&+ \frac{1}{2\alpha} [D^l u_t, (T-t-\varepsilon)^{N-1} e^{-\theta_1 t} D^l u_t]_{G_{t_1}} \\
&+ C_9 \sum_{|\gamma| \leqslant q} \{ [D^\gamma u_t, D^\gamma u_t (T-t-\varepsilon)^N e^{-\theta_1 t}]_{G^\varepsilon} \\
&+ [D^\gamma u, D^\gamma u (T-t-\varepsilon)^N e^{-\theta_1 t}]_{G^\varepsilon} \}.
\end{aligned}
$$

Here α is the constant in (3.2.30), the constant δ is positive and sufficiently small, and $C_9 = \text{const}$. It is easily seen that

$$
\begin{aligned}
&|[D^l (b_\varepsilon^0 u_t), D^l u_t (T-t-\varepsilon)^N e^{-\theta_1 t}]_{G^\varepsilon}| \\
&+ |[D^l (c_\varepsilon u), D^l u_t (T-t-\varepsilon)^N e^{-\theta_1 t}]_{G^\varepsilon}| \\
&\leqslant C_{10} \Big\{ \sum_{|\gamma| \leqslant q} [D^\gamma u_t, D^\gamma u_t (T-t-\varepsilon)^N e^{-\theta_1 t}]_{G^\varepsilon} \\
&+ \sum_{|\gamma| \leqslant q} [D^\gamma u, D^\gamma u (T-t-\varepsilon)^N e^{-\theta_1 t}]_{G^\varepsilon} \Big\}.
\end{aligned}
$$

The constant C_{10} depends on the coefficients of the equation (3.2.3) and on their derivatives with respect to x up to order q. Bearing in mind the estimates so obtained, inequality (3.2.33), condition (3.2.30), and also choosing the constant θ_1 sufficiently large, we obtain from (3.2.35) that the estimate (3.2.34) is also valid for $|l| = q \leqslant s + p$.

Expressions for those derivatives of u_ε involving more than one differentiation with respect to t are obtained from (3.2.3) and from the equations obtained from the latter by differentiating with respect to t and x. These derivatives are then estimated by use of (3.2.34) for $|l| \leqslant s + p$. Hence we obtain that

$$
\begin{aligned}
&\left[D^l \frac{\partial^p u}{\partial t^p}, D^l \frac{\partial^p u}{\partial t^p} (T-t-\varepsilon)^{N-1} \right]_{G^\varepsilon} \\
&\leqslant C_{11} \{ \| f_\varepsilon \|^2_{G^\varepsilon; s+p-2} + \| f_\varepsilon \|^2_{G^\varepsilon; 0, s+p, s+p} + \| \varphi_\varepsilon \|^2_{0; s+p+1} + \| \psi_\varepsilon \|^2_{0; s+p} \} \qquad (3.2.37)
\end{aligned}
$$

for $|l| + \rho \leqslant s + p$. From inequality (3.2.32) and estimate (3.2.37) it follows that if $N = 2p + 1$, then $u_\epsilon(t, x)$ satisfies inequality (3.2.31).

The following theorem follows easily from this lemma.

THEOREM 3.2.2. *Assume that the coefficients of the equation* (3.2.1) *satisfy the inequality*

$$\alpha (T - t)(b^k \xi_k)^2 \leqslant A a^{kj} \xi_k \xi_j - a_t^{kj} \xi_k \xi_j + \frac{a^{kj} \xi_k \xi_j}{\alpha (T - t)} \tag{3.2.38}$$

for every ξ *and for* t *in the range* $t_1 \leqslant t \leqslant T$, t_1 *being some number less than* T, *where* α *is a constant satsifying* $\alpha > (2p + 1)^{-1}$ (p *is a nonnegative integer*), *and* A *is a constant. Assume also that* (3.2.38) *is satisfied in the range* $0 \leqslant t \leqslant t_1$ *for some (possibly different) constants* $\alpha > 0$ *and* $A > 0$. *Suppose the coefficients* $a^{kj}, a^{kj}_{x_k}, b^k, b^0$ *and* c *have bounded derivatives up to order* $s + p - 2$ *with respect to* x *and* t ($s \geqslant 2$), *and up to order* $s + p$ *with respect to* x. *Then there exists a unique solution* $u(t, x)$ *of the problem* (3.2.1), (3.2.2) *in the class* $\mathcal{H}^s(G_T)$, *and for this solution the estimate*

$$\|u\|^2_{G_T;s} \leqslant C_{12} \{\|f\|^2_{G_T;s+p-2} + \|f\|^2_{G_T;0,s+p,s+p} + \|\varphi\|^2_{0;s+p+1} + \|\psi\|^2_{0;s+p}\} \tag{3.2.39}$$

is valid, if f, ϕ *and* ψ *have compact support in* x *and have bounded norms appearing in* (3.2.39).

PROOF. As in the proof of Theorem 3.2.1, we mollify the coefficients of the equation (3.2.1), as well as the functions f, ϕ and ψ, and consider the problem (3.2.3), (3.2.4), where the coefficients of (3.2.3) and the functions f_ϵ, ϕ_ϵ and ψ_ϵ are defined by (3.2.26). We verify that condition (3.2.30) of Lemma 3.2.2 is satisfied for the problem (3.2.3), (3.2.4). For this purpose we apply the operator $\widetilde{P}_\epsilon$ to the right and left sides of (3.2.38).

For $0 \leqslant t \leqslant T - \epsilon$ we obtain the inequality

$$\alpha \widetilde{P}_\varepsilon [(T - t)(b^k \xi_k)^2] \leqslant A \widetilde{P}_\varepsilon [a^{kj} \xi_k \xi_j] - \widetilde{P}_\varepsilon [a_t^{kj} \xi_k \xi_j] + \frac{1}{\alpha} \widetilde{P}_\varepsilon \left[\frac{a^{kj} \xi_k \xi_j}{T - t}\right]. \tag{3.2.40}$$

It is easily seen that

$$\widetilde{P}_\varepsilon [(T - t)(b^k \xi_k)^2] \geqslant (T - t - \varepsilon) \widetilde{P}_\varepsilon [(b^k \xi_k)^2] \geqslant (T - t - \varepsilon)(b^k_\varepsilon \xi_k)^2,$$

$$\widetilde{P}_\varepsilon [a_t^{kj} \xi_k \xi_j] = a_{\varepsilon t}^{kj} \xi_k \xi_j;$$

$$\frac{1}{\alpha} \widetilde{P}_\varepsilon \left[\frac{a^{kj} \xi_k \xi_j}{T - t}\right] \leqslant \frac{1}{\alpha} \frac{1}{(T - t - \varepsilon)} a_\varepsilon^{kj} \xi_k \xi_j.$$

Condition (3.2.30) follows from these estimates and from (3.2.40). We obtain the solution of (3.2.1), (3.2.2) as the limit as $\epsilon \to 0$ of the solution of the problem (3.2.3),

(3.2.4). It is clear that inequality (3.2.39) is valid for the limit function $u(t, x)$.

REMARK 4. It follows from condition (3.2.38) that if $a^{kj}(t_0, x_0)\xi_k\xi_j = 0$ for some vector ξ at a point (t_0, x_0), then for this same ξ the equation $a^{kj}(t, x_0)\xi_k\xi_j = 0$ is satisfied for all $T \geqslant t \geqslant t_0$.

Theorems 3.2.1 and 3.2.2 easily yield a theorem allowing for the degeneration of equation (3.2.1) on a set of more general structure; in particular, degeneration merely at individual points.

THEOREM 3.2.3. *Suppose the domain* G_T *may be divided into a finite number of domains* $G^j\{T_{j-1} \leqslant t \leqslant T_j, x \in R^m\}$ *such that in each of the* G^j, *either the condition*

$$\alpha(t - T_{j-1})(b^k\xi_k)^2 \leqslant Aa^{kj}\xi_k\xi_j + a_t^{kj}\xi_k\xi_j$$

of Theorem 3.2.1 *is satisfied, or the condition*

$$\alpha(T_j - t)(b^k\xi_k)^2 \leqslant Aa^{kj}\xi_k\xi_j + \frac{a^{kj}\xi_k\xi_j}{\alpha(T_j - t)} - a_t^{kj}\xi_k\xi_j$$

of Theorem 3.2.2, *the constants* α *and* A *satisfying the requirements stated in those respective theorems. Then in* G_T *there exists a unique solution* $u(t, x)$ *of the problem* (3.2.1), (3.2.2) *in the class* $\mathcal{H}^s(G_T)$, *if the coefficients of the equation* (3.2.1) *and the functions* f, ϕ *and* ψ *(having compact support) are smooth enough. Namely, we require smoothness sufficient to provide for the satisfaction, in each domain* G^j, *of those conditions of Theorem* 3.2.1 *or Theorem* 3.2.2 *which will guarantee the existence of a solution in class* $\mathcal{H}^s(G^j)$ *of* (3.2.1) *in* G^j *with initial functions* $u|_{t=T_{j-1}}$ *and* $u_t|_{t=T_{j-1}}$.

Theorem 3.2.3 is proved by successively applying Theorems 3.2.1 and 3.2.2 in the domain G^j.

It follows from Theorem 3.2.3, for example, that the Cauchy problem for the equation

$$u_{tt} - t^2(2 - t)^2[(1 - t)^2 + x^2]u_{xx} + au_x + bu_t + cu = f$$

with conditions (3.2.2) is always correctly posed (i. e a solution of the Cauchy problem exists, is unique, and depends on the initial data continuously with respect to a certain norm) if the coefficients of this equation, and the functions f, ϕ and ψ, are sufficiently smooth in G_T.

§3. Necessary conditions for correctness of the Cauchy problem for second order equations *

The question of necessary conditions for correctness of the Cauchy problem for differential equations of any order has been considered under various assumptions by E. E. Levi [74], A. Lax [72], P. Lax [73], Mizohata and Ohya [82], Strang and Flaschka [127], V. Ja. Ivriĭ [62], V. Petkov [116], and others. The following exposition uses ideas found in these papers.

Let Ω be a domain in $R^{m+1} = \{x_0, \cdots, x_m\}$ whose boundary contains a domain ω lying in the hyperplane $x_0 = h$, and such that all points in Ω satisfy $x_0 > h$. In particular, Ω might be a domain of the type $\{h < x_0 < T, (x_1, \cdots, x_m) \in R^m\}$, $T = \text{const} \leqslant \infty$.

In $\bar{\Omega}$ we consider the equation

$$L(u) = \sum_{|\alpha| \leqslant 2} a_\alpha(x) D^\alpha u = f(x), \tag{3.3.1}$$

where $a_\alpha(x)$ and $f(x)$ are given complex valued functions, infinitely differentiable in $\bar{\Omega}$. Here and in the following,

$$D_j = -i \frac{\partial}{\partial x_j}, \qquad j = 0, 1, \cdots, m; \quad i^2 = -1; \qquad D^\alpha = D_0^{\alpha_0} \cdots D_m^{\alpha_m};$$

$$|\alpha| = \alpha_0 + \cdots + \alpha_m; \qquad \alpha! = \alpha_0! \cdots \alpha_m!.$$

In the domain Ω we consider the Cauchy problem for equation (3.3.1) with initial conditions

$$u|_{x\in\omega} = g_0, \quad D_0 u|_{x \in \omega} = g_1, \tag{3.3.2}$$

where g_0 and g_1 are given functions in ω. The Cauchy problem (3.3.1), (3.3.2) is called correct in the domain Ω if for any functions $f(x) \in C_0^\infty(R^{m+1})$ with support in $\Omega \cup \omega$ and any functions $g_0, g_1 \in C_0^\infty(\omega)$ there exists in $\bar{\Omega}$ a solution $u(x) \in C^{(2)}(\bar{\Omega})$ of equation (3.3.1) satisfying conditions (3.3.2); and furthermore the condition $f(x) \equiv 0$ in $\bar{\Omega} \cap \{x_0 \leqslant h^1\}$ for some $h^1 \geqslant h$, together with $g_0(x) \equiv 0$ and $g_1(x) \equiv 0$ in ω, imply that $u(x) \equiv 0$ in $\bar{\Omega} \cap \{x_0 \leqslant h^1\}$.

LEMMA 3.3.1. *Suppose the Cauchy problem* (3.3.1), (3.3.2) *is correct in the domain* Ω. *Then for each compact set* $K \subset \Omega \cup \omega$ *such that* $K \cap \omega$ *is nonempty there exists an integer* $k \geqslant 0$ *and a constant* $C > 0$ *such that for any function* $f \in C_0^\infty(K)$ *and any* $g_0, g_1 \in C_0^\infty(K \cap \omega)$ *the solution* $u(x)$ *of the problem* (3.3.1), (3.3.2) *satisfies the inequality*

---

* Added by the authors to the translation.

$$\sup_{\bar{\Omega}} |u| \leqslant C\left\{\sup_K \sum_{|\alpha|\leqslant k} |D^\alpha f| + \sup_{K\cap\omega} \sum_{|\alpha|\leqslant k} (|D^\alpha g_0| + |D^\alpha g_1|)\right\}. \tag{3.3.3}$$

PROOF. In the linear space M_K of functions (f, g_0, g_1), where $f \in C_0^\infty(R^{m+1})$, $\operatorname{supp} f \subset K$, $g_0, g_1 \in C_0^\infty(\omega)$, $\operatorname{supp} g_0 \subset K \cap \omega$, and $\operatorname{supp} g_1 \subset K \cap \omega$, we introduce the countable system of seminorms

$$p_j(f, g_0, g_1) = \sup_K \sum_{|\alpha|\leqslant j} |D^\alpha f| + \sup_{K\cap\omega} \sum_{|\alpha|\leqslant j} (|D^\alpha g_0| + |D^\alpha g_1|).$$

To each element (f, g_0, g_1) of the F-space so obtained, we associate the solution $u(x) \in C^{(2)}(\bar{\Omega})$ of the Cauchy problem (3.3.1), (3.3.2), thus defining an operator T on M_K. Since the Cauchy problem (3.3.1), (3.3.2) by assumption is correct in Ω, it is easy to see that the graph of the operator T is closed. Hence, by the closed graph theorem for F-spaces, the operator T is continuous. This means that there exists an integer $k \geqslant 0$ and $C = \text{const} > 0$ such that for any $(f, g_0, g_1) \in M_K$ the inequality (3.3.3) is valid for the solution $u(x)$ of the Cauchy problem (3.3.1), (3.3.2). The lemma is proved.

Let $0 \leqslant l < m$. We introduce the notation

$$x' = (x_0, \cdots, x_l, 0, \cdots, 0), \qquad x'' = (0, \cdots, 0, x_{l+1}, \cdots, x_m)$$
$$\xi' = (\xi_0, \cdots, \xi_l, 0, \cdots, 0). \qquad \xi'' = (0, \cdots, 0, \xi_{l+1}, \cdots, \xi_m).$$

Then $x = x' + x''$, $\xi = \xi' + \xi''$. For multi-indices we also introduce the corresponding notation

$$\alpha' = (\alpha_0, \cdots, \alpha_l, 0, \cdots, 0), \qquad \alpha'' = (0, \cdots, 0, \alpha_{l+1}, \cdots, \alpha_m),$$

$\alpha = \alpha' + \alpha''$, and, in the same way, $\beta = \beta' + \beta''$.

We write equation (3.3.1) in the form

$$L(u) = \sum_{s=0}^{2} L_s(x, D)u = f(x), \tag{3.3.4}$$

where

$$L_s(x, D)u = \sum_{|\alpha|=s} a_\alpha(x) D^\alpha u.$$

Let

$$L_s(x, \xi) = \sum_{|\alpha|=s} a_\alpha(x)\xi^\alpha; \qquad L_{s,\beta}^{(\alpha)}(x, \xi) = \left(\frac{\partial}{\partial x}\right)^\beta \left(\frac{\partial}{\partial \xi}\right)^\alpha L_s(x, \xi),$$

$$\left(\frac{\partial}{\partial x}\right)^\beta = \left(\frac{\partial}{\partial x_0}\right)^{\beta_0} \cdots \left(\frac{\partial}{\partial x_m}\right)^{\beta_m}, \qquad \left(\frac{\partial}{\partial \xi}\right)^\alpha = \left(\frac{\partial}{\partial \xi_0}\right)^{\alpha_0} \cdots \left(\frac{\partial}{\partial \xi_m}\right)^{\alpha_m}.$$

THEOREM 3.3.1. *Suppose, for some point* $\hat{x} = (\hat{x}_0, \cdots, \hat{x}_m) \in \Omega \cup \omega$ *such that* $a_{(2,0,\cdots,0)}(\hat{x}) \neq 0$, *there exist positive rational numbers* p_1 *and* p_2 *such that*

$$1 - p_1 + p_2 > 0 \tag{3.3.5}$$

and such that for any ξ''

$$L_{2,\beta}^{(\alpha')}(\hat{x}, \xi'') = 0, \tag{3.3.6}$$

provided that

$$|\alpha'| + p_1 |\beta'| + p_2 |\beta''| < 2. \tag{3.3.7}$$

Then, if the Cauchy problem (3.3.1), (3.3.2) *is correct in the domain* Ω, *it follows that for any* ξ''

$$L_{1,\beta}(\hat{x}, \xi'') = 0, \tag{3.3.8}$$

provided that

$$p_1(1 + |\beta'|) + p_2 |\beta''| < 1. \tag{3.3.9}$$

PROOF. We assume the contrary. Let the Cauchy problem (3.3.1), (3.3.2) be correct in Ω and suppose there exist β such that for some $\hat{\xi}''$ we have $L_{1,\beta}(\hat{x}, \hat{\xi}'') \neq 0$ and $p_1(1 + |\beta'|) + p_2|\beta''| < 1$.

We shall show then that there exists a compact set $K \subset \Omega \cup \dot{\omega}$ and functions f, g_0, g_1 in M_K such that the corresponding solution $u(x)$ of the Cauchy problem (3.3.1), (3.3.2) does not satisfy inequality (3.3.3). In order to construct such a solution of problem (3.3.1), (3.3.2) we proceed as follows. Let t be a rational number with $t \geqslant p_1$. We denote by $\mathfrak{M}_t$ the set of multi-indices $\beta = \beta' + \beta''$ such that

$$R_\beta(t) = t(1 + |\beta'|) + p_2 |\beta''| < 1 \tag{3.3.10}$$

and $L_{1,\beta}(\hat{x}, \hat{\xi}'') \neq 0$ for some $\hat{\xi}'' \neq 0$. It is clear that $\mathfrak{M}_{t_1} \subset \mathfrak{M}_{t_2}$ if $t_1 > t_2$. By E_t we denote the set of those β in $\mathfrak{M}_{p_1}$ for which the inequality

$$R_\beta(t) > 2(p_2 - t) - 1 \tag{3.3.11}$$

is satisfied.

We may suppose that the numbers p_1 and p_2 entering into the conditions of Theorem 3.3.1 are such that for any multi-index $\beta \in \mathfrak{M}_{p_1}$ we have

$$R_\beta(p_1) > 2(p_2 - p_1) - 1, \tag{3.3.12}$$

since if for some $\beta_* \in \mathfrak{M}_{p_1}$ the inequality (3.3.12) is not fulfilled, we can show that there exists a rational number $t > p_1$ such that $\mathfrak{M}_t$ is nonempty and for any $\beta \in \mathfrak{M}_t$,

$$R_\beta(t) > 2(p_2 - t) - 1;$$

and we take this value t as the new value of p_1.

In fact, suppose there exists a $\beta_* \in \mathfrak{M}_{p_1}$ such that $R_{\beta_*}(p_1) \leqslant 2(p_2 - p_1) - 1$. It is easy to see that

$$R_{\beta_*}(t_1) = R_{\beta_*}(p_1) + (t_1 - p_1)(1 + |\beta'_*|) > 2(p_2 - t_1) - 1$$

if

$$t_1 > (2p_2 + p_1(1 + |\beta'_*|) - R_{\beta_*}(p_1) - 1)(3 + |\beta'_*|)^{-1} = K_1.$$

On the other hand, $R_{\beta_*}(t_1) < 1$ if

$$t_1 < (1 - R_{\beta_*}(p_1) + p_1(1 + |\beta'_*|))(1 + |\beta'_*|)^{-1} = K_2.$$

It is clear that $K_2 > 0$, since $R_{\beta_*}(p_1) < 1$ and

$$K_2 - K_1 = [2(1 - R_{\beta_*}(p_1)) + 2(1 - p_2 + p_1)(1 + |\beta'_*|)](1 + |\beta'_*|)^{-1}(3 + |\beta'_*|)^{-1} > 0,$$

since by assumption $1 - p_2 + p_1 > 0$ and $1 - R_{\beta_*}(p_1) > 0$. Therefore there exist rational positive numbers t_1 such that

$$1 > R_{\beta_*}(t_1) > 2(p_2 - t_1) - 1.$$

We take $t_1 = K_2 - \epsilon$, where $\epsilon = \text{const} > 0$, and $\epsilon < (1 - R_{\beta_*}(p_1))(1 + |\beta'_*|)^{-1}$. Then $t_1 > p_1$; hence $\mathfrak{M}_{t_1} \subset \mathfrak{M}_{p_1}$, with $\mathfrak{M}_{t_1}$ nonempty. It is easy to see that if $\beta \in E_{p_1}$, then $\beta \in E_{t_1}$, since

$$\begin{aligned} R_\beta(t_1) &= R_\beta(p_1) + (t_1 - p_1)(1 + |\beta'|) \\ &> 2(p_2 - t_1) - 1 + 2(t_1 - p_1) + (t_1 - p_1)(1 + |\beta'|) > 2(p_2 - t_1) - 1. \end{aligned}$$

Furthermore, if there exists a multi-index $\beta \in \mathfrak{M}_{t_1}$ such that $R_\beta(t_1) \leqslant 2(p_2 - t_1) - 1$, then, as above, we may also find a rational number $t_2 > t_1$ such that

$$1 > R_\beta(t_2) > 2(p_2 - t_2) - 1.$$

Since $E_{t_1} \subset E_{t_2}$ and E_{t_2} differs from E_{t_1} by at most one element, and the set $\mathfrak{M}_{p_1}$ is finite, we may continue in such a manner and, after a finite number of steps, obtain a rational positive number t such that $t > p_1$, $\mathfrak{M}_t$ is nonempty, and $\mathfrak{M}_t \subset E_t$. We take this t as a new value for p_1. Then (3.3.12) is clearly satisfied for all $\beta \in \mathfrak{M}_{p_1}$.

With no loss of generality, we may assume under the conditions of the theorem that $\hat{x} = 0$. We make a change of variables x:

$$y' = \rho^\kappa x', \qquad y'' = \rho^\tau x''; \qquad \kappa, \tau = \text{const} > 0, \qquad \rho = \text{const} > 1, \tag{3.3.13}$$

with the corresponding change of the variables ξ:

$$\eta' = \rho^{-\kappa}\xi', \qquad \eta'' = \rho^{-\tau}\xi''.$$

If estimate (3.3.3) is fulfilled, then in the new variables y the estimate

$$\sup_{\bar{\Omega}_1} |u| \leqslant C\rho^{k\delta}\left\{\sup_{K_1} \sum_{|\alpha| \leqslant k} |D^\alpha f| + \sup_{K_1 \cap \omega_1} \sum_{|\alpha| \leqslant k} (|D^\alpha g_0| + |D^\alpha g_1|)\right\} \tag{3.3.14}$$

holds, where Ω_1, ω_1 and K_1 are the images of Ω, ω and K respectively in the space y, and $\delta = \max\{\tau, \kappa\}$. We represent the symbol of the operator L_s in the form of a portion of a Taylor's series in some neighborhood $\hat{G}$ of the point $x = 0$. We thus obtain

$$L_s(x, \xi) = \sum_{\alpha',\beta,\lambda_s(\alpha',\beta)>-N} \frac{x^\beta}{\alpha'!\beta!} L_{s,\beta}^{(\alpha')}(0, \xi'')\xi'^{\alpha'} + R_N(x, \xi) \tag{3.3.15}$$

$$= \sum_{\alpha',\beta,\lambda_s(\alpha',\beta)>-N} \rho^{\lambda_s(\alpha',\beta)} \frac{y^\beta}{\alpha'!\beta!} L_{s,\beta}^{(\alpha')}(0, \eta'')\eta'^{\alpha'} + \rho^{-N} \sum_{|\alpha|=s} b_\alpha \eta^\alpha,$$

where b_α are functions in $\hat{G}$, uniformly bounded with respect to ρ, $L_{s,\beta}^{(\alpha')}$ are the corresponding derivatives of the symbol $L_s(x, \xi)$ with respect to the old variables, N is some positive integer to be defined later, and

$$\lambda_s(\alpha', \beta) = \tau(s - |\alpha'| - |\beta''|) + \kappa(|\alpha'| - |\beta'|). \tag{3.3.16}$$

We assumed that the conditions (3.3.6) and (3.3.7) are satisfied at the point $x = 0$, but that conditions (3.3.8) and (3.3.9) are not. We consider the set $\mathfrak{M}_{p_1}$ and let

$$R = \min_{\beta \in \mathfrak{M}_{p_1}} R_\beta(p_1). \tag{3.3.17}$$

Let μ be the set of those $\beta \in \mathfrak{M}_{p_1}$ for which $R_\beta(p_1) = R$. Thus there exists $\eta'' = \hat{\eta}'' \neq 0$ and a multi-index $\hat{\beta} = (\hat{\beta}', \hat{\beta}'') \in \mu$ such that $L_{1,\hat{\beta}}(0, \hat{\eta}'') \neq 0$.

In a neighborhood of the point $x = 0$ we construct an asymptotic solution u_ρ of the equation $L(u) = 0$ in the form

$$u_\rho(y) = \sum_{k=0}^{N_1} v^k(y)\rho^{-\frac{k}{n}} \exp i\left[(y'', \hat{\eta}'')\gamma\rho + \sum_{j=0}^{N_2} \phi^j(y)\rho^{\sigma_j}\right], \tag{3.3.18}$$

where σ_j are certain rational numbers, $1 > \sigma_0 > \cdots > \sigma_{N_2} > 0$, n is a positive integer, γ is a real number, $\gamma \neq 0$, $(y'', \hat{\eta}'') = \Sigma_{l+1}^m y_j \hat{\eta}_j$; N_1 and N_2 are integers; and the functions $v^k(y)$, $\phi^j(y)$ and numbers σ_j, n, N_1, N_2, γ and ρ will be defined below.

We set

$$W = \exp i[(y'', \hat{\eta}'')\gamma\rho + \phi^0(y)\rho^{\sigma_0}],$$

$$w = \exp i\left[\sum_{j=1}^{N_2} \phi^j(y)\rho^{\sigma_j}\right], \qquad v = \sum_{k=0}^{N_1} v^k(y)\rho^{-\frac{k}{n}},$$

$$L_{[\rho,N]} \equiv \sum_{s,\lambda_s>-N} \rho^{\lambda_s} \frac{y^\beta}{\alpha'!\beta!} L_{s,\beta}^{(\alpha')}(0, D'')D'^{\alpha'}. \tag{3.3.19}$$

By Leibniz' formula we have

$$L_{[\rho,N]}(u_\rho) = L_{[\rho,N]}(W)wv + \sum_{1\leqslant|\alpha|\leqslant 2} \frac{1}{\alpha!} L^{(\alpha)}_{[\rho,N]}(W)D^\alpha(vw). \tag{3.3.20}$$

It is easy to see that

$$L_{[\rho,N]}(W) = W\Bigg[\sum_{s,\lambda_s>-N} \frac{\rho^{\lambda_s}y^\beta}{\alpha'!\beta!}\sum_{\alpha''}\frac{1}{\alpha''!} L^{(\alpha'+\alpha'')}_{s,\beta}(0,\gamma\hat{\eta}'')$$
$$\times \rho^{s-|\alpha'+\alpha''|(1-\sigma_0)}((\phi^0_{y_0})^{\alpha_0}\cdots(\phi^0_{y_m})^{\alpha_m} + Q_\alpha(\rho, y))\Bigg], \tag{3.3.21}$$

where $Q_\alpha(\rho, y)$ consists of terms containing second derivatives of ϕ^0. We shall show that for a definite choice of κ, τ and σ_0 the highest power of ρ occurring in the expansion (3.3.21) is equal to $\nu = 2(\kappa + \sigma_0)$, and this coincides with the power of ρ in the term containing $(\phi^0_{y_0})^2$. We set $q^{-1} = 1 - p_2 + p_1$. By virtue of the condition (3.3.5) of the theorem, the number $q > 0$. We now choose

$$\sigma_0 = \frac{q}{2}(1-R) = \frac{q}{2}\,[1 - p_1(1+|\hat{\beta}'|) - p_2|\hat{\beta}''|],$$
$$\kappa = qp_1, \qquad \tau = qp_2. \tag{3.3.22}$$

Clearly $0 < \sigma_0 < 1$, since $2(p_2 - p_1) - 1 < R < 1$. The exponent of ρ in the term of (3.3.21) containing the factor $L^{(\alpha)}_{s,\beta}(0, \gamma\hat{\eta}'')\,(\phi^0_y)^\alpha$ will be denoted by $J(s, \alpha, \beta)$. We shall show that always $\nu - J(s, \alpha, \beta) \geqslant 0$, provided $L^{(\alpha)}_{s,\beta}(0, \gamma\hat{\eta}'') \neq 0$.

We have

$$\begin{aligned}\nu - J(s,\alpha,\beta) &= \nu - [\lambda_s + s - |\alpha' + \alpha''|(1-\sigma_0)]\\ &= |\alpha''|(1-\sigma_0) + q[(2-s)(1+p_1) + p_1|\beta'| + p_2|\beta''| - (2-|\alpha'|)(1-q^{-1}\sigma_0)]\\ &= |\alpha''|(1-\sigma_0) + q\Big[(2-s)(1+p_1) + p_1|\beta'|\\ &\qquad + p_2|\beta''| - \frac{(2-|\alpha'|)}{2}(1 + p_1(1+|\hat{\beta}'|) + p_2|\hat{\beta}''|)\Big].\end{aligned} \tag{3.3.23}$$

If $s = 1, \beta \in \mathfrak{M}_{p_1}$, or $s = 0$, then clearly $\nu - J(s, \alpha, \beta) \geqslant 0$ for every α, by virtue of the definition of $\hat{\beta}$ and the facts that $\sigma_0 < 1, p_1(1 + |\hat{\beta}'|) + p_2|\hat{\beta}''| < 1$ and $R_{\hat{\beta}}(p_1) \leqslant R_\beta(p_1)$ for every $\beta \in \mathfrak{M}_{p_1}$. We note that in this case $\nu - J(s, \alpha, \beta) = 0$ if $s = 1, \alpha' = 0, \alpha'' = 0, \beta \in \mu$. Suppose $s = 1$ and β does not belong to $\mathfrak{M}_{p_1}$. Then, if $|\alpha'| = 1$, we have $\nu - J(s, \alpha, \beta) > 0$, since

$$\frac{(2-|\alpha'|)}{2}(1 + p_1(1+|\hat{\beta}'|) + p_2|\hat{\beta}''|) < 1.$$

But if $|\alpha''| = 1$ and $L^{(\alpha'')}_{1,\beta}(0, \gamma\hat{\eta}'') \neq 0$, then, clearly, $L_{1,\beta}(0, \gamma\eta'') \neq 0$ for some $\eta'' \neq 0$. Consequently $(1 + |\beta'|)p_1 + p_2|\beta''| > 1$, and it follows from the relation (3.3.23) that in this case $\nu - J(s, \alpha, \beta) > 0$.

If $s = 2$ and $L^{(\alpha'+\alpha'')}_{2,\beta}(0, \gamma\hat{\eta}'') \neq 0$, then there exists a point η'' such that

$L_{2,\beta}^{(\alpha')}(0, \gamma\eta'') \neq 0$, and therefore, by virtue of conditions (3.3.6) and (3.3.7), we have $p_1|\beta'| + p_2|\beta''| \geqslant 2 - |\alpha'|$. Consequently in this case $\nu - J(s, \alpha, \beta) \geqslant 0$ and $\nu - J(s, \alpha, \beta) = 0$ only when $\alpha'' = 0$, $|\alpha'| = 2$, and $\beta = 0$.

If n is a common denominator of the rational numbers σ_0, qp_1, qp_2 and q, then each exponent of ρ in the expansion (3.3.21) has the form j/n, where j is an integer. Let $\nu = \nu_0/n$. It then follows from the preceding considerations and from (3.3.21) that

$$L_{[\rho,N]}(W) = W \sum_{j=0}^{N_3} A_j \rho^{\frac{\nu_0 - j}{n}}, \tag{3.3.24}$$

where $A_j(y, \phi^0)$ are polynomials in the derivatives of ϕ^0 with analytic coefficients, and the number N_3 depends on N.

We have shown above that

$$A_0(y, \phi^0) \equiv \sum_{|\alpha'|=2} \frac{1}{\alpha'!} L_2^{(\alpha')}(0, \gamma\hat{\eta}'')(\phi^0_{y_0})^{\alpha_0} \cdots (\phi^0_{y_l})^{\alpha_l} + \gamma \sum_{\beta \in \mu} \frac{y^\beta}{\beta!} L_{1,\beta}(0, \hat{\eta}''). \tag{3.3.25}$$

We further obtain an expansion in powers of ρ for $L^{(\alpha)}_{[\rho,N]}(W)$ with $|\alpha| = 1$. It is easy to see that for $|\alpha'| = 1$ we have

$$\begin{aligned} L^{(\alpha')}_{[\rho,N]}(W) &= \frac{\partial}{\partial(D^{\alpha'}\phi^0)} (L_{[\rho,N]}(W))\rho^{-\sigma_0} \\ &= \rho^{\nu-\sigma_0} W \left\{ \frac{\partial A_0}{\partial(D^{\alpha'}\phi^0)} + \sum_{j=1}^{N_3} \rho^{-\frac{j}{n}} B_{j,\alpha'}(y, \phi^0) \right\}. \end{aligned} \tag{3.3.26}$$

If $|\alpha''| = 1$ and $D^{\alpha''} = -i\partial/\partial y_k$, $l + 1 \leqslant k \leqslant m$, then it is easy to see that

$$\begin{aligned} L^{(\alpha'')}_{[\rho,N]}(W) &= \frac{\partial}{\partial(\gamma\hat{\eta}_k)} (L_{[\rho,N]}(W))\rho^{-1} \\ &= W\rho^{\nu-\sigma_0} \sum_{j=1}^{N_4} \rho^{-\frac{j}{n}} C_{j,\alpha''}(y, \phi^0). \end{aligned} \tag{3.3.27}$$

Here $B_{j,\alpha'}$ and $C_{j,\alpha''}$ are linear functions of the derivatives of ϕ^0 with coefficients which are analytic in y, and N_4 is an integer depending on N.

We now find an expansion for the terms on the right of (3.3.20) of the form

$$H \equiv \sum_{|\alpha|=2} \frac{1}{\alpha!} L^{(\alpha)}_{[\rho,N]}(W) D^\alpha(vw) = W \sum_{|\alpha|=2,\beta,\lambda_2 > -N} \rho^{\lambda_2(\alpha',\beta)} \frac{y^\beta}{\beta!} a_{\alpha,\beta} D^\alpha(vw). \tag{3.3.28}$$

Here $a_{\alpha,\beta}$ are certain constants.

For $|\alpha| = 2$ we have

$$D^\alpha(wv) = w\Bigg\{v\Bigg[\Bigg(\sum_{j=1}^{N_2} \rho^{\sigma_j}\phi^j_{y_0}\Bigg)^{\alpha_0} \cdots \Bigg(\sum_{j=1}^{N_2} \rho^{\sigma_j}\phi^j_{y_m}\Bigg)^{\alpha_m} + i\sum_{j=1}^{N_2} \rho^{\sigma_j}D^\alpha\phi^j\Bigg]$$
$$+ D^\alpha v + \sum_{k,s=0}^{m} b^\alpha_{k,s} D_s\Bigg(\sum_{j=1}^{N_2} \rho^{\sigma_j}\phi^j\Bigg) D_k v\Bigg\},$$

where $b^\alpha_{k,s} = \text{const}$.

From this we obtain that

$$H = Ww \sum_{|\alpha|=2,\beta,\lambda_2>-N} \rho^{\lambda_2(\alpha',\beta)}\frac{y^\beta}{\beta!}a_{\alpha,\beta}$$
$$\times\Bigg[v\sum_{j,k=1}^{N_2} \rho^{\sigma_j+\sigma_k}\Theta^\alpha_{kj} + \sum_{j=1}^{N_2}\rho^{\sigma_j}\Bigg(v\Theta^\alpha_j + \sum_{k=0}^{m} h^\alpha_{kj}D_k v\Bigg) + D^\alpha v\Bigg], \tag{3.3.29}$$

where Θ^α_{kj} depend on the derivatives of ϕ^k and ϕ^j, and Θ^α_j and h^α_{kj} depend on the derivatives of ϕ^j. We shall find an estimate for $\lambda_2(\alpha', \beta)$ with $|\alpha| = 2$. Considering that $p_1|\beta'| + p_2|\beta''| \geqslant 2 - |\alpha'|$, if $L^{(\alpha'+\alpha'')}_{2,\beta}(0, \hat{\eta}'') \neq 0$ for some $\hat{\eta}''$, we obtain from relation (3.3.23) that

$$\nu - \lambda_2(\alpha', \beta) - (2 - |\alpha' + \alpha''|(1 - \sigma_0))$$
$$\geqslant \frac{(2 - |\alpha'|)}{2} q(1 - p_1(1 + |\hat{\beta}'|) - p_2|\hat{\beta}''|) = (2 - |\alpha'|)\sigma_0.$$

From this it follows that

$$\lambda_2(\alpha', \beta) \leqslant \nu - [(2 - |\alpha'|)\sigma_0 + 2 - |\alpha' + \alpha''|(1 - \sigma_0)] \leqslant \nu - 2\sigma_0,$$

since $|\alpha' + \alpha''| = 2$. Considering the expressions (3.3.24)–(3.3.29) and using formula (3.3.20), we obtain that

$$L_{[\rho,N]}(u_\rho) = Ww\Bigg\{v\rho^\nu \sum_{j=0}^{N_3} A_j\rho^{-\frac{j}{n}}$$
$$+ \rho^{\nu-\sigma_0}\Bigg[\sum_{|\alpha'|=1}\Bigg(\frac{\partial A_0}{\partial(D^{\alpha'}\phi^0)} + \sum_{j=1}^{N_3}\rho^{-\frac{j}{n}}B_{j,\alpha'}\Bigg)\Bigg(D^{\alpha'}v + vi\sum_{j=1}^{N_2}\rho^{\sigma_j}D^{\alpha'}\phi^j\Bigg)$$
$$+ \sum_{|\alpha''|=1}\sum_{j=1}^{N_4}\rho^{-\frac{j}{n}}C_{j,\alpha''}\Bigg(D^{\alpha''}v + vi\sum_{j=1}^{N_2}\rho^{\sigma_j}D^{\alpha''}\phi^j\Bigg)\Bigg]$$
$$+ \rho^{\nu-2\sigma_0}\sum_{|\alpha|=2}\Bigg[v\sum_{j,k=1}^{N_2}\rho^{\sigma_j+\sigma_k}\Theta^\alpha_{kj} \tag{3.3.30}$$
$$+ \sum_{j=1}^{N_2}\rho^{\sigma_j}\Bigg(\Theta^\alpha_j v + \sum_{k=0}^{m} h^\alpha_{kj}D_k v\Bigg) + D^\alpha v\Bigg]\Bigg(\sum_{k=0}^{N_5} r_{\alpha,k}\rho^{-\frac{k}{n}}\Bigg)\Bigg\},$$

where $r_{\alpha,k}$ are certain functions independent of v, ϕ^k, and ρ.

In order to determine the function ϕ^0, we consider the differential equation

$$A_0(y, \phi^0) = 0, \tag{3.3.31}$$

where $A_0(y, \phi^0)$ is determined by equation (3.3.25).

Since $L_{1,\hat{\beta}}(0, \hat{\eta}'') \neq 0$ by virtue of the choice of $\hat{\eta}''$ and $\hat{\beta}$, it follows that the polynomial in y

$$b(y) \equiv \sum_{\beta \in \mu} \frac{y^\beta}{\beta!} L_{1,\beta}(0, \hat{\eta}'')$$

has at least one coefficient different from zero. Hence in a neighborhood of the origin there is a point $y^* = (y_0^*, \cdots, y_m^*)$ such that $y_0^* > 0$ and $b(y^*) \neq 0$.

We shall show that equation (3.3.31) may be solved for $\phi^0_{y_0}$. For this purpose we consider

$$A_0(y^*, \zeta_0, \cdots, \zeta_l) = \sum_{|\alpha'|=2} \frac{1}{\alpha'!} L_2^{(\alpha')}(0, \gamma\hat{\eta}'')\zeta_0^{\alpha_0} \cdots \zeta_l^{\alpha_l} + \gamma b(y^*).$$

The equation

$$A_0(y^*, \zeta_0, 0, \cdots, 0) \equiv a_{(2,0,\cdots,0)}(0)\zeta_0^2 + \gamma b(y^*) = 0$$

for ζ_0 has a simple root $F(y^*)$ such that $\operatorname{Im} F(y^*) < 0$, if the sign of the number γ is chosen appropriately. Therefore in some neighborhood G_1 of the point y^* and in a neighborhood of the origin in the space $(\zeta_1, \cdots, \zeta_l)$ there is defined an analytic function $\zeta_0 = F(y, \zeta_1, \cdots, \zeta_l)$, which satisfied the equation $A_0(y, \zeta_0, \cdots, \zeta_l) = 0$ and $\operatorname{Im} F(y, \zeta_1, \cdots, \zeta_l) < 0$. By the Cauchy-Kowalewski Theorem, in a neighborhood $G_2 \subset G_1$ of the point y^* there exists an analytic solution $\phi^0(y)$ of the equation

$$\phi^0_{y_0} = F(y, \phi^0_{y_1}, \cdots, \phi^0_{y_l}) \tag{3.3.32}$$

with the condition

$$\phi^0 \Big|_{y_0 = y_0^*} = i \sum_{k,j=1}^{m} b^{kj}(y_k - y_k^*)(y_j - y_j^*), \tag{3.3.33}$$

where $\| b^{kj} \|$ is a real symmetric positive definite matrix.

Expanding the solution $\phi^0(y)$ of the problem (3.3.32), (3.3.33) in a Taylor series at the point $y = y^*$, we obtain

$$\phi^0(y) = \left[\phi^0_{y_0}(y^*) + \frac{1}{2}\sum_{j=0}^{m} \frac{\partial^2 \phi^0(y^*)}{\partial y^0 \partial y_j}(y_j - y_j^*)\right](y_0 - y_0^*)$$

$$+ \sum_{k,j=1}^{m} i b^{kj}(y_k - y_k^*)(y_j - y_j^*) + O(|y - y^*|^3). \tag{3.3.34}$$

Since $\operatorname{Im} \phi^0_{y_0}(y^*) < 0$, we may choose a sufficiently small neighborhood G_3 of the point y^* so that

$$\operatorname{Im} \phi^0(y) \geqslant C_0\left(|y_0 - y_0^*| + \sum_{j=1}^{m} |y_j - y_j^*|^2\right) \tag{3.3.35}$$

holds in $G_3^- = G_3 \cap \{y_0 \leqslant y_0^*\}$, where $C_0 = \text{const} > 0$.

Thus for the function ϕ^0 in (3.3.18) we choose the solution $\phi^0(y)$ of the problem (3.3.32), (3.3.33) constructed above. We now set

$$\sigma_j = \sigma_0 - \frac{j}{n} = \frac{n\sigma_0 - j}{n}, \qquad j = 1, \cdots, n\sigma_0 - 1 = N_2.$$

The functions ϕ^j in (3.3.18) are specified in succession, equating to zero the sum of all terms in the expansion (3.3.30) containing the factor $\rho^{\nu - j/n}$, i.e. we define the ϕ^j successively as solutions of the equations

$$\sum_{|\alpha'|=1} i\left(\frac{\partial A_0}{\partial(D^{\alpha'}\phi^0)}\right) D^{\alpha'}\phi^j + \Phi_j(y, \phi^0, \cdots, \phi^{j-1}) = 0 \tag{3.3.36}$$

with conditions

$$\phi^j\Big|_{y_0 = y_0^*} = 0, \qquad j = 1, \cdots, n\sigma_0 - 1, \tag{3.3.37}$$

where Φ_j are known analytic functions. Since $\partial A_0/\partial D_0\phi^0 \neq 0$ for $y = y^*$, it follows that solutions ϕ^j of the problem (3.3.36), (3.3.37) exist in a sufficiently small neighborhood G_4 of the point y^*. Substituting into (3.3.30) the functions ϕ^0 and ϕ^j, $j = 1, \cdots, n\sigma_0 - 1$ constructed in this way, we obtain that in G_4

$$\begin{aligned} L_{[\rho,N]}(u_\rho) = Ww\rho^{\nu-\sigma_0}\Bigg[&\sum_{j=0}^{N_6} \rho^{-\frac{j}{n}} \widetilde{Q}_j v + \sum_{|\alpha'|=1}\left(\frac{\partial A_0}{\partial(D^{\alpha'}\phi^0)} + \sum_{j=1}^{N_7} \rho^{-\frac{j}{n}} \widetilde{B}_{j,\alpha'}\right) D^{\alpha'} v \\ &+ \sum_{|\alpha''|=1} \sum_{j=1}^{N_8} \rho^{-\frac{j}{n}} \widetilde{C}_{j,\alpha''} D^{\alpha''} v + \rho^{-\sigma_0} \sum_{|\alpha|=2} \sum_{j=0}^{N_5} \rho^{-\frac{j}{n}} r_{\alpha,j} D^\alpha v\Bigg], \end{aligned} \tag{3.3.38}$$

where $\widetilde{Q}_j$, $\widetilde{B}_{j,\alpha'}$ and $\widetilde{C}_{j,\alpha''}$ are certain known functions, and N_6, N_7 and N_8 certain numbers depending on N.

We shall seek a function v in the form

$$v = \sum_{j=0}^{N_1} \rho^{-\frac{j}{n}} v^j(y) \tag{3.3.39}$$

and substitute it in the equation (3.3.38). Equating to zero the sum of terms on the right side of (3.3.38) containing the factor $\rho^{\nu - \sigma_0 - j/n}$, $j = 0, 1, \cdots, N_1$, we obtain equations for the successive determination of the functions v^j of the form

$$\sum_{|\alpha'|=1}\left(\frac{\partial A_0}{\partial(D^{\alpha'}\phi^0)}\right)D^{\alpha'}v^j+\widetilde{Q}_0 v^j+\psi_j(y,v^0,\cdots,v^{j-1})=0, \tag{3.3.40}$$

where ψ_j are known analytic functions. For $v^j(y)$ in the expansion (3.3.39) we take the solutions of the equation (3.3.40) with initial conditions

$$v^0\big|_{y_0=y_0^*}=1, \tag{3.3.41}$$

$$v^j\big|_{y_0=y_0^*}=0, \qquad j=1,\cdots,N_1, \tag{3.3.42}$$

which clearly exist in some sufficiently small neighborhood of the point y^*. With such a construction of the function u_ρ in some neighborhood $G\subset\Omega_1$ of the point y^*, we have

$$L_{[\rho,N]}(u_\rho)=WwQ(y,\rho), \tag{3.3.43}$$

where $|Q(y,\rho)|\leqslant C_1\rho^{\nu-\sigma_0-(N_1+1)/n}$, and the constant C_1 does not depend on ρ.

Let the function $\phi(y)\in C_0^\infty(G)$, $\phi\geqslant 0$ in G, and $\phi\equiv 1$ in some neighborhood G_0 of the point y^*. We denote $G^-=G\cap\{y_0\leqslant y_0^*\}$, and $G_0^-=G_0\cap\{y_0\leqslant y_0^*\}$. There clearly exists a constant $C_2>0$ such that

$$|y_0-y_0^*|+\sum_{j=1}^{m}|y_j-y_j^*|^2>C_2 \tag{3.3.44}$$

for points of the set $G^-\backslash G_0^-$. Considering the representation (3.3.15) of the symbol of the operator L_s, equation (3.3.43), and the estimates (3.3.35), (3.3.44), we obtain that for large ρ

$$\sup_{G^-}\sum_{|\alpha|\leqslant k}|D^\alpha L(\phi u_\rho)|$$

$$\leqslant C_3\sup_{G^-}|Ww|\left(\rho^{\nu-\sigma_0-\frac{N_1+1}{n}+k}+\rho^{-N+k+2}\right)+C_4\rho^M\exp\left(-\tfrac{1}{2}C_0C_2\rho^{\sigma_0}\right),$$

where C_3, C_4 and M are constants independent of ρ.

We choose N and N_1 such that $-N+k+2\leqslant -k\delta-1$ and $\nu-\sigma_0-(N_1+1)/n+k\leqslant -k\delta-1$. Then for sufficiently large ρ

$$\rho^{k\delta}\sup_{G^-}\sum_{|\alpha|\leqslant k}|D^\alpha L(\phi u_\rho)|\leqslant C_5\rho^{-1}\left(\sup_{G^-}|Ww|+1\right),$$

where the constant C_5 does not depend on ρ.

Since estimate (3.3.35) is valid for ϕ^0 in G_3^-, and $\phi^j=0$ for $y_0=y_0^*$ and $j\geqslant 1$, it follows that for sufficiently large ρ the estimate

$$\sup_{G^-}|Ww|=\left|\exp\left(i\sum_{j=0}^{N_2}\phi^j\rho^{\sigma_j}\right)\right|\leqslant 1$$

holds. We choose $\rho=\rho_0$ so large that we have, in addition,

$$C\rho_0^{k\delta} \sup_{G^-} \sum_{|\alpha|\leqslant k} |D^\alpha L(\phi u_{\rho_0})| < \frac{1}{2}, \tag{3.3.45}$$

where C is the constant entering into the inequality (3.3.14) for some compact set K_1 containing G. We may suppose that $\phi u_\rho = 0$ in a neighborhood of the plane $x_0 = h$, where initial conditions are given, since $\hat{x}_0 \geqslant h$ and $y_0^* > 0$. We therefore have

$$\phi u_{\rho_0}\big|_{x_0=h} = 0, \quad \frac{\partial}{\partial x_0}(\phi u_{\rho_0})\big|_{x_0=h} = 0.$$

We consider the function $f(y)$, equal to $L(\phi u_{\rho_0})$ in G^- and such that $f(y) \in C_0^\infty(G)$ and

$$C\rho_0^{k\delta} \sup_G \sum_{|\alpha|\leqslant k} |D^\alpha f| \leqslant \frac{1}{2}.$$

Since by our assumption the Cauchy problem (3.3.1), (3.3.2) is correct in Ω, it follows that there exists in Ω a solution u' of the problem

$$L(u') = f, \quad u'\big|_{x_0=h} = \frac{\partial u'}{\partial x_0}\bigg|_{x_0=h} = 0,$$

with $u' = \phi u_{\rho_0}$ for $y_0 \leqslant y_0^*$.

By construction $|\phi u_\rho(y^*)| = 1$, and consequently $\sup_{\Omega_1} |u'| \geqslant 1$. We thus obtain that estimate (3.3.14) is not satisfied for u', which contradicts the assumption that the Cauchy problem (3.3.1), (3.3.2) is correct and that condition (3.3.8), (3.3.9) is not fulfilled.

The theorem is proved.

It is interesting to compare the necessary conditions for correctness of the Cauchy problem for equation (3.3.1) obtained in Theorem 3.3.1, with the sufficient conditions obtained in §2.

Let

$$L(u) \equiv u_{x_0x_0} - x_0^{2k} u_{x_1x_1} + a x_0^s u_{x_1} = 0, \tag{3.3.46}$$

where $a = \text{const}$, and k and s are nonnegative integers.

According to Theorem 3.2.1, the Cauchy problem in the domain $\Omega = \{0 \leqslant x_0 < T, -\infty < x_1 < +\infty\}$ with initial conditions at $x_0 = 0$ is correct if $s \geqslant k - 1$. From Theorem 3.3.1 we obtain that the condition $s \geqslant k - 1$ is also a necessary condition for the correctness of the Cauchy problem.

Let

$$L(u) \equiv u_{x_0x_0} - x_1^{2k} u_{x_1x_1} + a x_1^s u_{x_1} = 0, \tag{3.3.47}$$

where $a = \text{const}$, and k and s are nonnegative integers. According to Theorem 3.2.1, the Cauchy problem for equation (3.3.47) is correct in the domain

$$\Omega\,\{0 < x_0 < T,\ -\infty < x_1 < +\infty\}$$

with initial conditions at $x_0 = 0$, if $s \geqslant k$. Theorem 3.3.1 shows that this condition is also necessary.

A theorem analogous to Theorem 3.3.1 may be proved for equations of any order.

BIBLIOGRAPHY

1. A. D. Aleksandrov, *Investigations on the maximum principle.* I, Izv. Vysš. Učebn. Zaved. Matematika **1958**, no. 5 (6), 126–157. (Russian) MR **24** # A3400a.

2. ———, *Investigations on the maximum principle.* II, Izv. Vysš. Učebn. Zaved. Matematika **1959**, no. 3 (10), 3–12. (Russian) MR **24** # A3400b.

3. ———, *Investigations on the maximum principle.* III, Izv. Vysš. Učebn. Zaved. Matematika **1959**, no. 5 (12), 16–32. (Russian) MR **24** # A3400c.

4. ———, *Investigations on the maximum principle.* IV, Izv. Vysš. Učebn. Zaved. Matematika **1960**, no. 3 (16), 3–15. (Russian) MR **24** # A3400d.

5. ———, *Investigations on the maximum principle.* V, Izv. Vysš. Učebn. Zaved. Matematika **1960**, no. 5 (18), 16–26. (Russian) MR **24** # A3400e.

6. M. S. Baouendi, *Sur une classe d'opérateurs elliptiques dégénérant au bord*, C. R. Acad. Sci. Paris Sér. A–B **262** (1966), A3337–A340. MR **33** # 2950.

7. ———, *Sur une classe d'opérateurs elliptiques dégénérés*, Bull. Soc. Math. France **95** (1967), 45–87. MR **37** #4398.

8. M. S. Baouendi and P. Grisvard, *Sur une équation d'évolution changeant de type*, J. Functional Analysis **2** (1968), 352–367. MR **40** # 6034.

9. ———, *Sur une équation d'évolution changeant de type*, C. R. Acad. Sci. Paris Sér. A–B **265** (1967), A556–A558. MR **36** # 4158.

10. I. S. Berezin, *On Cauchy's problem for linear equations of the second order with initial conditions on a parabolic line*, Mat. Sb. **24 (66)** (1949), 301–320; English transl., Amer. Math. Soc. Transl. (1) **4** (1962), 415–439. MR **11**, 112; **13**, 559.

11. S. N. Bernšteĭn, *Sur une généralisation des théorèmes de Liouville et de M. Picard*, C. R. Acad. Sci. Paris **151** (1910), 635–638.

12. L. Bers, *Mathematical aspects of subsonic and transonic gas dynamics*, Surveys in Appl. Math., vol. 3, Wiley, New York; Chapman and Hall, London, 1958. MR **20** # 2960.

13. L. Bers, F. John and M. Schechter, *Partial differential equations*, Lectures in Appl. Math., vol. 3, Interscience, New York, 1964. MR **29** # 346.

14. P. Bolley and J. Camus, *Études de la régularité de certains problèmes elliptiques dégénérés dans des ouverts non réguliers, par la méthode de réflexion*, C. R. Acad. Sci. Paris Sér. A–B **268** (1969), A1462–A1464. MR **39** #7275.

15. J.-M. Bony, *Principe du maximum, inégalité de Harnack et unicité du problème de Cauchy pour les opérateurs elliptiques dégénérés*, Ann. Inst. Fourier (Grenoble) **19** (1969), fasc. 1, 277–304. MR **41** # 7486.

16. J.-M. Bony, *Problème de Dirichlet et inégalité de Harnack pour une classe d'opérateurs elliptiques dégénérés du second ordre*, C. R. Acad. Sci. Paris Sér. A–B **266** (1968), A830–A833. MR **37** # 6579.

17. ———, *Sur la propagation des maximums et l'unicité du problème de Cauchy pour les opérateurs elliptiques dégénérés du second ordre*, C. R. Acad. Sci. Paris Sér. A–B **266** (1968), A763–A765. MR **37** # 6578.

18. Chi Min'-yu, *The Cauchy problem for a class of hyperbolic equations with initial data on a line of parabolic degeneracy*, Acta Math. Sinica **8** (1958), 521–530 = Chinese Math. Acta **9** (1967), 246–254. MR **21** # 5815.

19. R. Courant, *Methods of mathematical physics*. Vol. 2: *Partial differential equations*, Interscience, New York, 1962. MR **25** # 4216.

20. M. Derridj, *Sur une classe d'opérateurs différentiels hypoelliptiques à coefficients analytiques*, Séminaire Goulaouic-Schwartz 1970/71, Exposé no. 12, École Polytechnique, Centre de Mathématiques, Paris.

21. Ju. V. Egorov, *Hypoelliptic pseudodifferential operators*, Dokl. Akad. Nauk SSSR **168** (1966), 1242–1244 = Soviet Math. Dokl. **7** (1966), 808–810. MR **34** # 3078.

22. ———, *On subelliptic pseudoddifferential operators*, Dokl. Akad. Nauk SSSR **188** (1969), 20–22 = Soviet Math. Dokl. **10** (1969), 1056–1059. MR **41** # 630.

23. ———, *The canonical transformations of pseudodifferential operators*, Uspehi Mat. Nauk **24** (1969), no. 5 (149), 235–236. (Russian) MR **42** # 657.

24. Ju. V. Egorov and V. A. Kondrat'ev, *The oblique derivative problem*, Mat. Sb. **78** (**120**) (1969), 148–176 = Math. USSR Sb. **7** (1969), 139–166. MR **38** # 6230.

25. L. P. Eisenhart, *Continuous groups of transformations*, Princeton Univ. Press, Princeton, N. J., 1933; reprint, Dover, New York, 1961. MR **23** # A1328.

26. G. M. Fateeva, *The Cauchy problem and boundary value problem for linear and quasilinear degenerate second-order hyperbolic equations*. Dokl. Akad. Nauk SSSR **172** (1967), 1278–1281 = Soviet Math. Dokl. **8** (1967), 281–284. MR **35** # 569

27. ———, *Boundary value problems for degenerate quasilinear parabolic equations*, Mat. Sb. **76** (**118**) (1968), 537–565 = Math. USSR Sb. **5** (1968), 509–532. MR **37** # 4419.

28. V. S. Fediĭ, *Estimates in $H_{(s)}$ norms and hypoellipticity*, Dokl. Akad. Nauk SSSR **193** (1970), 301–303 = Soviet Math. Dokl. **11** (1970), 940–942. MR **42** # 6419.

29. G. Fichera, *Sulle equazioni differenziali lineari ellittico-paraboliche del secondo ordine*, Atti Accad. Naz. Lincei. Mem. Cl. Sci. Fis. Mat. Nat. Sez. I (8) **5** (1956), 1–30. MR **19**, 658; 1432.

30. ———, "On a unified theory of boundary value problems for elliptic-parabolic equations of second order," in *Boundary problems. Differential equations*, Univ. of Wisconsin Press, Madison, Wis., 1960, pp. 97–120. MR **22** # 2789.

31. J. N. Franklin and E. R. Rodemich, *Numerical analysis of an elliptic-parabolic partial differential equation*, SIAM J. Numer. Anal. **5** (1968), 680–716. MR **39** # 6522.

32. M. I. Freĭdlin, *Markov processes and differential equations*, Theory of Probability. Mathematical Statistics. Theoretical Cybernetics 1966, Akad. Nauk SSSR Inst. Naučn. Informaciĭ, Moscow, 1967, pp. 7–58 = Progress in Math. **3** (1969), 1–55. MR **38** # 3618.

33. M. I. Freĭdlin, *The first boundary value problem for degenerating elliptic differential equations*, Uspehi Mat. Nauk **15** (1960), no. 2 (92), 204–206. (Russian)

34. ———, *On the formulation of boundary value problems for degenerating elliptic equations*, Dokl. Akad. Nauk SSSR **170** (1966), 282–285 = Soviet Math. Dokl. **7** (1966), 1204–1207. MR **35** # 1908.

35. ———, *The stabilization of the solutions of certain parabolic equations and systems*, Mat. Zametki **3** (1968), 85–92 = Math. Notes **3** (1968), 50–54. MR **36** #5536.

36. ———, *Quasilinear parabolic equations, and measures on a function space*, Funkcional. Anal. i Priložen. **1** (1967), no. 3, 74–82 = Functional Anal. Appl. **1** (1967), 234–240. MR **37** # 584.

37. A. Friedman, *Partial differential equations of parabolic type*, Prentice-Hall, Englewood Cliffs, N. J., 1964. MR **31** # 6062.

38. K. O. Friedrichs, *Pseudo-differential operators. An introduction*, Courant Inst. Math. Sci., New York University, 1970. MR **44** #859.

39. L. Gårding, *Dirichlet's problem for linear elliptic partial differential equations*, Math. Scand. **1** (1953), 55–72. MR **16**, 366.

40. I. M. Gel'fand and G. E. Šilov, *Generalized functions*. Vol. 1: *Operations on them*, Fizmatgiz, Moscow, 1958; English transl., Academic Press, New York, 1964. MR **20** # 4182; **29** # 3869.

41. ———, *Generalized functions*. Vol. 2: *Spaces of fundamental functions*, Fizmatgiz, Moscow, 1958; English transl., Academic Press, New York, 1964. MR **21** # 5142a; **37** # 5693.

42. S. Gellerstedt, *Sur une équation linéaire aux dérivées partielles de type mixte*, Ark. Mat. Astr. Fys. **25A** (1937).

43. T. G. Genčev, *Ultraparabolic equations*, Dokl. Akad. Nauk SSSR **151** (1963), 265–268 = Soviet Math. Dokl. **4** (1963), 979–982. MR **27** # 1715.

44. V. P. Gluško, *Coerciveness in L_2 of general boundary value problems for a degenerate second order elliptic equation*, Funkcional. Anal. i Priložen. **2** (1968), no. 3, 87–88 = Functional Anal. Appl. **2** (1968), no. 3, 261–263. MR **38** # 3585.

45. B. Hanouzet, *Régularité pour une classe d'opérateurs elliptiques dégénérés du deuxième ordre*, C. R. Acad. Sci. Paris Sér. A–B **268** (1969), A1177–A1179. MR **39** # 3147.

46. G. Hellwig, *Anfangs- und Randwertprobleme bei partiellen Differentialgleichungen von wechselndem Typus auf den Rändern*, Math. Z. **58** (1953), 337–357. MR **15**, 130.

47. D. Hilbert, *Grundlagen der Geometrie*, 7th ed., Teubner, Leipzig, 1930.

48. ———, *Über die Darstellung definiter Formen als Summen von Formenquadraten*, Math. Ann. **32** (1888), 342–350.

49. E. Hopf, *Elementare Bemerkungen über die Lösungen partiellen Differentialgleichungen zweiter Ordnung vom elliptischen Typus*, S.-B. Preuss. Akad. Wiss. **19** (1927), 147–152.

50. L. Hörmander, *On the theory of general partial differential operators*, Acta Math. **94** (1955), 161–248. MR **17**, 853.

51. ———, *On interior regularity of the solutions of partial differential equations*, Comm. Pure Appl. Math. **11** (1958), 197–218. MR **21** # 5064.

52. L. Hörmander, *Hypoelliptic differential operators*, Ann. Inst. Fourier (Grenoble) **11** (1961), 477–492. MR **23** # A3368

53. ———, *Pseudo-differential operators and hypoelliptic equations*, Proc. Sympos. Pure Math., vol. 10, Amer. Math. Soc., Providence, R. I., 1967, pp. 138–183.

54. ———, *Pseudo-differential operators*, Comm. Pure Appl. Math. **18** (1965), 501–517. MR **31** # 4970.

55. ———, *Hypoelliptic second order differential equations*, Acta Math. **119** (1967), 147–171. MR **36** # 5526.

56. ———, *Linear partial differential operators*, Die Grundlehren der math. Wissenschaften, Band 116, Academic Press, New York; Springer-Verlag, Berlin, 1963. MR **28** #4221.

57. A. M. Il'in, *On a class of ultraparabolic equations*, Dokl. Akad. Nauk SSSR **159** (1964), 1214–1217 = Soviet Math. Dokl. **5** (1964), 1673–1676. MR **30** # 1315.

58. ———, *On Dirichlet's problem for an equation of elliptic type degenerating on some set of interior points of a region*, Dokl. Akad. Nauk SSSR **102** (1955), 9–12. (Russian) MR **17**, 269.

59. ———, *Degenerate elliptic and parabolic equations*, Mat. Sb. **50 (92)** (1960), 443–498. (Russian) MR **22** # 2788.

60. ———, *Degenerating elliptic and parabolic equations*, Naučn. Dokl. Vysš. Školy. Fiz.-Mat. Nauki **1958**, no. 2, 48–54. (Russian) RŽ Mat. **1960** # 1669.

61. A. M. Il'in, A. S. Kalašnikov and O. A. Oleĭnik, *Second-order linear equations of parabolic type*, Uspehi Mat. Nauk 17 (1962), no. 3 (105), 3–146 = Russian Math. Surveys **17** (1962), no. 3, 1–143. MR **25** # 2328.

62. V. Ja. Ivriĭ, *The Cauchy problem for nonstrictly hyperbolic equations*, Dokl. Akad. Nauk SSSR **197** (1971), 517–519 = Soviet Math. Dokl. **12** (1971), 483–486.

63. M. V. Keldyš, *On certain cases of degeneration of equations of elliptic type on the boundary of a domain*, Dokl. Akad. Nauk SSSR **77** (1951), 181–183; English transl., Amer. Math. Soc. Transl. (2) (to appear). MR **13**, 41.

64. J. J. Kohn and L. Nirenberg, *Degenerate elliptic-parabolic equations of second order*, Comm. Pure Appl. Math. **20** (1967), 797–872. MR **38** # 2437.

65. ———, *An algebra of pseudodifferential operators*, Comm. Pure Appl. Math. **18** (1965), 269–305. MR **31** # 636.

66. J. J. Kohn, *Pseudo-differential operators and non-elliptic problems*, Pseudo-Differential Operators (C. I. M. E., Streza, 1968), Edizioni Cremonese, Rome, 1969, pp. 157–165. MR **41** # 3972.

67. A. N. Kolmogorov, *Zufällige Bewegungen*, Ann. of Math. **35** (1934), 116–117.

68. V.A. Kondrat'ev, *Boundary value problems for parabolic equations in closed regions*, Trudy Moskov. Mat. Obšč. **15** (1966), 400–451 = Trans. Moscow Math. Soc. **1966**, 450–504. MR **35** # 579.

69. S. N. Kružkov, *Boundary value problems for second order elliptic equations*, Mat. Sb. **77 (119)** (1968), 299–334 = Math. USSR Sb. **6** (1968), 275–308. MR **39** # 1794.

70. L. D. Kudrjavcev, *On the solution by the variational method of elliptic equations which degenerate on the boundary of the region*, Dokl. Akad. Nauk SSSR **108** (1956), 16–19. (Russian) MR **19**, 283.

71. L. D. Kudrjavcev, *Direct and inverse imbedding theorems. Applications to the solution of elliptic equations by variational methods*, Trudy Mat. Inst. Steklov. **55** (1959); English Transl., Transl. Math. Monographs, vol. 00, Amer. Math. Soc., Providence, R. I. (to appear). MR **33** # 7838.

72. A. Lax, *On Cauchy's problem for partial differential equations with multiple characteristics*, Comm. Pure Appl. Math. **9** (1956), 135–169. MR **18**, 397.

73. P. D. Lax, *Asymptotic solutions of oscillatory initial value problems*, Duke Math. J. **24** (1957), 627–546. MR **20** # 4096.

74. E. E. Levi, *Opere*, A cura dell'unione Matematica Italiana e col contributo del Consiglio Nazional delle Ricerche, 2 vols., Edizioni Cremonese, Rome, 1959, 1960, papers 15–18. MR **23** # A790.

75. J. L. Lions, *Quelques méthodes de résolution des problèmes aux limites non linéaires*, Dunod; Gauthier-Villars, Paris, 1969. MR **41** # 4326.

76. B. Malgrange, *Sur une classe d'opérateurs différentiels hypoelliptiques*, Bull. Soc. Math. France **85** (1957), 283–306. MR **21** # 5063.

77. V. G. Maz'ja, *The degenerate problem with oblique derivative*, Uspehi Mat. Nauk **25** (1970), no. 2 (152), 275–276. (Russian) MR **41** # 4031.

78. V. G. Maz'ja and B. P. Panejah, *Degenerate elliptic pseudodifferential operators on a smooth manifold without boundary*, Funkcional. Anal. i Priložen. **3** (1969), no. 2, 91–92 = Functional Anal. Appl. **3** (1969), 159–160. MR **40** # 3372.

79. V. P. Mihaĭlov, *An existence and uniqueness theorem for the solution of a certain boundary value problem for a parabolic equation in a domain with singular points on the boundary*, Trudy Mat. Inst. Steklov. **91** (1967), 47–58 = Proc. Steklov Inst. Math. **91** (1967), 47–60. MR **36** # 6804.

80. S. G. Mihlin, *Degenerate elliptic equations*, Vestnik Leningrad, Univ. **9** (1954), no. 8, 19–48. (Russian) MR **17**, 493.

81. C. Miranda, *Equazioni alle derivate parziali di tipo ellittico*, Ergebnisse der Mathematik und ihrer Grenzgebiete, Heft 2, Springer-Verlag, Berlin, 1955; English transl., Springer-Verlag, Berlin, 1970. MR **19**, 421.

82. S. Mizohata and Y. Ohya, *Sur la condition de E. E. Levi concernant des équations hyperboliques*, Publ. Res. Inst. Math. Sci. Ser. A **4** (1968), 511–526. MR **43** # 2349b.

83. A. B. Nersesjan, *The Cauchy problem for a second-order hyperbolic equation degenerating on the initial hyperplane*, Dokl. Akad. Nauk SSSR **181** (1968), 798–801 = Soviet Math. Dokl. **9** (1968), 934–938. MR **38** # 4814.

84. S. M. Nikol'skiĭ, *Approximation of functions of several variables, and imbedding theorems*, "Nauka", Moscow, 1969; English transl., Springer-Verlag, New York (to appear)

85. L. Nirenberg, *A strong maximum principle for parabolic equations*, Comm. Pure Appl. Math. **6** (1953), 167–177. MR **14**, 1089. **16**, 1336.

86. O. A. Oleĭnik, *Alcuni risultati sulle equazioni lineari e quasi lineari ellittico-paraboliche a derivate parziali del secondo ordine*, Atti Accad. Naz. Lincei. Rend. Cl. Sci. Fis. Mat. Natur. (8) **40** (1966), 775–784. MR **37** # 5542.

87. O. A. Oleĭnik, *A boundary value problem for linear elliptic-parabolic equations,* Lecture Ser., no. 46, University of Maryland Inst. Fluid Dynamics and Appl. Math., 1965.

88. ———, *The Cauchy problem and the boundary value problem for second-order hyperbolic equations degenerating in a domain and on its boundary,* Dokl. Akad. Nauk SSSR **169** (1966), 525–528 = Soviet Math. Dokl. **7** (1966), 969–973. MR **34** # 3120.

89. ———, *On second order hyperbolic equations degenerating in the interior of a region and on its boundary,* Uspehi Mat. Nauk **24** (1969), no. 2 (146), 229–230. (Russian) MR **40** # 548.

90. ———, *Mathematical problems of boundary layer theory,* Uspehi Mat. Nauk **23** (1968), no. 3 (141), 3–65 = Russian Math. Surveys **23** (1968), no. 3, 1–66. MR **37** # 4428.

91. ———, *A problem of Fichera,* Dokl. Akad. Nauk SSSR **157** (1964), 1297–1300 = Soviet Math. Dokl. **5** (1964), 1129–1133. MR **30** # 1293.

92. ———, *On the smoothness of solutions of degenerate elliptic and parabolic equations,* Dokl. Akad. Nauk SSSR **163** (1965), 577–580 = Soviet Math. Dokl. **6** (1965), 972–975. MR **34** # 486.

93. ———, *Linear equations of second order with nonnegative characteristic form,* Mat. Sb. **69** (111) (1966), 111–140; English transl., Amer. Math. Soc. Transl. (2) **65** (1967), 167–199. MR **33** # 1603.

94. ———, *On equations of elliptic type degenerating on the boundary of a region,* Dokl. Akad. Nauk SSSR **87** (1952), 885–888. (Russian) MR **15**, 366.

95. ———, *On properties of solutions of certain boundary problems for equations of elliptic type,* Mat. Sb. **30** (**72**) (1952), 695–702. (Russian) MR **14**, 280.

96. ———, *Discontinuous solutions of non-linear differential equations,* Uspehi Mat. Nauk **12** (1957), no. 3 (75), 3–73; English transl., Amer. Math. Soc. Transl. (2) **26** (1963), 95–172. MR **20** # 1055; **27** # 1721.

97. O. A. Oleĭnik and T. D. Ventcel′, *The first boundary problem and the Cauchy problem for quasi-linear equations of parabolic type,* Mat. Sb. **41** (83), (1957), 105–128. (Russian) MR **19**, 149.

98. O. A. Oleĭnik and E. V. Radkevič, *On the local smoothness of weak solutions and hypoellipticity of differential equations of second order,* Uspehi Mat. Nauk **26** (1971), no. 2 (158), 265–281 = Russian Math. Surveys **26** (1971), no. 2, 139–156.

99. O. A. Oleĭnik, *On the equations of unsteady filtration,* Dokl. Akad. Nauk SSSR **113** (1957), 1210–1213. (Russian) MR **20** # 1056.

100. I. G. Petrovskiĭ, *Lectures on partial differential equations,* 3rd aug. ed., Fizmatgiz, Moscow, 1961; English transl., Iliffe, London, Distributed by Saunders, Philadelphia, Pa., 1967. MR **25** # 2308; **35** # 1906.

101. ———, *Lectures on the theory of ordinary differential equations,* 5th ed., "Nauka", Moscow, 1964; English transl., Prentice-Hall, Englewood Cliffs, N. J., 1966. MR **30** # 5055; **33** # 1518.

102. R. S. Phillips and L. Sarason, *Elliptic-parabolic equations of the second order,* J. Math. Mech. **17** (1967/68), 891–917. MR **36** # 2942.

103. R. S. Phillips and L. Sarason, *Singular symmetric positive first order differential operators*, J. Math. Mech. **15** (1966), 235–271. MR **32** # 4357.

104. M. Picone, *Some forgotten almost sixty years old Lincean notes on the theory of second order linear partial differential equations of the elliptic-parabolic type*, Atti Accad. Naz. Lincei. Rend. Cl. Sci. Fis. Mat. Natur. (1968),

105. ———, *Teoremi di unicità nei problemi dei valori al contorno per le equazioni ellittiche e paraboliche*, Atti Accad. Naz. Lincei. Rend. Cl. Sci. Fis. Mat. Natur. **22** (1913), No. 2, 275–282.

106. M. H. Protter, *The Cauchy problem for a hyperbolic second order equation with data on the parabolic line*, Canad. J. Math. **6** (1954), 542–553. MR **16**, 255.

107. M. H. Protter and H. F. Weinberger, *Maximum principles in differential equations*, Prentice-Hall, Englewood Cliffs, N. J., 1967. MR **36** # 2935.

108. C. Pucci, *Proprietà di massimo e minimo delle soluzioni di equazioni a derivate parziali del secondo ordine di tipo ellittico e parabolico*. I, Atti Accad. Naz. Lincei. Rend. Cl. Sci. Fis. Mat. Nat. (8) **23** (1957), 370–375. MR **21** # 6467.

109. ———, *Proprietà di massimo e minimo delle soluzioni di equazioni a derivate parziali del secondo ordine di tipo ellittico e parabolico*. II, Atti Accad. Naz. Lincei. Rend. Cl. Sci. Fis. Mat. Nat. (8) **24** (1958), 3–6. MR **21** # 6467.

110. E. V. Radkevič, *The second boundary value problem for a second order equation with non-negative characteristic form*, Vestnik Moskov. Univ. Ser. I Mat. Meh. 22 (1967), no. 4, 3–11. (Russian) MR **37** # 5531.

111. ———, *A Schauder type estimate for a certain class of pseudo-differential operators*, Uspehi Mat. Nauk **24** (1969), no. 1 (145), 199–200. (Russian) MR **40** # 525.

112. ———, *On a theorem of L. Hörmander*, Uspehi Mat. Nauk **24** (1969), no. 2 (146), 233–234. (Russian) MR **39** # 7286.

113. ———, *A priori estimates and hypoelliptic operators with multiple characteristics*, Dokl. Akad. Nauk SSSR **187** (1969), 274–277 = Soviet Math. Dokl. **10** (1969), 849–852. MR **40** # 4590.

114. ———, *Hypoelliptic operators with multiple characteristics*, Mat. Sb. **79** (121) (1969), 193–216 = Math. USSR Sb. **8** (1969), 181–205. MR **41** # 5763.

115. P. K. Raševskiĭ, *On the joinability of any two points of a completely nonholonomic space by an admissible line*, Uč. Zap. Moskov. Gos. Ped. Inst. im. Libknehta Ser. Fiz.-Mat. **2** (1938), 83–94. (Russian) Fiz.-Mat. RŽ **1** (1939), # 2719.

116. V. M. Petkov, *Necessary conditions for correctness of the Cauchy problem for hyperbolic systems with multiple characteristics*, Uspehi Mat. Nauk **27** (1972), no. 4 (166), 221–222. (Russian)

117. F. Riesz and B. Sz.-Nagy, *Leçons d'analyse fonctionnelle*, Akad. Kiadó, Budapest, 1953; English transl., *Functional analysis*, Ungar, New York, 1955. MR **15**, 132; **17**, 175.

118. J. Schauder, *Über lineare elliptische Differentialgleichungen zweiter Ordnung*, Math. Z. **38** (1934), 257–282.

119. M. Schechter, *On the Dirichlet problem for second order elliptic equations with coefficients singular at the boundary*, Comm. Pure Appl. Math. **13** (1960), 321–328. MR **22** #3872.

120. L. Schwartz, *Théorie des distributions.* Tomes I, II, Actualités Sci. Indust., nos. 1091, 1122, Hermann, Paris, 1950, 1951. MR **12**, 31; 833.

121. ———, *Méthodes mathématiques pour les sciences physiques,* Hermann, Paris, 1961. MR **26** # 919.

122. G. E. Šilov, *Local properties of solutions of partial differential equations with constant coefficients,* Uspehi Mat. Nauk **14** (1959), no. 5 (89), 3–44; English transl., Amer. Math. Soc. Transl. (2) **42** (1964), 129–173. MR **22** # 9708.

123. M. M. Smirnov, *Degenerating elliptic and hyperbolic equations,* "Nauka", Moscow, 1966. (Russian) MR **36** # 1850.

124. G. N. Smirnova, *Linear parabolic equations which degenerate on the boundary of the region,* Sibirsk. Mat. Ž. **4** (1963), 343–358. (Russian) MR **26** # 5299.

125. S. L. Sobolev, *Applications of functional analysis in mathematical physics,* Izdat. Leningrad. Univ., Leningrad, 1950; English transl., Transl. Math. Monographs, vol. 7, Amer. Math. Soc., Providence, R. I., 1963. MR **14**, 565; **29** # 2624.

126. ———, *Méthode nouvelle à résoudre le problème de Cauchy pour les équations linéaires hyperboliques normales,* Mat. Sb. **1** (43) (1936), 39–72.

127. G. Strang and H. Flaschka, *The correctness of the Cauchy problem,* Advances in Math. **6** (1971), 347–379.

128. K. Suzuki, *The first boundary value problem and the first eigenvalue problem for the elliptic equations degenerate on the boundary,* Publ. Res. Inst. Math. Sci. Ser. A 3 (1967/68), 299–335. MR **38** # 412.

129. ———, *The first boundary value and eigenvalue problems for degenerate elliptic equations.* I, Publ. Res. Inst. Math. Sci. Ser. A **4** (1968/69), 179–200. MR **40** # 544.

130. F. Trèves, *Opérateurs différentiels hypoelliptiques,* Ann. Inst. Fourier (Grenoble) **9** (1959), 1–73. MR **22** # 4886.

131. F. Tricomi, *Sulle equazioni lineari alle derivate parziali di secondo ordine, di tipo misto,* Rend. Reale Accad. Lincei. (5) **14** (1923), 134–247.

132. M. I. Višik, *On the first boundary problem for elliptic equations degenerating on the boundary of a region,* Dokl. Akad. Nauk SSSR **93** (1953), 9–12. (Russian) MR **15**, 798.

133. ———, *Boundary-value problems for elliptic equations degenerating on the boundary of a region,* Mat. Sb. **35** (77) (1954), 513–568; English transl., Amer. Math. Soc. Transl. (2) **35** (1964), 15–78. MR **16**, 927.

134. M. I. Višik and V. V. Grušin, *On a class of degenerate elliptic equations,* Mat. Sb. **79** (121) (1969), 3–36 = Math. USSR Sb. **8** (1969), 1–32. MR **40** # 1706.

135. ———, *Boundary value problems for elliptic equations degenerate on the boundary of a domain,* Mat. Sb. **80** (122) (1969), 455–491 = Math. USSR Sb. **9** (1969), 423–454. MR **41** # 2212.

136. ———, *Elliptic pseudodifferential operators on a closed manifold which degenerate on a submanifold,* Dokl. Akad. Nauk SSSR **189** (1969), 16–19 = Soviet Math. Dokl. **10** (1969), 1316–1319.

137. M. I. Višik and G. I. Eskin, *Convolution equations in a bounded region,* Uspehi Mat. Nauk **20** (1965), no. 3 (123), 89–152 = Russian Math. Surveys **20** (1965), no. 3, 85–151. MR **32** # 2741.

138. L. G. Volevič, *Hypoelliptic equations in convolutions*, Dokl. Akad. Nauk SSSR **168** (1966), 1232–1235 = Soviet Math. Dokl. 7 (1966), 797–800. MR **34** # 3077.

139. L. G. Volevič and B. P. Panejah, *Some spaces of generalized functions and embedding theorems*, Uspehi Mat. Nauk **20** (1965), no. 1 (121), 3–74 = Russian Math. Surveys **20** (1965), no. 1, 1–73. MR **30** # 5160.

140. A. I. Vol'pert and S. I. Hudjaev, *Cauchy's problem for degenerate second order quasilinear degenerate parabolic equations*, Mat. Sb. **78** (120) (1969), 374–396 = Math. USSR Sb. 7 (1969), 365–387. MR **41** # 8828.

141. N. D. Vvedenskaja, *On a boundary problem for equations of elliptic type degenerating on the boundary of a region*, Dokl. Akad. Nauk SSSR **91** (1953), 711–714. (Russian) MR **15**, 711.

142. M. Weber, *The fundamental solution of a degenerate partial differential equation of parabolic type*, Trans. Amer. Math. Soc. 71 (1951), 24–37. MR **13**, 41.

143. A. Weinstein, *Generalized axially symmetric potential theory*, Bull. Amer. Math. Soc. 59 (1953), 20–38. MR **14**, 749.

144. K. Yosida, *Functional analysis*, Die Grundlehren der math. Wissenschaften, Band 123, Academic Press, New York; Springer-Verlag, Berlin, 1965. MR 31 # 5054.

145. E. C. Zachmanoglou, *Propagation of zeros and uniqueness in the Cauchy problem for first order partial differential equations*, Arch. Rational Mech. Anal. **38** (1970), 178–188. MR **41** # 5769.

146. T. Nagano, *Linear differential systems with singularities and an application to transitive Lie algebras*, J. Math. Soc. Japan **18** (1966), 398–404. MR **33** # 8005.

147. J. J. Kohn and L. Nirenberg, *Non-coercive boundary value problems*, Comm. Pure Appl. Math. **18** (1965), 443–492. MR **31** # 6041.

148. O. A. Oleĭnik and E. V. Radkevič, *On the analyticity of solutions of linear partial differential equations*, Mat. Sb. **90** (132) (1973), 592–606 = Math. USSR Sb. 19 (1973) (to appear).

149. V. M. Petkov, *Necessary conditions for correctness of the Cauchy problem for nonstrictly hyperbolic equations*, Dokl. Akad. Nauk SSSR 206 (1972), 287–290 = Soviet Math. Dokl. 13 (1972), 1218–1219.